Integrated Assessment of Scale Impacts of Watershed Intervention

Integrated Assessment of Scale Impacts of Watershed Intervention

Assessing Hydrogeological and Bio-physical Influences on Livelihoods

Edited by

V. Ratna Reddy
Livelihoods and Natural Resource Management Institute, Hyderabad, India

Geoffrey J. Syme
Edith Cowan University, Perth, Australia

ELSEVIER

AMSTERDAM   BOSTON   HEIDELBERG   LONDON   NEW YORK   OXFORD
PARIS   SAN DIEGO   SAN FRANCISCO   SINGAPORE   SYDNEY   TOKYO

Elsevier
Radarweg 29, PO Box 211, 1000 AE Amsterdam, The Netherlands
The Boulevard, Langford Lane, Kidlington, Oxford OX5 1GB, UK
225 Wyman Street, Waltham, MA 02451, USA

Notices
Knowledge and best practice in this field are constantly changing. As new research and experience broaden our understanding, changes in research methods, professional practices, or medical treatment may become necessary.

Practitioners and researchers must always rely on their own experience and knowledge in evaluating and using any information, methods, compounds, or experiments described herein. In using such information or methods they should be mindful of their own safety and the safety of others, including parties for whom they have a professional responsibility.

To the fullest extent of the law, neither the Publisher nor the authors, contributors, or editors, assume any liability for any injury and/or damage to persons or property as a matter of products liability, negligence or otherwise, or from any use or operation of any methods, products, instructions, or ideas contained in the material herein.

ISBN: 978-0-12-800067-0

Library of Congress Cataloging-in-Publication Data

Integrated assessment of scale impacts of watershed intervention: assessing hydrogeological and bio-physical influences on livelihoods / edited by V. Ratna Reddy, Geoffrey J. Syme.
      pages cm
   Includes index.
   ISBN 978-0-12-800067-0
   1. Watershed management. 2. Watershed management–Economic aspects. 3. Soil conservation.
   I. Reddy, V. Ratna, editor. II. Syme, Geoffrey J., editor.
   TC413.I58 2015
   333.73–dc23
                          2014026080

British Library Cataloguing in Publication Data
A catalogue record for this book is available from the British Library

For information on all Elsevier publications visit
our web site at http://store.elsevier.com/

Working together
to grow libraries in
developing countries

www.elsevier.com • www.bookaid.org

Contents

Part II
Hydro-geological and Bio-physical Aspects of the Watersheds

3.　Investigating Geophysical and Hydrogeological Variabilities and Their Impact on Water Resources in the Context of Meso-Watersheds

P.D. Sreedevi, S. Sarah, Fakhre Alam, Shakeel Ahmed, S. Chandra and Paul Pavelic

4. **Application of a Simple Integrated Surface Water
and Groundwater Model to Assess Mesoscale
Watershed Development**
*Paul Pavelic, Jian Xie, P.D. Sreedevi, Shakeel Ahmed
and Daniel Bernet*

5. **Modeling the Impact of Watershed Development
on Water Resources in India**
Barry Croke, Peter Cornish and Adlul Islam

6. Sustainable Watershed Development Design Methodology

K.V. Rao, Pratyusha Kranti, Hamsa Sandeep, P.D. Sreedevi and Shakeel Ahmed

Part III
Socio-economic and Livelihood Impacts of Watersheds

7. Assessing Livelihood Impacts of Watersheds at Scale: An Integrated Approach
V. Ratna Reddy, T. Chiranjeevi, Sanjit Kumar Rout and M. Sreenivasa Reddy

8. Evaluating the Determinants of Perceived Drought Resilience: An Empirical Analysis of Farmers' Survival Capabilities in Drought-Prone Regions of South India
Ram Ranjan, Deepa Pradhan, V. Ratna Reddy and Geoffrey J. Syme

Part IV
Integrating Science into Policy and Practice

11.   High Stakes—Engagement with a Purpose
T. Chiranjeevi, Geoffrey J. Syme and V. Ratna Reddy

12.   Exploring Implications of Climate, Land Use, and Policy
      Intervention Scenarios on Water Resources,
      Livelihoods, and Resilience
*Wendy Merritt, K.V. Rao, Brendan Patch, V. Ratna Reddy,
G. Syme and P.D. Sreedevi*

List of Contributors

Shakeel Ahmed, CSIR-National Geophysical Research Institute, Hyderabad, India

Fakhre Alam, CSIR-National Geophysical Research Institute, Hyderabad, India; Presently at CGWB (MER), Patna, India

Daniel Bernet, Institute of Geography & Oeschger Centre for Climate Change Research, University of Bern, Bern, Switzerland

S. Chandra, CSIR-National Geophysical Research Institute, Hyderabad, India

T. Chiranjeevi, Livelihoods and Natural Resource Management Institute, Hyderabad, India

Peter Cornish, University of Western Sydney, Hawkesbury Campus, Australia

Barry Croke, Australian National University, Canberra, Australia

Adlul Islam, Natural Resources Management Division (ICAR), New Delhi, India

Pratyusha Kranti, Central Research Institute for Dryland Agriculture, Hyderabad, India

Wendy Merritt, Fenner School of Environment and Society, The Australian National University, Canberra, ACT, Australia

Brendan Patch, Fenner School of Environment and Society, The Australian National University, Canberra, ACT, Australia

Paul Pavelic, International Water Management Institute, Vientiane, Lao PDR

Deepa Pradhan, Department of Environment and Geography, Faculty of Science, Macquarie University, Sydney, Australia

Ram Ranjan, Department of Environment and Geography, Faculty of Science, Macquarie University, Sydney, Australia

K.V. Rao, Central Research Institute for Dryland Agriculture, Hyderabad, India

V. Ratna Reddy, Livelihoods and Natural Resource Management Institute, Hyderabad, India

M. Sreenivasa Reddy, Centre for Economic and Social Studies, Hyderabad, India

Sanjit Kumar Rout, Livelihoods and Natural Resource Management Institute, Hyderabad, India

Hamsa Sandeep, Central Research Institute for Dryland Agriculture, Hyderabad, India

S. Sarah, CSIR-National Geophysical Research Institute, Hyderabad, India

P.D. Sreedevi, CSIR-National Geophysical Research Institute, Hyderabad, India

Geoffrey J. Syme, Edith Cowan University, Perth, Australia

Jian Xie, International Water Management Institute, Vientiane, Lao PDR; Beijing Normal University, College of Water Sciences, Beijing, PR China

Foreword

Watershed development (WSD) is one of the most effective interventions used to stabilize rainfed agriculture by providing sources of water for small-scale irrigation. In India, WSD is one of the government's flagship programs with substantial budget allocations during the last two decades. While these allocations have increased the area of WSD programs, their effectiveness has been limited in terms of achieving the stated objectives of strengthening the natural resource base and improving agricultural productivity of rainfed regions. The reasons for this include: (1) adopting a narrow approach by treating a micro watershed (500 ha) rather than a complete watershed (hundreds of hectares) or a whole river basin; (2) little understanding of the spatial variability of hydrological processes and the effects of WSD and its likely socioeconomic impacts; and (3) a lack of easy-to-use tools to analyze watershed hydrology and its potential socioeconomic impacts. To address some of these concerns the Integrated Watershed Management Program (2009) has increased the scale of WSD implementation from 500 to 5000−10000 ha; however, the other aspects have not been addressed.

The increased scale of intervention has facilitated a more comprehensive and integrated approach to WSD, providing scope for assessing the spatial variability in hydrogeological and biophysical aspects—mainly the upstream/ downstream differences of access to water resources and the variability in socioeconomic or livelihood impacts. The planning and design of WSD interventions ought to be in tune with these upstream/downstream differences and the impacts predicted with close attention to these locational factors. Such an integrated design will enhance the effectiveness of the interventions and help sustainable resource management in general and water resources in particular. To achieve this, an interdisciplinary approach is necessary. However, to follow such an integrated approach and reap the resultant benefits is difficult without a clear understanding of the processes and methodologies required for integration.

This volume is the outcome of a collaborative research effort between Australian and Indian researchers carried out over a period of 5 years in Andhra Pradesh, India, and Australia. The research reported in this book has transcended discipline boundaries to evolve into an approach integrating hydrogeological, biophysical, and socioeconomic disciplines. These include modeling techniques, Bayesian networks, and classical econometric techniques to analyze resilience, equity, and scale economics. This collection of research covers a wide range of issues including aquifer characteristics and their linkages to groundwater potential and storage; groundwater demand; linkages between surface and subsurface hydrology; rainfall, soils, land use, and recharge linkages; upstream/downstream variations in the groundwater potential; the impact of such variations on household livelihoods and resilience; factors influencing household resilience; assessing the relative importance of hydrogeological, biophysical, and socioeconomic aspects in determining household resilience (integration); assessing the inter- and intra-location equity aspects of watershed interventions; and the processes for successful stakeholder engagement.

This book demonstrates the need for an interdisciplinary research to make investments in WSD effective and efficient. It is shown that hydrogeology varies widely within a mesoscale watershed affecting the groundwater potential and potential for access to irrigation water. Soils, rainfall, and land use influence the runoff and recharge coefficients. These variations in turn influence the socioeconomic aspects (livelihoods or resilience) depending upon location in a watershed. The current accepted view that there are increasing economic benefits as one moves from upstream to downstream may not hold in all situations due to hydrogeological variations. Also enhancing equity, although a more complex issue requiring a number of policy instruments, can be improved by appropriate watershed design.

Planning and design of watershed interventions should consider hydrogeological and biophysical aspects to make interventions effective. It is demonstrated in this volume that the present approach of blanket solutions and allocations of watershed interventions is not achieving the greatest potential benefits. Modifying the design of WSD works according to hydrogeological and biophysical aspects, and will not only help achieve sustainable livelihoods but also improve the long-term management of the water resources of watersheds.

Long-term impact indicators like household resilience (instead of the traditional indicators such as crop yields, irrigation, income, etc.) are more appropriate for watershed impact assessments, especially as there is always a time lag between implementation and assessment.

For this volume of work to bring about the much needed changes in the approach to WSD in India, it is important that the implementing agencies are aware of the advances in understanding and tools developed for better implementation of WSD. The Department of Rural Development of Andhra Pradesh has taken steps in this direction, converting science into practice by taking these scientific research outputs and guiding the adaptation of their WSD implementation programs.

We hope that this book will initiate debate and change in WSD implementation and improve the outcomes for India's farmers and increased the value from WSD for the governments of India.

Dr. Evan W. Christen (ACIAR)
Dr. C. Suvarna (DRD, Andhra Pradesh)

Acknowledgments

Multidisciplinary, interdisciplinary, or transdisciplinary approaches to research are widely believed to be more desirable than the traditional disciplinary approach. This is evident in natural resource research where a multiplicity of sciences is involved. Although such a philosophy is well established, its implementation is rather limited. This is mainly attributed to the complexities involved with coordinating and communicating the activities of different disciplines that use different methodologies and approaches.

This process can be time-consuming and demanding on resources. Fully-fledged collaboration across disciplines needs a concerted effort and systematic approach to understand and appreciate the linkages among different disciplines. Such a systematic approach needs a commitment of resources and time to bring the teams on board and ensure collaboration in creating a collective understanding of the subject and the problem at hand. Apart from the close collaboration within the core research team (scientists), support and understanding of other stakeholders, such as funders and policy makers, is critical for the success of integrated research projects. This volume is the outcome of such a multifaceted research program carried out over a period of 5 years involving a variety of disciplines and subdisciplines including hydrogeology, surface hydrology, biophysical sciences, mathematical modeling, economics, psychology, management, and geography.

This was made possible with the support from the Australian Center of International Agricultural Research (ACIAR) and the proactive role and involvement of the Department of Rural Development (DRD) and the Government of Andhra Pradesh (AP). Promotion of science through integrated research is at the heart of ACIAR's approach, which funded this research and provided ample encouragement to carry the research through periods of unforeseen initial delays due to multiple factors. The support and cooperation from ACIAR is gratefully acknowledged. The research team has gained substantially from the research, which was a new experience for many of us.

The DRD and the Government of AP were very supportive from the design stage of the research through to the end. As a formal partner in the project, the DRD not only supported and participated in the selection of sample sites, but also provided the much needed feedback to make the research policy relevant. More importantly, the DRD agreed to mainstream the results through capacity building of their staff. Adapting the research at the implementation level is one of the major achievements of this research. Our sincere thanks go to the DRD and the Government of AP.

A number of individuals have contributed to the completion of this research. Given that the project lasted 5 years, it has naturally experienced a degree of turnover of team members. It has gone through the rigorous scrutiny of four ACIAR Program Managers. Dr. Christian Roth initiated the project under the Land and Water Cluster Research Program in AP. He saw it through a number of hurdles at the beginning, and without his persistent efforts the project would not have taken off. We gratefully acknowledge his strong support at the initial stages of the project. Our thanks go out to his successors at the ACIAR, the late Dr. Mirko Stauffacher, Dr. Andrew Nobel, and Dr. Evan Christen who were equally supportive and provided constructive feedback at various stages of the research. Special thanks go to Dr. Evan Christen who has overseen the merits of the research and provided the much needed support to carry it forward. Dr. Kuhu Chatterjee and Ms. Simrat Labana, from the ACIAR India office, provided valuable feedback and excellent administrative support at various stages of the project. We gratefully acknowledge their support and cooperation.

The project benefited immensely from Dr. Tirupathaiah, then Special Commissioner, DRD, who provided constructive feedback in setting the objectives and designing the research at the initial stages. His valuable insights were instrumental in making the research relevant to policy. His successor, Dr. C. Suvarna, has ably carried this collaboration forward through her constructive suggestions and support in converting research outputs into inputs for implementation and policy through capacity-building activities. Our sincere thanks go to both of them. At the national level Dr. Alok K. Sikka, DDG (Natural Resource Management), ICAR, and Technical Expert, National Rainfed Area Authority (NRAA), provided much needed recognition to the research by ensuring his support to spread the recommendations through capacity-building activities to the implementing agencies at the national level.

Apart from the contributors to this volume, a number of people have provided valuable input to the research at various stages. Our thanks go to Dr. Mac Kirby, Dr. K.A.S. Mani, Dr. Madar Samad, and Mr. Sreedhar Nallan for their respective contributions. The research has benefited from the critical comments from Dr. Jeff Bennett in the initial stages and the external reviewers Dr. Dennis Wichelnsand and Dr. M. Dinesh Kumar for their constructive feedback.

But for the time and patience of the rural households in the sample villages, the field research, which forms the core of this research, would have been impossible. The information provided by the sample households and the participants of formal and informal group discussions at the time of data collection and their feedback at the time of validation of the results has been invaluable. Their cooperation, despite a long and tiring schedule filling out and answering numerous detailed questionnaires, cannot be compensated. We only hope that this research will help enhance their living conditions. The field research was ably supported by two nongovernmental organizations—Development Initiatives for Peoples Action (DIPA) and Star Youth association (SYA)—operating under the Bharatiya Integrated Rural Development Society (BIRDS) and field investigators who completed the hard task of collecting needed information from the farmers. We sincerely appreciate and thank them for their efforts.

The heads of all the partner institutions and their supporting staff [Edith Cowan University (ECU), Australian National University (ANU), Macquarie University, CSIRO, IWMI, CSIR-NGRI, CRIDA, and LNRMI] have provided the indirect but valuable support and cooperation facilitating the smooth running of this project. We thankfully acknowledge their support. We are particularly grateful to Dr. V.P. Dimri; Dr. Mrinal K. Sen, Director; Dr. Y.J.B. Rao, Acting Director CSIR-NGRI; B. Venkateswarlu, Director, CRIDA; Dr. J.V.N.S. Prasad, Pr. Scientist (Agronomy), CRIDA; Dr. Maheswari, Director, CRIDA; and Dr. A.K. Singh, Former DDG (Natural Resource Management), ICAR, for their cooperation and support throughout the project.

V. Ratna Reddy
Geoffrey J. Syme

Part I

Setting

Chapter 1

Introduction

V. Ratna Reddy* and Geoffrey J. Syme[§]
*Livelihoods and Natural Resource Management Institute, Hyderabad, India, [§]Edith Cowan
University, Perth, Australia

Chapter Outline

1.1 BACKGROUND

Rainfed agriculture accounts for more than 75% of the cropped area in the world. One-third of the developing world's population lives in the less-favored rainfed regions [1]. In India, rainfed agriculture accounts for 60% of the cropped area, and is the food basket for the poor, with a millet-dominant crop pattern. About 70% of India's population is dependent on rainfed agriculture. Therefore it holds promise for future food security because of the saturation of productivity in the green revolution regions.

Rainfed regions house the largest proportion of poor people in India. Further, these regions are expected to be the worst affected in the context of climate variability (e.g., natural disasters like frequent droughts, floods, etc.) and, as a result, productivity. In this context, watershed technology is seen as one of the best alternatives for improving land productivity in terms of reducing soil degradation, runoff, improved *in situ* soil moisture, access to irrigation, and so on, which in turn improves the resilience of the system. The resilience of the farming community in the context of watershed development (WSD) and livelihood strategies at the household level is closely linked to hydrogeological and biophysical attributes of the ecosystem. However, these aspects have not been integral to watershed assessments.

In the recent years, the WSD program in India has transformed from a soil and water conservation initiative to a comprehensive rural development and

Integrated Assessment of Scale Impacts of Watershed Intervention
http://dx.doi.org/10.1016/B978-0-12-800067-0.00001-3. Copyright © 2015 Elsevier Inc. All rights reserved.

3

livelihoods program; although soil and water conservation remains the core. Recent changes in the scale of watersheds from micro (500 ha) to meso (5000−10000 ha) under the Integrated Watershed Management Program (IWMP) facilitates the integration of hydrogeological and biophysical aspects. Comprehensive impact assessments at the meso level can be demanding in terms of data and methods of assessment. The larger scale of watershed should assist in capturing the externalities relating to groundwater and surface water flows in comparison with the micro approach. Mesoscale evaluation accounts for the impact of positive and negative externalities across the streams while assessing watershed impacts.

Impact measurements of developmental initiatives are more often used to correct the type and nature of interventions and implementation modalities. Often the objective is to improve allocative efficiency of resources and improve the value for money. This assumes specific programs, such as WSD in India, are important since they receive huge budgetary allocations (Rs.25,000 crores per year, i.e., $4.545 million per year). Measuring the watershed impacts becomes more complex as watershed interventions consider how hydrogeological and biophysical aspects affect livelihoods. Integration of hydrogeological and biophysical aspects into watershed interventions makes resilience an important attribute, especially in the context of climate change impacts. Integrated assessments of watersheds from a resilience perspective are either rare or absent, and there are several reasons for this.

Until now, watershed impact assessment studies focused on the socioeconomic and natural resource impacts [2,3]. Such assessments are also used to estimate the benefit−cost ratios of the program [3]. With the introduction of a livelihood component along with a participatory approach to implementation during the late 1990s, impact studies have started to use the sustainable livelihoods (SL) framework to assess impacts [4,5]. The SL framework is a more comprehensive approach that looks beyond the income and employment aspects of poverty, assessing the impacts using the five capitals financial: natural, social, human, and physical dimensions of poverty. These dimensions of poverty are more long term in nature. Despite the fact that the prime objective of WSD is soil and water conservation and thus improved productivity and environmental sustainability of the system, not much attention has been paid to assessing the societal resilience aspects of WSD.

In most cases, watershed impact studies do not have the backing of valid baseline information. This limits the appropriate interpretation of the perceived impacts, as the data generated from the households suffer from memory lapse when "before and after" methods are used. In addition, getting a perfectly matching sample becomes a limitation when "with and without" methods are used. Hence, adopting a "double difference" method where both approaches are combined is expected to provide the best proxy in the absence of baseline [4] information. Of late, methods like propensity matching have been used to overcome the baseline deficiencies.

Impact assessments are also influenced by the timing of the study. While impacts are clearly captured in the immediate post-implementation phase, attribution of impacts can get blurred by potential exogenous influences as the gap between implementation and assessment increases. In this context, using resilience as a robust impact indicator would help to address the current limitations of impact assessment to a significant extent. In a way, resilience is directly linked to watershed interventions; if there are more water resources available, then production should also be more reliable. Resilience is also more long term in nature and hence addresses the sustainability aspects of WSD. When resilience is linked to the five capitals, it becomes a robust and comprehensive concept in understanding the IWMP impacts in the absence of baseline information.

This book outlines an integrated approach derived to provide insights into appropriate designs of watershed interventions in the hydrogeological and biophysical context. The hypothesis is that specific watershed interventions are required that suit the technical attributes of the location rather than a blanket approach of uniform interventions. While advanced hydrogeological and biophysical models are used to assess the water and land use impacts, a sustainable rural livelihood framework is implemented to assess the community-level impact. Finally, a Bayesian network (BN) is used to integrate the dimensions. This network approach is also used to develop scenarios of climate and land use changes, while providing a generalizable evaluation tool for policy analysis, including the scale at which watershed interventions should be delivered.

1.2 RAINFED AGRICULTURE AND WSD IN INDIA

While the policy bias, resulting in intensive agricultural practices, has paid off in terms of meeting the country's food demands in the short run, it has proved to be unsustainable, economically as well as environmentally, in the long run. This, coupled with the limited scope for expanding irrigation (through traditional methods of damming the rivers), has prompted the policy shift toward rainfed agriculture. Although recent policies failed to address the problems of irrigated agriculture through improving the allocative efficiency of crucial inputs like water, concerted efforts have been made toward improving the conditions of rainfed farming. Development of such regions, in terms of enhancing the crop yield, holds the key for future food security. Also, these regions are increasingly confronted with environmental problems such as wind and soil erosion; it is feared that the intensity of resource degradation is reaching irreversible levels in some of these regions. Thus, promotion of appropriate technologies and development strategies in these regions would result in multiple benefits: (1) ensuring food security, (2) enhancing the viability of farming, and (3) restoring the ecological balance. Approximately 15% of India's 329 million hectares of geographical area is already degraded [6].

Rainfed regions account for more than 50% of the cultivable land and support 40% of India's population. For the government of India, WSD is one of the primary vehicles of water resource management used to assist in rural poverty reduction in the more marginal semi-arid, rainfed areas of the Central Plateau. These regions house a large share of the poor, food-insecure, and vulnerable populations in the country. Moreover, as productivity growth in the more favored green revolution areas is already showing signs of slowing down or stagnation [7], future growth in agricultural production and food security will depend on improving productivity in the semi-arid rainfed areas [8]. Accordingly, WSD in India has had significant investments over many years. More than $4 billion was spent by the Central Government alone since the beginning of the Indian Government's Eighth Plan (1992). These allocations are being doubled during the Eleventh Plan period with enhanced per hectare investments.[1]

A watershed is a topographically delineated area drained by a stream system. It is a hydrologic unit described and used both as a physical—biological unit and as a socioeconomic and sociopolitical unit for planning and implementing resource-management activities. WSD is a land-based tech-nology intended to help conserve and improve *in situ* soil moisture and to check soil erosion and improved water resources, especially groundwater, in the rainfed regions. WSD simply means improving the management of a watershed or rainfall catchment area, for instance, by building contour bunds, water-harvesting structures (check dams), field bunds (raised edges), etc. It facilitates higher land productivity through improved overall ecological con-ditions such as moisture and water availability for agriculture. WSD deals with the adoption of watershed technologies in specific watersheds. In recognition of the socioeconomic and environmental benefits, India has one of the largest micro-WSD programs in the world.

However, the cost-effectiveness of these allocations and the sustainability of the WSD program as well as the cost-effectiveness at individual, household, and community levels are widely questioned [9]. There is also an increasing concern about the impact of WSD on downstream water flows, raising equity issues between upstream (rainfed farmers supported by WSD) and downstream water users (irrigation, urban, and industrial water use). Earlier studies have observed that declined discharge levels into the downstream surface water bodies can be attributed to WSD in its upper reaches [10].

At the institutional level, the impact of WSD on water availability at the mesoscale are not considered as different by the government departments responsible for managing rainfed areas (Department of Rural Development) and downstream irrigation areas (Irrigation and Command Area Department). Recent discussions with departmental representatives indicate that considering issues at the mesoscale when investing WSD funds will lead to better outcomes, but guidance is needed.

---

1. The per hectare allocation under IWMP is Rs.12,000 ($300).

1.3 WATERSHED POLICIES IN INDIA

WSD has particularly taken off during the last decade, with the 1994 National Guidelines providing the framework within which this expansion has taken place. The recent Government of India (GoI) Working Group on Rainfed Agriculture [11] estimates that the total area covered by watershed programs was about 45.58 million hectares by March 2005, which is about 40% of the total potential area, at an investment of Rs.170,370 crores. Annual expenditure on WSD during the Tenth Plan is about Rs.2300 crores [12]. Although this represents a substantial achievement, given that this progress has taken over 40 years, the speed of implementation of the watershed programs and the ability to scale up successful experiences is clearly a major policy issue.

In particular, the need for effective on-the-ground implementation capacity is recognized as an important constraint in many areas, both for the government agencies and the nongovernmental organizations intended to be the main project implementation agencies. This constraint is also reflected in the quality of watershed implementation, which at its best can be a flexible and empowering process that can transform the livelihoods and resource base of poor communities. In particular, the guideline figures for both the size (500 ha) of a watershed as well as the amount spent per hectare have been rigidly applied in many places, regardless of the local needs or conditions. In some cases, the actual implementation is far from satisfactory, with little effort to engage the local communities or implement appropriate interventions to an adequate quality.

Although there are exceptions, much of the WSD is concentrated on physical interventions, such as contour bunding and check dams, which are intended to improve groundwater recharge and reduce land and soil degradation. These physical interventions are often not balanced against nonstructural measures, measures to improve the production process, or measures to open up new livelihood opportunities. Such measures include policy changes that bring in cropping pattern shifts and changes in livelihood patterns. The need to widen the scope of activities in watershed programs is reflected in the "watershed plus" approach in which a wider range of interventions is considered. However, this is still recognized as limited and there are active strategies to develop the approach further. The development of a process to widen the scope of possible interventions and make them more effective in local conditions is critical for further evolution of the watershed policy.

Equity is seen as a major policy issue, since the watershed programs in the past have often failed to reach the poorest households and have disproportionately benefited the better sections of the community. This is reflected, above all, in the pattern of expenditure on different activities in watershed programs, where an estimated 70% of the funds were used for land and water management interventions that predominantly benefit larger farmers, while only 7.5% was used to support the livelihoods of the poor and landless families [9]. Hence, there is a need to more effectively target the needs and potential of the landless and land-poor (especially those with rainfed lands on upper slopes) families and of women if WSD is truly to become the catalyst for a wider process of local-level development and poverty reduction.

Several approaches are considered for addressing the equity issues, although these approaches are mainly technical in nature and hence their impact is limited to physical coverage rather than actual benefits. Moreover, the focus of these approaches continues to be on landed households. One such approach is the "ridge to valley" treatment of the watershed area. This approach gives preference to small and marginal farmers who are located on the degraded slopes of the higher reaches of the watershed. Another approach is to treat the entire land in the village rather than restricting it to 500 ha. This approach facilitates the coverage of all the sections of the landed households and ensures better participation and cooperation. Also, focusing on the landless households in the community through initiating specific programs for them is crucial for enhanced livelihoods of the poor. These aspects are incorporated at the national level in the recently initiated IWMP.

There are also concerns over the sustainability of many of the interventions and the benefits gained, with the mid-term appraisal of the Ninth Plan program by the Planning Commission [13] stating that, for watersheds surveyed in Maharashtra and Andhra Pradesh, "increase in agricultural production did not last for more than two years. Structures were abandoned because of lack of maintenance and there was no mechanism for looking after common lands." This situation has not changed much, even during the Eleventh Plan. This picture is verified by other studies, with the root cause of poor sustainability seen as a failure to engage effectively local people in the process. The issue of participation gets complicated under IWMP as the size of the watershed increased 10-fold, with the transcending number of villages. At the same time, the size of the watershed (5000 ha and above) facilitates the integration of hydrogeological and biophysical aspects that can facilitate sustainable watershed management. However, these aspects of scale have not received due attention at either the research or policy level.

1.4 WSD AND IMPORTANCE OF SCALE

The watershed management approach has evolved to deal with the complex challenges of natural resource management by adopting watershed as an appropriate unit of implementation. Watersheds consist of areas of any size, because small watersheds are part of large watersheds that can be located within larger watersheds up to entire river basins. The size of watersheds ranges from 2 ha [14] to 30,000 ha [15]. Although a watershed can be defined at different levels, international practice reveals that the micro-watershed is usually the chosen scale of implementation for watershed management. This scale facilitates a program to act in response to human needs and natural resource problems at the local level.

Watershed management at the micro level is ecologically and institutionally sustainable as well as capable, under the right conditions, of empowering vulnerable segments of the society [16]. The micro-watershed approach enables amicable integration of land, water, and infrastructure development, particularly because of the homogenous nature of soil, water, and overall physical conditions within the micro-watershed. Theory and experience have shown that facilitating collective action in small, village-level watersheds has fewer constraints compared to more complex communities. Moreover, organizing collective action at the micro-watershed level generally proved to result in lower costs and in improved use of financial and human resources, particularly for the management of common resources.

The recent generation of watershed management projects has been mostly successful in its integrated and participatory approach to sustainable conservation and development in upstream areas. This has provided some impetus for scaling up. However, the micro-watershed approach encounters adversity when it comes to scaling up. Operating at the micro-watershed scale does not necessarily aggregate up or capture upstream/downstream interactions. A mix of upstream interventions would only have a considerable impact downstream if prioritized and planned within the larger watershed perspective and with understanding of the spatial and hydrological links between the perceived externalities and their underlying factors (for example, land and water use).

Watershed management projects are generally anticipated to provide local on-site benefits at the micro-watershed level and to offer positive externalities in the form of valuable environmental services downstream as well as to provide a way to correct downstream negative externalities within the larger watershed. Therefore, investment in services upstream cannot be justified by their on-site benefits alone and can only pass economic reasoning when downstream benefits are embodied.

However, watershed management programs have usually paid attention only to on-site interventions and their benefits. Whether these actions also benefited the downstream locations or were the best possible approach to minimizing negative externalities was often not ascertained. Similarly, stakeholder involvement and participation normally covered on-site requirements of local farmers, and the spatial dimension was tackled through community-based planning of their region. The institutional approach only focused on the micro-watershed, with limited or no cooperation across the watersheds or between upstream and downstream populations.

The success of the project was assessed on-site, and the individual-level outcomes (income increase, land area treated, and yield increase) were in general aggregated across the watershed area. There is hardly any evidence that can prove the improved conditions in the wider watershed results as a consequence of micro-level activities and institutions at the upstream level, or even that the activities were optimal or cost-effective ways to improve conditions in the watershed.

Despite their apparent objective of improving natural resource conditions in a watershed, the WSD programs may prove detrimental to downstream areas. Research has revealed that the micro-watershed approach may be producing hydrological problems that would be best addressed by operating at a macro-watershed scale. For example, in India, recent hydrological research cautions that watershed projects may be aggravating the very water scarcity they intend to overcome. The study by Batchelor et al. [10] noticed that successful water harvesting in upper watersheds came at the expense of lower watershed areas. On the basis of the data from the macro-watershed level (covering many villages), they documented cases where water harvesting in upper watersheds reduced water availability downstream. With the worsening of the groundwater table downstream, more intensive drilling of wells is needed, which the poor often cannot afford, leading to inequitable distribution and use of water [17]. Calder et al. [18] cited this as "catchment closure," whereby water harvesting upstream accumulates groundwater locally and then intensive pumping depletes the shallow aquifer. In this case, WSD checks the movements of both surface runoff and groundwater toward downstream locations. This indicates two adverse project outcomes: (1) what is good for one micro-watershed can be bad for others in the downstream locations and (2) what is good for a watershed in the short term can be bad in the long term. Thus, while addressing socioeconomic considerations favors small micro-watersheds as the unit of operation, approaching this hydrological problem calls for working in large macro-watersheds, and the two may be inconsistent.

From the biophysical context, it is observed that as the size of the watershed increases, the influence of land use on the upstream/downstream hydrology reduces, while the influence of precipitation increases [19].

While successful watershed projects have overcome the inherent constraints to collective action, they have not contravened two outstanding barriers: (1) projects with high investment in social organization may not be replicable beyond a small number of cases and (2) operating on the basis of a feasible social unit (a village micro-watershed instead of a macro-watershed that crosses administrative boundaries) trades one set of problems for another. This would involve working simultaneously to promote watershed governance capacity both within and between micro-watersheds. However, research implies potentially severe trade-offs between these two approaches.

Resolving these trade-offs is necessary for the widespread success of the WSD program, but no obvious solutions exist. The difficulty of managing watershed interventions at diverse scales to achieve the larger scale objectives of downstream impacts is further complicated because of participatory approaches, which basically give the option of interventions to the communities rather than to the planners.

One of the most important characteristics of watershed management is the ability to improve the management of externalities, which generally emerge because of land and water interactions. There exist a number of approaches to "internalizing externalities"—compensating those who generate positive externalities and taxing those who cause negative ones. These approaches include attaching the adoption of conservation practices to other benefits, such as access to credit [20], and practices like cost sharing—full subsidy to the cost of adoption or partial subsidy. Investment subsidies, particularly cost sharing, have been the most frequently applied procedures.

However, one study on the Indian experience observed investment subsidies to be the least effective mechanism [21]. Experience also suggests that subsidies, if not sustained, do not realize long-term changes in conservation practices. It is observed that once the projects end and the subsidies cease, land users often resume their previous land uses disregarding the conservation measures they had adopted; they may even actively destroy them [22].

To avoid the apparent problems of a "compensation" approach, watershed management programs resort to a variety of nonfinancial approaches to persuade stakeholders to adopt the recommended conservation practices. While some have recommended alternative income-generation activities to compensate for lost income because of conservation practices, some have relied on a hoped-for "demonstration effect"—assuming that conservation practices would eventually demonstrate their usefulness to stakeholders who would then approve them once their benefits had been established [20]. Others have used approaches such as awareness generation, moral suasion, and regulatory limits and fines. Generally, these approaches have not proved to be effective [23,24]. While the alternative income-generating activities approach has had mixed results with the demonstration effect often failing because the assumption that conservation practices were lucrative to upland stakeholders was often not the case, regulatory approaches are often very difficult to implement and may entail high costs on poor land users by forcing them to adopt land uses that generate lower returns.

Apart from the previously mentioned approaches, market-based contracting approaches—payment for environmental services (PES)—have also been used in some cases, particularly, in Latin America in small-scale initiatives involving water services; several countries are already experimenting with such systems [25]. The basic principle behind such approaches is that those who supply environmental services should be compensated for their service, and that those who receive the services should pay for their provision. This approach has the added advantage of providing supplementary income sources for poor upstream land users, thus helping them to improve their livelihoods. However, in most cases, although a PES approach is apparently attractive, putting it into practice is far from simple; therefore, application of these approaches requires the presence of several building blocks [26].

It could be argued that upstream/downstream hydrological relationships within watersheds are just externalities and can be managed through approaches such as Coasian bargaining and command-and-control or taxes and subsidies. However, in the context of a developing country characterized by dense population and small holdings, approaches such as command-and-control, taxes, and land use restrictions are not viable [27]. Coasian bargaining is seldom feasible because of the high transaction costs of generating and implementing agreement among various dispersed actors.

Harnessing upstream activities to management objectives at the broader watershed level is obviously a major challenge, as upstream/downstream linkages are multifaceted and the information essential to understanding the interactions has until recently proved to be complicated and costly to accumulate. However, development of dynamic modeling at the basin level, coupled with more affordable monitoring tools such as remote sensing, allows for enhanced understanding of watershed properties better capable of defining upstream/downstream relations, functions, and management impacts. Moreover, if watershed management is to be justified by its beneficial impact on the downstream environment, institutional arrangement is needed to endorse interaction among the micro-watershed groups within a large macro-watershed, and to determine and monitor outcomes and impacts. This could involve specific mechanisms to facilitate the interaction such as a new legislation or new arrangements for sharing upstream/downstream costs and benefits.

A variety of institutional mechanisms exist, ranging from simply maintaining an information system to identifying externalities, through the formation of platforms for dialog between upstream and downstream communities, to building higher level watershed planning institutions. Preferably, the institutional framework should be capable of incorporating the micro-watershed management plans into the broader scale of the watershed as a whole. This would involve developing something like a "nested platforms" approach at the macro-watershed scale.

An important question regarding the trade-off between operating at an optimal hydrological unit versus an optimal social unit is its severity. During the early days of watershed projects, disregarding the optimal social unit resulted in the failure of the projects as they could not accomplish effective watershed governance. Of late, the pendulum has swung in the opposite direction and now, most projects operate at the village level, disregarding hydrological linkages between micro-watersheds. Catchment closure has appeared in part by overlooking these linkages, and it illustrates the need to deal with them by working at a meso-watershed scale.

1.5 NEED FOR AN INTEGRATED APPROACH

Thus, we see that the increased scale of watersheds has its advantages as well as disadvantages as far as the effectiveness of the program is concerned. As discussed previously, the IWMP at the 5000 ha scale should help internalize the externalities associated with hydrogeological and biophysical aspects. On the other hand, it could hinder the institutional aspects pertaining to collective strategies. Hence, it is necessary to assess the impacts of watershed interventions using an integrated approach.

In this book, the integration is mainly in terms of biophysical and socio-economic models (Figure 1.1) of the watersheds at a scale of 5000 ha and

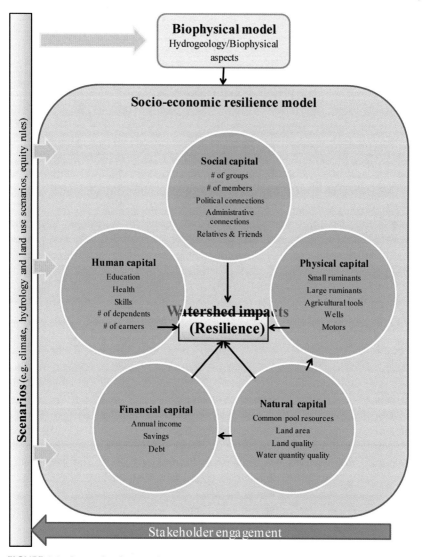

FIGURE 1.1   Integration framework.

above. The biophysical model consists of hydrogeology, rainfall, soil type, and land use, while the socioeconomic model incorporates household resilience in relation to its livelihood capitals. The integrated model provides insight into the interactions between the hydrogeological and biophysical aspects of a watershed and the resulting influence on how the quality and quantity of the watershed impacts the livelihoods of the local communities. It also explores the potential of WSD in the context of increasing climate variability as a mitigation or adaptation strategy for improved resilience of the farming communities.

The model highlights the importance of understanding these complex interactions specifically in the context of scale and their importance in achieving not only sustainable soil and water management, but also economic and livelihood outcomes. The model also simulates different scenarios such as climate (rainfall), hydrology, and land use. Further, it identifies the attendant equity issues and the need for stakeholder engagement at all levels if the integrative approach is to be useful for evaluating meaningful alternative IWMP programs at different levels.

The integrated model is primarily driven by the socioeconomic model. Within this model, watershed impacts are assessed in terms of household resilience to changes in climate, especially droughts. The level or degree of resilience (number of droughts a household can withstand) varies across households. The degree of household resilience is linked to the household's assets and capabilities. The SL (five capitals) framework is used to assess the household assets and capabilities.

The biophysical model influences household assets and capabilities through its natural capital, especially the quantity and quality of water and land. Biophysical attributes, including hydrogeology, rainfall, and soil type, are exogenous or given to the household and need to be taken into account while assessing the watershed impacts. These attributes are critical in determining the extent of impacts, and should be considered while designing interventions to optimize the impacts. Of these, hydrogeology and soil type are highly variable and instrumental in creating inequity in access to resources, assets, and capability. However, some households could substitute the lacuna in these attributes with other capabilities (capitals) such as human or social capital to enhance their resilience.

The integrated model is based on the research that has adopted a clear analytical framework and scientific approach for assessing the watershed impacts. The aim of this approach is to provide design inputs for sustainable watershed interventions that enhance livelihood outcomes. The biophysical model uses appropriate modeling techniques that include groundwater, surface—subsurface water modeling, and land use modeling. These models are used to arrive at appropriate watershed intervention designs that are location specific. The nature and density of the interventions are determined by exogenous factors including rainfall, soil quality, slope, aquifer structure, and land use (forests, wastelands, etc.).

The cropping pattern in a specific area influences the groundwater use and balance. Crop patterns are sustainable when crops are grown according to these biophysical attributes—when crops are chosen according to the soil type and groundwater potential (sustainable groundwater yields), this is called a sustainable crop pattern.

Community livelihoods are determined by the biophysical potential of the region that can support farm systems. While agricultural or farm systems could enhance livelihoods in terms of financial capital, there are other forms of household assets and capabilities (human, physical, and social) that could do the same. Watershed interventions might directly or indirectly influence these capitals.

Hence, watershed impact assessments should look beyond natural and financial capital on which watershed has a direct bearing. The socioeconomic model adopted here looks at the five capitals and the capabilities of the household, along with a number of indicators of these five capitals, including the biophysical aspects, to explain the variations in watershed impacts (resilience) between upstream/downstream and control situations.

Equity is assessed in terms of horizontal and vertical distribution of benefits. Horizontal equity is assessed by comparing upstream/downstream impacts, while vertical equity is assessed in terms of distribution of benefits within upstream/downstream locations. The integrated model helps in assessing whether the distribution of benefits is optimum, given the biophysical attributes of the specific location. This model helps in arriving at alternative and appropriate design interventions that could optimize the benefits. Equity would be optimum when benefits are maximized across locations. Maximal equity is not necessarily the absolute equity, which is ideal and desirable, and appropriate policies (compensation, subsidies, incentives, payments for environmental services, etc.) could help improve equity to a large extent.

While the integrated model is built using the actual data generated at different levels, viz., village, household, etc., it is also capable of developing alternative scenarios that pertain to climate change predications, groundwater (hydrology), land use changes, and so on. Generation of these scenarios is based on the perceptions of various stakeholders and implementing agencies through the stakeholder engagement process.

1.6 ABOUT THIS BOOK

This volume is a collection of research work performed during the last 5 years, following an integrated approach in assessing watershed impacts at scale. The chapters explore the generality of the approach taken in Andhra Pradesh, India, with a comparative case study in Australia where sustainable groundwater management is an issue within a catchment or hydrological unit context. In particular, the relevance of the livelihood approach and the concept of resilience in a country with differing socioeconomic and demographic conditions and alternative water and land management institutions are examined. This volume is organized into 13 chapters (including an introduction and conclusions).

The multidisciplinary nature and the integrated approach adopted in the research include various methods and tools cutting across hydrology, biophysical, and socioeconomic methodologies. An overview of the overall framework of hydrogeological modeling, biophysical modeling, and socioeconomic analyses in relation to the chapters is provided in the following paragraphs. A discussion on this framework and other details, such as the sampling design and profile of sample sites, are presented in Chapter 2.

Groundwater is the dominant source of irrigation in the rainfed regions and plays a key role in the socioeconomic development in agrarian economies. Since groundwater remains hidden in a complex system of rocks, its precise assessment is difficult and has resulted in a large mismatch between groundwater demand and availability. While watershed interventions are expected to improve groundwater recharge through better soil and water conservation practices, the actual availability of groundwater for final use depends on the suitability of interventions to the aquifer system.

In the context of watershed interventions, it is often presumed that groundwater recharge improves as one moves from upstream to downstream locations. However, these observations are not based on scientific information on aquifers and drainage systems and often efforts on WSD go waste. Thus, hydrogeological investigation through geophysical methods provides a clear link to the socioeconomics since a precise knowledge of the subsurface is helpful in two ways: ensuring efficacy and suitability of the type of WSD and to plan for its sustainable use. Due to the high variability of the system parameters as well as the complex heterogeneous nature of the system, new and sophisticated techniques of geophysical logging and electrical resistivity imaging have been deployed to characterize the system. Furthermore, differences in the nature and type of aquifers across the locations within a hydrological unit could result in contradictory evidence; in the absence of such scientific information, watershed impacts have been attributed to physical interventions. Hence, there is a need to understand the role of hydrogeology in terms of water resource potential and its use in the context of watershed interventions (Chapter 3).

Coupled hydrological modeling emerges as an increasingly important pursuit in water-scarce environments where decision making must consider the limited availability of surface water and groundwater in an integrated manner to evaluate the trade-offs that emerge under alternative development scenarios. In the drought-prone, rainfed regions of India, it has become increasingly apparent that programs implemented on small scales do not always have the intended hydrological impacts and the issue of the optimal scale has been seriously questioned. Hence, an integrated hydrologic model was developed for this study. The simulations for current and potential WSD interventions indicated strong adverse effects on the availability of water resources in the downstream areas in spite of improved local water usage for agriculture and livelihood (Chapter 4).

Hydrological modeling forms the backbone of an integrated model investigating water resources. In such a situation, a hydrological model needs to be an appropriately simplified representation of a catchment, providing the necessary inputs to enable an assessment of impacts on water resources from key drivers (e.g., policy options, institutions, climate change, and land use). Based on an intensively studied catchment in West Bengal, an appropriate hydrological model has been built (Chapter 5) that not only captures the impact of WSD on surface water storage, but also captures the subsequent impact on recharge as well as surface and groundwater use. The model runs on a catchment scale, with upstream/downstream impacts of WSD work investigated through a surface—subsurface routing model. The spatial scale is driven by the requirements of the integrated model, and the temporal scale is driven mostly by the resolution of the key datasets—most importantly rainfall, which is generally available on a daily timescale.

A key ensuring sustainable WSD lies in choosing the optimum watershed interventions by considering both the existing requirements of ecosystems and agricultural systems as well as the available water infrastructure. For this, the biophysical resources such as rainfall, soil type, slope, and land use, along with topography and aquifer characteristics, play an important role. While watershed interventions are based on the biophysical resources existing in the watershed, they also influence the availability of surface water flows and groundwater in a spatial context; any excess diversion leading to changes in the available water downstream may cause a conflict between the upstream and downstream users. Such upstream/downstream conflicts have been demonstrated in large-scale water storage systems (Chapter 6). Any evidence of similar impacts of watershed interventions (if any) was thoroughly scrutinized and assessed prior to this study. Watershed interventions in terms of nature and density need to be planned keeping these issues in mind without causing large changes in the existing hydrological system. Furthermore, appropriate designing helps improve the water-holding capacity of the soil as well as the water storage capacity of the existing and new structures. Nevertheless, the new interventions need to be in harmony with the existing storage structures.

Evaluation of WSD, which has the inherent potential to enhance the resilience of the system, needs to integrate hydrogeological and biophysical attributes of impact assessment at scale, since resilience is defined as the perceived ability to cope with drought in the future. This perception is examined along with retrospective thoughts about change in the absence of baseline data. Resilience is directly linked to average rainfall and downstream locations and is related to the socioeconomic position of the households. The significance of these relationships to water reallocation is examined in Chapter 7.

As far as the factors influencing resilience or drought survival are concerned, education and/or number of earning members in the household are observed to have a positive influence. In addition, households with better health rely more on government intervention programs such as the Mahatma Gandhi National Rural Employment Guarantee Act for drought survival (Chapter 8). Similarly, households with better health could draw more from common property resources for their income. This implies that health plays an important role in accessing livelihood opportunities outside farming and hence, drought survival. Further, it is speculated that watershed programs may also have led to increased inequality in these regions. Finally, when compared to untreated regions, watershed-treated areas show improvement in drought resilience in some regions, while no improvement was observed in others.

The BN methodology has been used in the mesoscale project to develop a socioeconomic model that relates the stocks of indicators for financial, human, natural, physical, and social capital, reported by households, to their capacity to survive consecutive droughts (resilience). BNs are probabilistic modeling approaches that have garnered popularity in the field of environmental modeling because they are well-suited to representing relationships between the biophysical and societal factors critical to the success of natural resource management programs. The development process used to construct the component capitals and resilience BNs is outlined in Chapter 9, followed by a demonstration of the model behavior and performance.

Watershed intervention policies are likely to lead to different social outcomes and formulations of policies. For example, in Australia water reforms began in the 1990s with the primary goal of environmental protection. However, as the program developed, environmental protection was seen in the separation of land and water resources, the introduction of concrete water entitlement policies, and in the introduction of markets. Further, social goals were muted and largely assessed in terms of the Western social impact methodology, which had the underlying assumption that there were no unacceptable social impacts. This approach has led to a community-wide discussion regarding the "rights" of irrigators vis-à-vis other interests and the presentation of a variety of equity and ethical arguments. However, these arguments have become confusing as water allocation issues have moved from local to state arenas.

In contrast, there has been a clear enunciation of social goals for Indian watershed interventions and concern for equity issues in terms of the distribution of benefits from WSD. Issues such as property rights and the role of markets, which have been so important in Australia, have been less evident in India. Further, different underlying issues associated with karma also exist in India but not in Australia. The social, ethical, and equity issues, as in Australia, have also been shown to change when the scale of intervention is considered. The empirical results of the case studies in Andhra Pradesh and the findings of a comparative study in two sites in Australia (Southern and Western Australia) are used to examine how the different social and equity premises of the two countries could lead to different outcomes (Chapter 10). The study also examines the issues regarding whether or not the move toward property rights and markets evidenced in several developed countries are the inevitable, and the most successful approach for all countries concerned with improving the public good nature of watershed interventions.

As the approach is new, sharing experiences with different stakeholders, from the field to the implementing, monitoring, and policy-making agencies, is crucial for effective policy formulations. Experience shows that the main hurdle in applying the scientific findings to the existing practices comes from the lack of sustained and continuous dialog between the scientific community and the policy makers or implementers. As a result, when scientific evidence is presented as an end product of long-drawn research work, it either loses practical relevance because of the time lag or is considered impractical because it lacks insights from the practitioners. A continuous dialog with the relevant stakeholders, on the other hand, ensures that the research not only sustains its relevance but also becomes practicable through understanding the constraints and practical issues faced at various levels of policy making and implementation. The key objective of an effective stakeholder engagement process should be to link the policy and practice to science through an empathetic appreciation of policy limitations and an acknowledgement of practical problems.

Most impact studies have a limited scope when evaluating the socioeconomic benefits, instead of taking an integrated approach to evaluating the socioeconomic impacts in the given context of hydrogeological and biophysical aspects of a given watershed. As a result, they also fail to highlight the importance of adopting a scientific approach to the entire process of selecting, designing, and implementing a watershed intervention. Thus, there is very little understanding among the different stakeholders regarding the need for this integrated approach; as the design, implementation, and impact assessments take place at higher levels, it is very important to create awareness in this direction. Likewise, it is even more important to create such awareness at the community level because the community needs to be convinced that the design and implementation aspects are beneficial. Moreover, balancing of the equity and efficiency aspects cannot be achieved without conscious collaboration among different socioeconomic groups. Thus, equal emphasis needs to be put on educating, creating awareness, and getting a buy-in for the integrated approach suggested by the study from stakeholders at the community level, in implementing agencies, and in government departments. It is believed that an educated participant will have a better commitment to sustain and maintain the project infrastructure and impacts. The process adopted, the lessons learned, and the next steps planned in the process of engaging the different stakeholders provide insights for future stakeholder engagement processes (Chapter 11).

Integrated modeling methodologies have greater potential compared with purely disciplinary approaches to support comprehensive assessment of social, economic, and biophysical aspects of complex natural resource issues such as the IWMP. Climate and recharge estimates drive predictions and assessment of the availability of surface and groundwater resources as impacted by IWMP, climate, and land use (i.e., water extractions). Scenario analysis of the likely impacts of climate, land use, and IWMP or other policy interventions on surface and groundwater resources, agricultural productivity, and people's livelihoods and resilience is performed using the results from different models (Chapter 12). Examples of biophysical scenarios and social policy scenarios are also used to demonstrate the value of the integrated and disciplinary models for assessing IWMP and other impacts on water resources and resilience. Chapter 13 provides some concluding remarks.

REFERENCES

[1] World Bank. World Development Report 2008. Washington DC: World Bank; 2008.

[2] Reddy V, Ratna. M, Reddy Gopinath, Soussan John. Political Economy of Watershed Management: Policies, Institutions, Implementation and Livelihoods. Jaipur: Rawat Publishers; 2010.

[3] Joshi PK, Jha AK, Wani SP, Joshi L, Shiyani RL. Meta-Analysis to Assess Impact of Watershed Programme and People's Participation, Comprehensive Assessment Research Report No.8. Andhra Pradesh, India: International Crops Research Institute for the Semi-Arid Tropics (ICRISAT), Patancheru 502 324; 2005.

[4] Reddy V, Ratna M, Gopinath Reddy, Galab S, Soussan John, Springate-Briganski Oliver. Participatory Watershed Development in India: Can it Sustain Rural Livelihoods? Dev Change April 2004;35(2):297−326.

[5] Reddy V, Ratna M, Gopinath Reddy YV, Reddy Malla, Soussan J. Sustaining Rural Livelihoods in Fragile Environments: Resource Endowments and Policy Interventions-A Study in the Context of Participatory Watershed Development in Andhra Pradesh. Indian J Agric Econ 2008;63(2):169−87.

[6] Reddy V, Ratna. Watershed Development for Sustainable Agricultural: Need for an Institutional Approach. Econ Political wkly 2000;35(38):3435−44.

[7] Pingali PL, Rosegrant M. Intensive food systems in Asia: Can the degradation problems be resolved? In: Lee DR, Barrett CB, editors. Tradeoffs or synergies? Agricultural intensification, economic development and the environment. CABI publishing; 2001.

[8] Fan S, Hazell P. Should developing countries invest more in less favoured areas? An empirical analysis of rural India. Econ political wkly 2000;35(17):1455−64.

[9] GoI. Approach Paper to the Tenth Five Year Plan (2002-07). New Delhi: Planning Commission; September 2001.

[10] Batchelor CAK, Singh CH, Rama Mohan Rao, Butterworth C. Watershed Development: a Solution to Water Shortages or Part of the Problem? Land Use Water Resources Res 2003;3:1−10.

[11] GoI. Report of the Working Group on Natural Resource Management: Eleventh Five Year Plan (2007-2011), vol I. New Delhi, India: Planning Commission, Government of India; Synthesis 2007.

[12] GoI. From Haryali to Neeranchal: Report of the Technical committee on Watershed Programmes in India, Department of Land Resources. Ministry of Rural Development, Government of India; January 2006.

[13] GoI. Mid-Term appraisal of the Ninth Plan (1997−2002)- Chapter 10. New Delhi: Planning Commission, Government of India; October 2000.

[14] White TA, Runge CF. The Emergence and Evolution of Collective Action: Lessons from Watershed Management in Haiti. World Dev 1995;23(10):1683−98.

[15] World Bank. Watershed Management Approaches, Policies and Operations: Lessons For Scaling-Up (draft report). Washington, DC: Agriculture and Rural Development Department, World Bank; 2007.

[16] Farrington J, Turton C, James AJ. Participatory Watershed Development: Challenges for the Twenty-First Century. New Delhi, India: Oxford University Press; 1999.

[17] Calder IR. Blue Revolution, Integrated Land and Water Resource Management. 2nd ed. VA: Earthscan: London and Sterling; 2005.

[18] Calder I, Gosain A, Rama Mohan Rao MS, Batchelor C, Snehlatha M, Bishop E. Planning Rainwater Harvesting in India − 1) Biophysical and Societal Impacts. In: National Workshop on Priorities for Watershed Management in India. Bangalore, 22-23 May, Government of Karnataka and the World Bank; 2006.

[19] FAO. The new generation of watershed management projects. Rome, Italy: FAO Forestry Paper 150; 2006. http://www.fao.org/docrep/009/a0644e/a0644e00.htm.

[20] Pagiola S. Paying for Water Services in Central America: Learning from Costa Rica. In: Pagiola Bishop, Landell-Mills, editors. Selling Forest Environmental Services. London: Earthscan: Market-Based Mechanisms for Conservation and Development; 2002.

[21] Kerr J, Milne G, Chhotray V, Baumann P, James AJ. Managing Watershed Externalities in India: Theory and Practice. Environ, Dev Sustainability 2007;9(3).

[22] Lutz E, Pagiola S, Reiche C. Cost-benefit analysis of soil conservation: The farmers' viewpoint. World Bank Res Obs 1994:273−95.

[23] Enters T. The token line: adoption and non-adoption of soil conservation practices in the highlands of northern Thailand. In: Sombatpanit S, Zöbisch MA, Sanders DW, Cook MG, editors. Soil Conservation Extension: From Concepts to Adoption. Enfield, New Hampshire: Soil Publishers; 1997. pp. 417−27.

[24] Pagiola S. Economic Analysis of Incentives for Soil Conservation. In: Sanders DW, Huszar PC, Sombatpanit S, Enters T, editors. Incentives in Soil Conservation: From Theory to Practice. New Delhi: Oxford and IBH Publishing Co. Pvt. Ltd.; 1999. pp. 41−56.

[25] Landell-Mills N, Porras IT. Silver Bullet or Fools' Gold? A Global Review of Markets for Forest Environmental Services and Their Impact on the Poor. A component of the international collaborative research project steered by IIED. London: IIED: Instruments for Sustainable Private Sector Forestry; 2002.

[26] Pagiola S, Platais G. Payments for Environmental Services: From Theory to Practice; 2006. Draft, June 12.

[27] Pagiola S, Bishop J, Landell-Mills N, editors. Selling Forest Environmental Services. London: Earthscan: Market-based Mechanisms for Conservation and Development; 2002.

Chapter 2

Analytical Framework, Study Design, and Methodology

Geoffrey J. Syme *, V. Ratna Reddy [§], Shakeel Ahmed [¶], K.V. Rao [‖],
Paul Pevalic [#], Wendy Merritt ** and T. Chiranjeevi [§]
* Edith Cowan University, Perth, Australia, [§] Livelihoods and Natural Resource Management
Institute, Hyderabad, India, [¶] CSIR-National Geophysical Institute, Hyderabad, India, [‖] Central
Research Institute for Dryland, Agriculture, Hyderabad, India, [#] International Water management
Institute, Vientiane, Lao PDR, ** Fenner School of Environment and Society, The Australian
National University, Canberra, ACT, Australia

Chapter Outline

Integrated Assessment of Scale Impacts of Watershed Intervention
http://dx.doi.org/10.1016/B978-0-12-800067-0.00002-5. Copyright © 2015 Elsevier Inc. All rights reserved.

23

2.1 INTRODUCTION

As most of the chapters in this book are based on the research that has adopted a clear analytical framework and scientific approach for assessing the watershed impacts, it is necessary to illustrate the aims of this research, framework, and approach before going into the analytical details of these chapters. The objective of this research is to provide design inputs for sustainable watershed interventions that enhance livelihood outcomes. The multidisciplinary nature and the integrated approach include various methods and tools used in hydrogeological, biophysical, and socioeconomic methodologies. While the specific details of the methodologies used are discussed in their respective chapters, this chapter provides the overall framework of the hydrogeology modeling, biophysical modeling, and "sustainable livelihoods." This chapter also provides the framework of the Bayesian network (BN) used for integrating these three aspects. Also, details such as the sampling design and profile of sample sites are also presented in this chapter.

The study design and methods evolved out of preliminary workshops and consultations with stakeholders and the community who have experienced watershed development (WSD) in rainfed areas. The major goal of the research was to establish what the issues would be if WSD or integrated water resources management was shifted from a micro- to a meso-level application. In the following sections, we discuss the methods for each component together with its strengths and weaknesses.

2.2 STRATEGIC CONCEPTUAL AND METHODOLOGICAL ISSUES

First, there has to be an understanding of the basic requirements for effective evaluation at the meso rather than micro level [1]. For the purpose of this research (Chapter 1) and for the design of mesoscale WSD, it was generally considered that hydrological legibility is required. That is, WSD application should be modeled on hydrological units (HUNs) in which there was a good possibility for relating ground and surface water flows to land use and the ability to model the effects of this to potential users throughout the sub-catchment. It was also concluded that there was a need for a relatively simple catchment model to assist decision makers in deciding the most beneficial pattern of meso-WSD for the sustainability of water management as a whole. Using the hydrological legibility, two HUNs with high coverage of WSD along with control villages were selected. The selection of the HUNs was the basis for the social, economic, and survey data collection. It must be noted that this constrained the power of generalization of the survey data collected, because a randomized whole of catchment data was not collected. Nevertheless, given that WSD evaluation needs to relate to the relevant HUN with its unique hydrology and land use, situational circumstances always have significance for the delivery of WSD whether in hydrological, economic, or societal terms.

Having selected two HUNs with contrasting rainfall pattern, we needed to select a methodological vehicle for integration and appropriate indicators of socioeconomic WSD outcomes. The research objectives also required an integration methodology that could accommodate both quantitative and qualitative data along with expert opinions with be applied to scenario evaluation analysis. Further, this approach also needed to be able to accommodate biophysical, social, and economic data for the evaluation of future possible WSD designs and modes of application and explore the cause-and-effect relationships between them. The preferred choice for this tool was the development and application of a BN approach.

The holistic concept of sustainable livelihoods (SL) was chosen as an approach to understanding a range of five "capitals" that constituted the overall well-being of the beneficiaries (see Chapter 1). These included all aspects of factors that are considered to be influential in governing the overall well-being of an individual, family, or community.

Finally, an output criterion was required to provide an overall estimate of whether the WSD was meeting its overall requirement of a socially cohesive and sustainable rainfed agriculture sector. The variable chosen for this purpose was resilience, which has been defined as the number of drought years a farmer could survive without having to leave cultivation. There are a number of theoretical formulations for the concept of resilience, and the reason for using this formulation in this project will be discussed in later sections.

2.3 ASSESSING SCALE IMPACTS OF WSD: AN ANALYTICAL FRAMEWORK

In their discussion on the issues of scale in relation to WSD, Syme et al. [1] showed that there was scope for applying a top-down, whole-of-catchment approach for strategically assessing the availability of water resources (in the form of surface water, soil water, and groundwater), and that it was already reserved for the various anthropogenic uses to identify allocation strategies at the subcatchment level. As part of that study, a "checkerboard hydrology" approach was devised to illustrate the types of impacts of alternative levels and distribution of WSD activities on water resources on the broader scale.

The ease of understanding the checkerboard makes it a well-suited tool for facilitating discussions with planners and policy makers regarding the benefits and trade-offs of different configurations of WSD. However, its gross simplicity makes it unsuitable for science-based planning; therefore, an improved approach was sought.

The key criteria that for in the model are summarized as follows:

Credibility: A process-based approach wins favor with policy makers wishing to promote scientifically based planning and implementation of WSD projects.

Simplicity: Complex models are accessible only to specialist modelers with an interest in scientific research, but are highly unlikely to be taken up by practitioners. Hence, there is a need to "bring the model to the users" in a form that is understandable and relevant. However, what the simple approach gains in terms of utility can be lost in terms of absolute accuracy. In the data-scarce conditions where WSDs are implemented, catchments are universally ungauged, with limited or no monitoring wells. Hence, the data required to support sophisticated approaches are not available.

Accessibility: Models should be available at no cost and must be run with the most basic computing requirements.

Our review showed the existing models did not meet these criteria. The closest we could identify was the Exploratory Climate Land Assessment and Impact Management (EXCLAIM) tool developed by the Centre for Land Use and Water Resources Research (Newcastle University, UK), as reported by Calder et al. [2]. EXCLAIM is a Java-based tool designed for nonspecialists and is used to indicate the range of outcomes and trade-offs associated with changes in land use within a catchment by incorporating climate, hydrology, land use, and socioeconomic variables. It has been applied to a range of problems such as rainwater harvesting and forestry. However, it does not account explicitly for watershed interventions, surface water–groundwater interactions, and groundwater use; hence it cannot be applied in this study.

Thus, the simple integrated hydrologic modeling approach was conceived and developed to assess water availability under alternative land use, climate, and WSD scenarios to create more effective and equitable WSD projects, as presented in detail in Chapter 4. The approach developed only addresses water-resource availability, which is seen as the most important biophysical constraint from the context of the Indian WSD. Hence, the need to incorporate other elements into the analysis, such as agricultural production and economic benefits, is recognized as a limitation in this model, which could be improved upon in the future. The tool, which is still under development, has been assessed against more complex models and data at two sites; efforts to make the tool more accessible to users are currently underway.

2.4 BIOPHYSICAL MODELING

2.4.1 Hydrological and Hydrogeological Methods

A detailed knowledge of subsurface aquifer geometry and its properties are equally important at the watershed scale for implementing the watershed management decisions. Therefore, geophysical and hydrogeological investigations were performed to decipher the aquifer geometry and its extent to understand the groundwater resources and select suitable sites for rain water harvesting.

Ultimate groundwater availability in space and time is important for the end user to decide the developments and maintain their socioeconomics. Hence, in the present study, electrical resistivity tomography (ERT) and electrical logging were performed to determine the aquifer geometry based on the geophysical signature along with aquifer resistivity properties.

2.4.2 Surface Electrical Geophysical Surveys

This method is based on the electrical property of the earth's subsurface. ERT was performed at a few points covering the whole watershed using the Wenner—Schlumberger configuration at 480 m spread length and employing 48 electrodes at 10 m interelectrode spacing. By injecting an appropriate DC current through two electrodes, electrical potential differences were measured using the other two electrodes. Thus using Ohm's law, the resistance, and ultimately, the apparent resistivity, was determined in 2D space. The inversion of the electrical measurements provided the distribution of the resistivity along the profiles, from the surface down to a depth of about 92 m. This depth of investigation primarily depends on the electrode spacing, strength of the current injected, and resistivity of the overburden, or the top formation. However, the resistivity distribution thus obtained in 2D space is constrained by the known values obtained from the drilling of the wells and geophysical logging.

2.4.3 Geophysical Electrical Resistivity Logging

Geophysical survey can be performed at various scales: the well-known electrical survey when performed using a bore well such that one or more electrodes are lowered into it measures the resistivity distribution in one dimension. The most commonly used electrode arrangement for such a survey is normal or potential sonde in which one current electrode and two potential electrodes are located on the sonde, while the other current electrode is kept on the surface. The curves obtained for potential or normal resistivity logs are symmetrical in form in which the maximum indicates a layer with higher resistivity and the minimum indicates a layer with lower resistivity. However, the information obtained thus is confined to a well scale only. Further, it is observed that logging provides more continuous data on the vertical and lateral distribution of the well section and depends on the sensitivity of the sondes. Hence, most of the resistivity logging surveys were performed close to the ERT sites to understand the geologic sequences and different lithological information.

2.4.4 Lithologically Constrained Rainfall Method

Quantitative estimates of recharge to aquifer and changes in groundwater storage are important to manage the development of groundwater resources and determine the amount of groundwater that can be withdrawn without exceeding recharge. In hard rock areas, the most common methods for recharge estimation are groundwater balance, water table fluctuation, soil water balance, and chloride mass balance [3−7]. However, these methods require analysis of a huge volume of hydrological data such as precipitation, surface runoff, evaporation, and change in groundwater storage accumulated over a considerable time span, which is generally either inadequate or lacking/unreliable in many areas [8].

Hence, for this study, the lithologically constrained rainfall method was adopted to estimate the natural recharge in the study area [9]. This method needs three input parameters, i.e., soil resistivity (ρ_s), vadose zone thickness (H), and precipitation (P). Since lithological alterations take place very slowly in the geological timescale, they can be considered as almost constant (say for ±50 years). Hence, rainfall is the only parameter varying with time for the study period.

The advantages of this method include a reasonably good estimate with the input parameters, which can be obtained easily in the field with good accuracy, lesser time frame, and in a cost-effective manner. The rainfall data were collected from the adjacent rain gauge (RG) stations in and around the watersheds, soil resistivity was obtained using geophysical methods, and the water levels were directly measured in the study area.

2.4.5 Change in Groundwater Storage (ΔS)

Estimation of the value of ΔS is a basic prerequisite for efficient groundwater resource management. It is particularly important in regions with large demands for groundwater, where such resources are key to economic development. The value of ΔS here describes the volumetric loss/gain of groundwater from the aquifer system between two time periods. This value is assessed by multiplying the difference in groundwater levels for the two corresponding monitoring periods with the specific yield of the formation and the area overlying the groundwater basin; estimation of aquifer water storage variability is of great importance for the management of water resources.

2.4.6 Depth of Water Level

Groundwater levels were monitored during pre- and post-monsoon seasons from 2010 to 2013 and the monthly water levels were monitored for the year 2013 (January to December) to understand the ground water fluctuation behavior and seasonal variations. The groundwater level data for the period from 2005 to 2009 were collected from BIRD (nongovernmental organization; NGO) for understanding the long-term trend of water levels in the study areas;

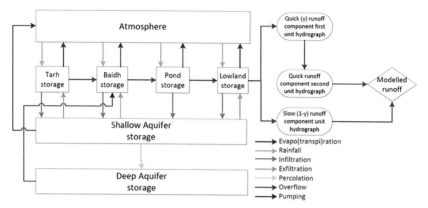

FIGURE 2.1   Structure of Andhra Pradesh model.

the rise and fall of the water table is observed to be a direct reflection of recharge and discharge conditions in the groundwater reservoir.

2.4.7 Modeling the Impact of Watershed Development on Water Resources

An intensive fieldwork campaign in the Purulia District of West Bengal has resulted in the development of a model designed to represent the impact of watershed development on a 2 km^2 catchment. This model has been adapted to be applied to larger scale catchments (of the order of 100 km^2) in Andhra Pradesh to investigate the upstream/downstream impacts of watershed development.

This model needed to be modified to include large in-stream dams. These large dams are distinct from the ponds used in the model developed for the West Bengal study site. Also, a deep aquifer has been added to the model. However, the climate in Andhra Pradesh is much dryer, with significantly less rainfall. Likewise, the shallow aquifer is much dryer and therefore the inhabitants pump water from the deep aquifer to irrigate their crops. The structure of the Andhra Pradesh model (Gooty site) is shown in Figure 2.1.

2.4.8 Strengths and Weaknesses

A spatial interpretation of the area resulted in a change of the model structure and the calculation sequence. Furthermore, some model processes have been changed because they were not included in the original model or were causing problems in the output generated by the model.

It is observed that the assumptions made during this research have great influence on the results generated by the model. The simple linear calibrated

---

2. Special Commissioner for Watershed Development, Department of Rural Development, Government of Andhra Pradesh.

percolation and exfiltration processes significantly influence the modeled runoff—a small change in parameter values results in largely changed amounts of runoff generated by the model. Unfortunately, no additional information is available to implement and underpin a more complex approach, or to estimate an order of magnitude. Hence, the parameters encapsulating these processes are calibrated during the research and are the major drivers of the uncertainty in the defined model processes. Therefore, further investigation during additional research to this catchment is highly recommended (see Section 7.3).

Nevertheless, the simple structure and processes of the model, based on visual interpretations of the catchment and study site using Google Earth and applying a simple approach of unknown model processes, gave a better representation of the catchment's hydrology compared with the original model.

2.4.9 WSD Design Methodology: Strengths and Weaknesses

The methodology for watershed assessment with and without interventions followed in the project includes a detailed rainfall assessment, resource conservation due to interventions (at on-stream and off-stream), and guidelines for the proper design of the watershed interventions. Hence, the data for this type of analysis must be detailed in terms of daily rainfall, temperature, land use information, and interventions made at the plot level. The data requirements are of a medium to high degree of complexity.

The methodology used for rainfall data collection is detailed and provides information on a monthly to annual scale on the quantum of rainfall, number of rainy days, etc., along with their variability, information about intense storms, and their contribution to the total rainfall in deficit, normal, and above normal years. Although the analysis is rigorous, it is simple and could be easily performed and interpreted using Microsoft Excel.

The methodology followed for watershed assessment includes a plot-level assessment for each land use and land parcel based on a water balance method, including runoff estimation, based on a soil moisture accounting process on a daily scale. Further, the intervention impacts are also assessed at each plot level by modifying the existing algorithm accounting for the augmentation of water within the plot on a daily scale. Although the algorithm requires daily data, it is considered to be essential to work on water balances in rainfed areas. This is a compromising methodology between the subdaily requirement of rainfall information needed by certain methods to the simpler methods with monthly runoff or for a 10 day interval.

The remaining datasets used such as the Digital Elevation Model or soil information are the publicly available domain datasets. These easily available datasets make use of the developed methodology by practitioners. When high-order resolution datasets are made available, the same could be used with this methodology. However, one of the lacunae in the methodology is the assessment of impacts on on-farm locations versus on-stream locations; it is observed that only in high rainfall areas do both on-farm and on-stream locations coexist and require an inclusion of on-stream interventions. Nevertheless, due to the net planning approach in watershed implementations, every land parcel is addressed for inclusion of watershed treatments and hence considered to be appropriate for inclusion of on-plot interventions.

The geographical information system (GIS) software used in the project is a commercial, open-source GIS system that is available with similar functionality.

2.5 ASSESSING SOCIOECONOMIC IMPACTS

Three different approaches are used to assess the socioeconomic impacts. A sustainable rural livelihoods (SRL) framework is used to assess the impact of WSD across locations. Resilience is used as an indicator of WSD impact and a resilience model is used to determine factors influencing household resilience. Finally, a BN model is adopted to integrate socioeconomic and hydro-geological and biophysical aspects.

2.5.1 SRLs Framework

The SRL approach is used widely as an analytical tool to facilitate poverty alleviation interventions. The recasting of households as the central focus for analysis helps prioritize interventions, which serve their developmental priorities. There are many different definitions of livelihoods. According to Carney [10] "a livelihood comprises the capabilities, assets (including both material and social resources) and activities required for a means of living. A livelihood is sustainable when it can cope with and recover from stresses and shocks and maintain or enhance its capabilities and assets both now and in the future, while not undermining the natural resource base" [11, p. 4].

In the aptly titled "Adaptable Livelihoods," Davies [11] provided a detailed understanding of the "dynamics" of the livelihoods of the poor in relation to food, as they respond to the highly variable conditions (natural as well as human) that confront them. Davies' conceptual framework is based on the following five key ideas, which can also be expanded to the broader issues of sustainable livelihoods:

- **Livelihood systems** and the security within them, encompassing a broader range of factors than household food systems and security to explain how and why producers pursue particular mixes of strategies to confront food insecurity
- **Entitlements** to explain different sources of food and the range of calls on them within the households and livelihood systems

- **Vulnerability** to explain the nature and intensity of food and livelihood insecurity
- **Resilience and sensitivity**, useful in analyzing changes in levels and intensity of vulnerability to food insecurity within different livelihood systems
- **Livelihood system diversity** to account for variation in the nature and intensity of vulnerability, depending on the different ways in which people acquire access to food [12, p. 15]

Rennie and Singh [12] provided an outline of the SRL approach for field project development. They stress that this should not be an esoteric exercise, but an analytically powerful contribution to policy for improving the position of the poor. They argue "livelihoods is a more tangible concept than 'development,' easier to discuss, observe, describe and even quantify" [13, p. 16]. They stress the importance of going beyond livelihoods at a conceptual level to identify robust research and implementation methodologies for field projects. They further argue: "Predominantly the poor of the world depend directly on natural resources, through cultivation, herding, collecting or hunting for their livelihoods. Therefore, for the livelihoods to be sustainable, the natural resources must be sustained" [13, p. 16].

Although not universal, this contention is undoubtedly true for many of the rural poor. Addressing the role of natural resources is therefore critical for any livelihoods model, and a watershed-centered approach of achieving SL and poverty alleviation is rather logical. In this context, particular attention may be paid to the issue of the sustainability and access the poor have to natural capital, as it is a key area not only in the lives of the poor but also in active policy development in rainfed regions.

This aspect of livelihoods cannot be considered in isolation; how access to and the use of natural capital is linked to other aspects of the livelihoods of the poor should also be included. We see that many policies concerning natural resources do not make these links, and focus instead on the management of the resources while excluding other issues (the same is true for policies concerned with other livelihood assets such as education or credit). The analysis of any one of these issues consequently needs to retain focus on the scope of the policy, as it exists while ensuring that it is in a context that allows linking to other aspects of livelihoods. Achieving this balance is one of the central goals of the model adopted here, which takes into account the basic dynamics of livelihoods, something that is inevitably complex, given the array of the factors that influence livelihood choices.

People draw on a set of "capital assets" as a basis for their livelihoods. Carney [10] identified five capitals: human, natural, financial, physical, and social. These capitals are defined as follows:

Human capital: Skills, knowledge, ability to labor, and good health and physical capabilities important for pursuing livelihoods; at the formal level these include health education, training, etc.

Natural capital: Natural resource stocks including soil, water, air, and genetic resources, as well as environmental services such as hydrological cycle and pollution sinks, which form the basis for deriving livelihoods.
Financial capital: The capital base that includes cash, credit/debit, savings, and other economic assets like basic infrastructure.
Physical capital: The basic and common infrastructure such as roads, connectivity, and other physical assets owned at the community and household level, viz., livestock, farm implements, machinery, etc.
Social capital: Social resources such as networks, social claims, social relations, political relations, administrative relations, and affiliations to local groups and associations, which help people overcome risks, uncertainties, shocks and vulnerabilities, and livelihood pursuits that require coordinated actions.

Of late, political capital is also gaining exclusive importance [13]; in this study, social capital is inclusive of political capital. The capitals available to individual households reflect their ability to gain access to systems (the resource base, the financial system, and society) through which these capitals are produced. As such, we can identify the "access profile" of the households, which defines their ability to gain access to capital assets.

2.5.2 The Livelihood Model

The conceptual framework presented here traces the interconnections between the different aspects of people's livelihoods and the factors that influence them (Figure 2.2). Recognizing and understanding the dynamics of the livelihoods process is fundamental for any analysis of the factors such as security, vulnerability, resilience, and sensitivity identified previously. These all relate to the processes of change in the conditions in which people's livelihoods operate and the response of livelihoods to these changes. The structure of people's livelihoods (and in particular, the strength and diversity of their livelihood assets) varies greatly, as do the effects of the external influences upon them. The key objective of the model is to provide a structure for understanding the dynamics and diversity.

Livelihoods are complex, especially in the developing countries. Rural livelihoods in the south of India are far more complicated than in the industrialized countries where one main income stream from formal employment, with fixed working hours and a known level of remuneration, is more the norm. This also has policy implications and suggests that pro-poor policy initiatives cannot be expected to have impacts that are predictable and easily aggregated across a diverse range of households and strategies. There is an increasing recognition that the livelihoods of people (and especially households) in the developing world are based around a wide range of activities: people are not just farmers, laborers, factory workers, or fisher folk [14].

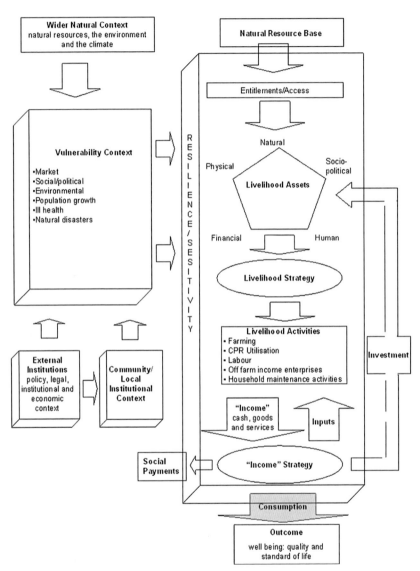

FIGURE 2.2 The livelihoods model.

Livelihoods are also influenced by a wide range of external forces—both within and outside the locality—that are beyond the control of the household. These include the social, economic, political, legal, environmental, and institutional dynamics of the locality, the wider region, the country, and, increasingly, the world as a whole. These factors are critical in defining the basic structure and operation of the livelihood systems. For example, land tenure laws are crucial in determining entitlements and, consequently, access to land for cultivation, which in turn is a critical determinant of the overall structure of livelihoods in rural areas, while prices and price variability is critical (for some crops) in determining what will be grown on that land in a particular season.

These external forces are not static. It is their dynamics, the processes of change in the wider economic, social, and natural environment, that creates the conditions in which livelihoods change. It was noted previously that these changes could be longer term trends (for example, changing attitudes to gender roles in a society or the gradual decline in groundwater stocks in a lake) or sudden shocks (the impact of a war, a drought, or a collapse of market prices for a key crop). Together, the threat of external shocks and trends directly affects the decision-making environment and the outcomes of livelihoods, and provides the "vulnerability context."

Rennie and Singh [12] also identified the responses of such threats as either adaptive strategies (where a household consciously adopts a process of change in response to long-term trends) or coping strategies (short-term responses to immediate shocks and stresses). In these, the household will seek to deploy its different assets to best effect within its often limited range of choices. This set of choices is again conditioned by the wider context within which the household lives, and in particular by the extent to which it can control the key decisions that affect the lives of its members. This is (or should be) why participation is widely advocated. This idea of people making conscious choices through deliberate strategies is fundamental to the approach to the livelihoods analysis presented here. This is integrated into the model at two stages: as a "livelihoods strategy," where a set of decisions is made on how to best employ the assets available, and as an "income strategy," where choices are made over the use of the products (cash, goods, and services) generated by the livelihood activities adopted.

The right-hand side of the model (Figure 2.2) represents the livelihood dynamics of a household. This submodel starts with the "entitlements" and "access" the members possess from the resource base in their locality. These in turn defines the "natural capital" available to the household. This natural capital is one form of "livelihood asset," represented by the pentagon, which can be deployed by the household for livelihood. When combined with the others (financial, social, physical, and human capitals), these capital assets represent the capabilities and assets—the "factors of production"—that the household can deploy to make a living. The "entitlements" box is consequently part of the "access profile" of the household. Similar access factors can also be identified for each of the other capitals; for example, the network of social and institutional relationships that a household possesses and the identity of the household in relation to factors such as caste, religion, clan, or other determinants of social structure are defined in terms of explaining the social capital that they possess.

A key aspect of any livelihoods approach is to understand how the access profile, and consequently, the assets available, changes over time and how increases or reductions in these values affect the livelihoods of the household. Together, these livelihood assets represent a potential: a set of possibilities for the household to secure a livelihood. But they do not automatically define that livelihood, as the extent to which their potential is realized will depend upon the way the assets are used. This is reflected in a set of decisions on what assets are to be utilized and when—decisions that together constitute the livelihood strategy of the household. There are always difficult choices to be made here; for example, what use of the assets will provide the best returns? What risks are involved in particular decisions? Which assents and what quantity should be held in reserve for the future? These and many other questions need to be considered in the livelihood strategy, and this strategy is at the heart of a livelihoods analysis.

The choices made in the strategy will in turn define the "livelihood activities" of the household: which activities are undertaken by whom and when. Land, labor, material inputs, social networks, and all the other capital assets available are used in different combinations to grow crops, raise livestock, gather common property resources, earn wages, make things, trade, provide services, and a multitude of different activities that the different members of the household engage in. Together these are their livelihood; the things that people do on a day-to-day basis to make a living. In some cases, there are one or two dominant activities, such as farming, fishing, or making pots, but for many households the pattern of livelihood activities is varied and no one activity dominates. Whatever the relative importance of the set of activities, the basis for understanding livelihoods is that they all need to be included in the analysis.

Households thus earn "income" (in cash or kind), which becomes part of the household budget. This income is in turn allocated through a second key set of decisions called the income strategy. Income can be allocated to savings or investments that enhance the value of the assets to pay for the production inputs (e.g., fertilizers, raw materials, labor), to repay loans or social payments (such as taxes) or, finally, for consumption that is part of "the outcome"—the total set of goods and services that constitute the material fabric of people's lives. Obviously, the greater the income, the more is left over after other obligations are met (inputs and social payments) either for consumption (meeting the daily needs) or investment (increasing the ability to meet tomorrow's needs). There are other factors that contribute to quality of life or well-being, however, the goal for which all strive is defined. This includes the social context within which one lives, a sense of freedom and security, and many other non-material factors.

Thus, we see that the core of the model reflects the internal dynamics of the process of gaining a livelihood on the part of individuals and the households to which they belong. It is clear that this process, however, does not operate in isolation from a wide range of influences that condition the flows through the livelihood, the choices available at any stage, and the overall outcomes of the livelihood:

The first of these is the "local community"—the social groupings, networks, and institutions within which the individual household is enmeshed. The social and institutional structures of local communities are locality specific, but reflect differing combinations of place (the locality or neighborhood) and people (kin, religious, ethnic, occupational grouping, or other social and economic characteristics) where an individual household lives.

The second conditioning factor is the "external institutional context"—the legal, political, social, economic, and institutional environment or those factors that link people and places into regional, national, and global systems. This includes the nature and operation of the government (which can have both direct effects, such as through agricultural subsidies or health services, and indirect impacts, such as through policy and macroeconomic frameworks and political climates), the structure and strength of the civil society (those non-state institutions and organizations that also regulate social and economic processes), the operation of markets, and so on.

The "wider natural environment" is also extremely important in the functioning of livelihoods. This can be through the character and variability of production conditions: the level and timing of rainfall, resource flows within an ecosystem, and its resilience in the face of management strategies, which can cause resource degradation. It can also reflect extreme events such as cyclones, earthquakes, or droughts.

In many ways these define the characteristics of the different parts of the livelihood model. For example, entitlements and access to common property resources (CPRs) in a watershed to gather products such as fuel wood and fodder can reflect both the legal and policy framework (which defines who owns the CPR and what form of external regulation exists) and local customs and traditions concerning who can gather what. This in turn defines a part of the natural capital in the livelihoods assets pentagon. Similarly, both external monetary policies as well as financial institutions and local moneylenders define the availability and cost of credit, which is crucial not only in determining how much income goes to repay past loans but also the credit available for investments and inputs into production.

These external factors are "filtered" through the vulnerability context, which was referred to previously. The vulnerability context describes the trends and variability in those factors that affect livelihood processes and, in particular, those that can materially disrupt different aspects of livelihoods. This can be specific—climate change directly affects the long-term characteristics of the resource base—with other consequences compounding through the system from there, while a devastating cyclone or drought will have massive immediate impacts and can cause structural change to the characteristics of a household's livelihood processes. The nature of vulnerabilities can also vary, depending on form or timing. For example, a sudden collapse in market prices for a dominant commercial crop can affect the assets available by making key assets of land and agricultural implements less valuable. It can affect the livelihood activity through affecting a decision to plant something different or affecting income if the price collapse happens after planting. Most vulnerabilities are not different in the local and external contexts described earlier (climate, markets); rather they reflect the dynamics and specific forms that those contexts take.

Finally, the fact that these forces affect different households differently has already been made. Some are more sensitive to the effects of vulnerability, while others are more resilient. This can be represented as a resilience filter, through which the flows of influence from the vulnerability context pass to define the specific impact of external forces on the livelihood system of particular households. The resilience of a household can be higher across the board. For example, secure access to credit or good financial reserves are important in relation to most forms of vulnerability, or it can be specific to particular vulnerabilities: owning higher land can be an advantage if there is a flood, but a disadvantage if there is a drought or erosion.

This model allows mapping of the consequences of specific changes, including changes brought about through external interventions intended to improve people's lives. For example, a dominant approach to natural resources management in recent years has been participatory mobilization to create community-based institutions to manage common property resources as well as private resources. Initiatives such as WSD or joint forest management in India typify this approach. The points of intervention and impact of this approach can be "mapped" on the livelihoods model.

2.5.3 Evaluating the Determinants of Perceived Drought Resilience

A combination of parametric and semiparametric approaches has been used to analyze the determinants of perceived drought-survival responses. Ordinary least-squares regression is performed to evaluate the factors that lead to drought-survival differences across watershed regions as well as across various socioeconomic categories. This conventional regression analysis provides results based upon the correlation between dependent and explanatory variables.

One of the drawbacks of such an approach is its inability to establish causality between the independent and dependent variables. Therefore, in Chapter 8 we make use of a semiparametric approach, namely the propensity score matching (PSM) method, to assess the effects of watershed intervention on enhancing perceived drought survival. The areas in the study region that have not seen watershed intervention are classified as "control regions," and the areas with watershed intervention are classified as "treated regions." Additionally, a distinction is made between various types of drought-survival responses that are associated with different types of capital ownerships of the farmers. The empirical analysis is performed in STATA.

The PSM method has been extensively used in situations where the effect of treatment on a parameter of interest needs to be assessed by separating its influence from other factors. By matching individuals with similar characteristics within the treated category to those in the control category or region, the PSM method allows for evaluation of the overall difference in the parameter of interest that could be solely ascribed to a particular treatment. The detailed procedure for performing PSM first involves the use of logit or probit methods to generate propensity scores; then a matching algorithm is used to generate the average treatment effect. The obvious advantage PSM offers over conventional regression analysis is that no functional form assumptions are needed to perform PSM analysis. However, PSM can only offer an average estimate of the impact and is prone to hidden biases.

2.6 MODEL OF INTEGRATION: THE BNs

BN submodels have been developed for the five SL capitals and linked to a measure of drought resilience. The component BNs have also been implemented within an integrated model that links key hydrogeological, biophysical, and social relationships. This is one of the first examples, to our knowledge, where the SLs framework has been operationalized within a modeling framework to explore the impact of WSD and other drivers on livelihoods and resilience of communities. The BN approach is well-suited to implementing the SL framework as it supports a relatively simple representation of cause-and-effect relationships and is flexible in terms of the data and information that can be used to define model relationships.

2.6.1 Description of BNs

BNs are a probabilistic modeling approach comprising:

- Network structure (or influence diagram) that represents cause-and-effect relationships between variables
- Probabilities that describe the strength and nature of relationships between variables

In the field of environmental science or management, BNs have been used for a range of purposes including data analysis, social learning, system understanding, decision making, and management [15].

2.6.2 Strengths

- Assigning probabilities to links between variable states allows explicit representation of uncertainty.
  - Complex systems can be modeled in a relatively simple way.
  - Cause-and-effect links can be described probabilistically.

- There is no need to represent complex processes mathematically.
- Integration across disciplines.
  - Hydrology, water quality, ecology
  - Economics, social, environmental
- Utilization of the best information available and guide data collection.
  - Expert assessments, monitoring data, simulation models, research data
  - Qualitative and quantitative data
- Well-suited to iterative, adaptive modeling and management processes.
- Models can be iteratively updated and used to improve system understanding as new information or evidence is acquired about links between variables.
  - The development process lends itself to the engagement of multiple people/groups promoting system and social learning.

2.6.3 Weaknesses

- Treatment of space
  - Most BNs are developed as nonspatial or lumped region models. However, BNs can be linked with GISs or integrated with other models to improve spatial representation.
- Treatment of time
  - Most BNs are developed as nontemporal or lumped temporal models. BNs are a directed acyclic graph, meaning that they cannot include feedback loops. However, some BN packages allow some representation of dynamics—dynamic Bayesian networks; also, BNs can be linked with other models to represent dynamics.
- Model structure and variables states
  - This includes decisions such as what is the right graph and how much detail to represent in variables as well as the balance between complexity and adequate representation of the output probability distribution.
  - Some BN packages support learning of model structure, but this requires considerable data.
- Populating BNs
  - Expert elicitation: can be an intensive process, especially with complex networks; balancing consensus with multiple views or models; and limitations in knowledge about interactions.
  - Learning algorithms: Large datasets are required to develop robust relationships between variables particularly with complex networks.

2.6.4 When are BNs Useful?

The decision tree in Figure 2.3 allows us to evaluate the selection of the BN approach as an integrating mechanism for the various components in the study. It was the team's assessment that the evaluation of WSD is a suitable application for the method.

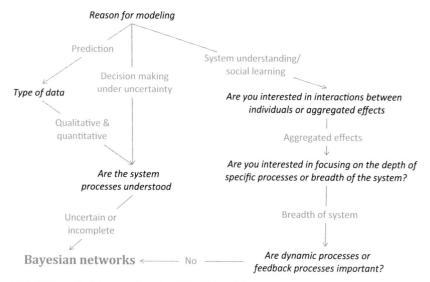

FIGURE 2.3 Decision tree for using BNs. *(Adapted from Kelly et al. [15])*

2.7 EQUITY AND JUSTICE ISSUES

These issues were largely addressed by an extensive literature review on the justice issues inherent in the development of water reforms in Australia and India. This review concluded that given the move to shift toward mesoscale implementation of WSD, there should be emphasis on the delivery of communal property rights and institutional arrangements that reflected procedural justice within WSD through appropriate institutional arrangements.

Data were collected on the current perceptions of communal decision making through community surveys and interviews with villagers. These results demonstrated that while there were potential equity or fairness issues that may be of concern in moving from micro- to meso-WSD, these are yet to be fully considered by the landholders and other stakeholders.

The strength of this approach is that it has broadly canvassed equity and justice issues, while its weakness is that these insights need to be applied during the planning of a new mesoscale WSD so that formative evaluation of the desired justice principles and the appropriate institutions can be undertaken.

2.8 STAKEHOLDER ENGAGEMENT

A systematic, three-stage stakeholder engagement was followed to achieve integration of research, policy, and implementation. As a first step, the state-level nodal agency for implementing the WSD program[2] has been made a formal partner in the process of developing the research and selection of sites.

The nodal agency has played a crucial role in redesigning the research proposal to the futuristic needs of the watershed implementation (IWMP). The policy-level stakeholder engagement is fostered even at the national level through the involvement of a national-level agency, the National Authority on Rainfed Areas. Although this agency is not a formal partner in the process, a continuous engagement with it in the process of research has helped to foster a strong understanding and support for the research at the national level. Engagement with the policy makers from the beginning has resulted in the articulated demand for research results in the form of training for the implementing agencies at the state as well as national level.

At the second level, we engaged with the watershed-implementing authorities and agencies through awareness creation and training. The main focus here is to bring awareness among the implementing agencies regarding the importance of integrating hydrogeological, biophysical, and socioeconomic aspects in the planning, designing, and implementation of the IWMP watersheds. The training programs were designed to highlight the role of hydrogeological and the biophysical aspects in realizing the socioeconomic impacts and the variations across upstream and downstream locations. These programs were designed in such a way that the participants were engaged in the learning process and were motivated to learn by doing (see Chapter 12).

At the third level, the engagement was with the farmers and the farming communities at the village level. Apart from collecting all the relevant information at the household, community, and watershed (village) level through participatory approaches, the results of the analysis were shared with them to validate and make them understand the hydrogeology in their locations. Researchers shared the visual attributes of the hydrogeological as well as the biophysical aspects with the communities and these were validated by the communities. The resulting socioeconomic impacts and resilience to droughts was also shared with the communities.

2.9 APPROACH AND SAMPLING DESIGN

The site selection was purposive because of the objective of the study, i.e., assessing the watershed impacts at scale. This was possible only because we have fully treated watersheds at scale that capture upstream/midstream/downstream variations. Given the fact that watersheds were small in size prior to the advent of IWMP, their implementation was not linked to hydrogeology. However, the purposive selection of the HUNs that are treated with watersheds to the maximum extent would substantially benefit (according to the Department of Rural Development; DRD) from the IWMP implementation in terms of scale issues for upstream/downstream impacts. While the purposive sampling limits the generalization of the findings in comparison to the randomized site selection, the latter was constrained by the absence of a substantial number of such sites from which a random sample could be drawn.

After looking at the watershed implementation data over the years from the department as well as the hydrogeology, two study sites (HUNs with highest coverage of watershed interventions in the state) were identified as best suited for the study. The sample villages were selected after visiting a number of villages within each HUN. Again, a simple random sample would not serve the purpose due to various considerations including the presence of watershed structures and land use. Within the sample villages, the households were selected randomly and the size of the sample was quite substantial. Also, qualitative research tools such as focus group discussions (FGDs) and case studies were used to infer insights. Together, the results are expected to be applicable in a broader context, although sweeping generalizations cannot be made.

Quantitative and qualitative research methods have been used to address the research questions in six watersheds spread over three HUNs located in the Kurnool/Anantapur and Prakasam districts of Andhra Pradesh. The sample watersheds are located at the upstream, midstream, and downstream locations of the HUNs. These HUNs are formed under the Andhra Pradesh Farm Managed Groundwater Systems (APFMGS) project in partnership with local NGOs and implemented in 650 villages spread over 63 HUNs across seven drought-prone districts of Andhra Pradesh, using hydrological boundaries as operational units.

Two broad criteria were adopted for selection of the field sites: (1) a technically demarcated HUN and (2) substantial coverage of area under the WSD program implemented by the DRD. Three HUNs were selected after an elaborate process of assessing the technical aspects of the HUNs under the APFMGS projects and coverage of area under the WSD through the DRD over the years (Table 2.1). The area covered under each HUN ranges between 5000

TABLE 2.1 Selected HUNs and coverage of watershed development program

Name	Vajralavanka	Maruvavanka	Peethuruvagu
District	Anantapur/ Kurnool	Anantapur/ Kurnool	Prakasam
Area (ha)	10,594	5025	9425
Villages covered	14	13	14
Watershed covered villages Approximate area (500 ha per village)	7 (3500 Ha)	4 (2000 Ha)	7 (3500 Ha)
Approximate percentage of coverage of DRD watershed to HUN Area	33	40	37

and 10000 ha, although the number of villages covered is ~13–14. The coverage of area under the watershed is between 33 and 40%.

Initially two districts, Anantapur and Prakasam, were identified after considering the variations in rainfall and hydrogeological formations. A few HUNs and villages were identified after assessing the cadastral maps of each HUN (Figures 2.4 and 2.5). After a rapid appraisal of these HUNs for ground truthing the upstream/downstream variations at the village/watershed level, three HUNs were identified for the study. The initially selected HUNs in the Anantapur District were dropped, since the natural upstream/downstream characteristics were not found due to water quality and hard rock pan issues. These were replaced by two other HUNs located in the Kurnool/Anantapur districts (Table 2.2). We have opted for two HUNs instead of one because the HUNs individually are not big enough to fulfill the criteria. These two HUNs are interconnected hydrologically as well as in terms of surface flow pattern and provide the upstream/downstream linkages between the HUNs. In the Prakasam District we could find a classic upstream/downstream case in a single HUN covered under the WSD program. These HUNs can be designated as hydrological sites.

From each hydrological site we have identified three villages: one each at the upstream, midstream, and downstream locations. The criteria for the village selection include: (1) location, (2) being covered under the watershed program, and (3) being covered under the APFMGS project. In both the sites, upstream villages are located at the mountain slopes and the downstream villages are located in the valley and drain into the major surface water bodies or streams. One of the main differences between the sample villages in the hydrological sites is that the sample villages in the Anantapur/Kurnool districts do not have any surface water body (tanks), while all of the three villages in the Prakasam District have surface water bodies. The Prakasam HUN drains into one of the biggest tanks (Kambam Cheruvu) in the state. All the sample villages are covered under the watershed program under different batches and programs. While watersheds in the Anantapur District are covered under the Desert Development Program (DDP), the other watersheds are covered under the Integrated Watershed Development Program (IWDP), Drought-Prone Area Development Program (DPAP), and the Andhra Pradesh Rural Livelihoods Program (APRLP; Table 2.3). These watersheds were implemented from 1995 to 1996 and from 2007to 2008. All the watersheds, except S. Rangapuram, were implemented by government agencies and have an average coverage area of 500 ha. The S. Rangapuram watershed covers more than 800 ha as it is extended to forest and hillocks outside of the village area. The size of the villages in terms of the number of households varies from 87 in S. Rangapuram to 425 in Basinepalle in the Anantapur/Kurnool districts.

Qualitative research tools such as FGDs, key informant discussions, case studies, and transect walks were used to elicit information. In each sample village, four FGDs were conducted covering different socioeconomic groups

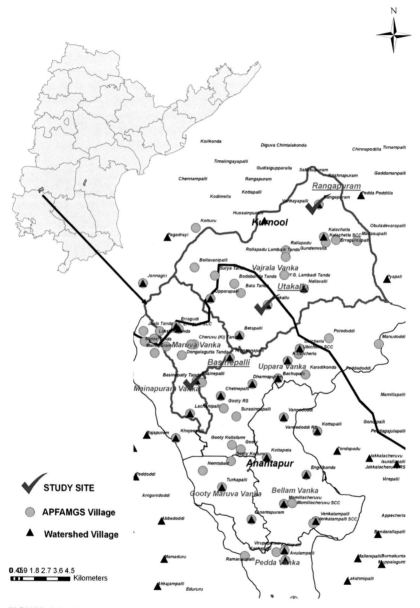

FIGURE 2.4   Location of mesoscale watershed project study sites in the Anantapur/Kurnool districts of Andhra Pradesh, India.

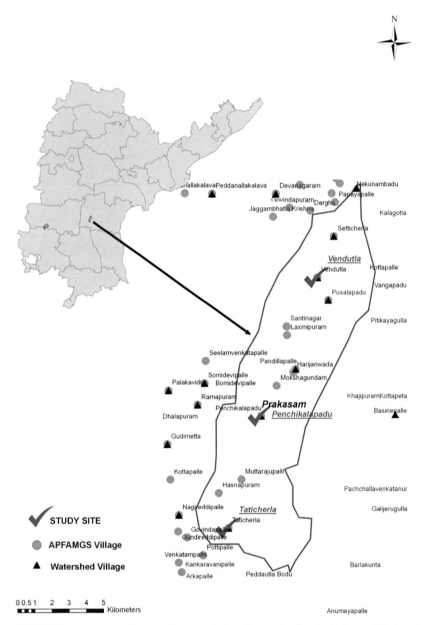

FIGURE 2.5 Location of meso-scale watershed project study sites in Prakasam District of Andhra Pradesh, India.

TABLE 2.2 Villages visited and the selected sample hydrological units

HUN	District	Village	Location in HUN	Program Status
Uppara Vanka	Anantapur	Vennedoddi	Upstream	WSD and APFMGS
Bellam Vanka	Anantapur	Mamilla Cheruvu Kothur Kottapet	Downstream Downstream/midstream Midstream	WSD and APFMGS WSD and APFMGS WSD and APFMGS WSD and APFMGS
Pedda Vanka	Anantapur	Dimmaguda Kottapalle	Downstream Downstream	WSD and APFMGS APFMGS
Maruvavanka	Anantapur	Lachanapalli Basinepalle[a]	Downstream Downstream	WSD and APFMGS WSD and APFMGS
Vajralavanka	Anantapur Kurnool	Utakallu[a] S. Rangapuram[a]	Midstream Upstream	WSD and APFMGS WSD and APFMGS
Peethuruvagu	Prakasam	Vendutia[a] Penchikalapadu[a] Thaticherla[a]	Downstream Midstream Upstream	WSD and APFMGS WSD and APFMGS WSD and APFMGS

[a]The selected villages.

as well as upstream/downstream households. The FGDs would provide insights into the community perceptions on the WSD as well as the APFMGS programs. The externality impacts of these programs can be captured better at the broader community level rather than at the individual household level. Key informant discussions with watershed and APFMGS committee members and village elders were conducted, case histories were collected from specific households, quantitative information was collected from the secondary as well as primary sources, and information pertaining to the WSD was collected from the implementing agency and the watershed committee at the village level.

TABLE 2.3 Basic features and household sample selection in the sample villages

Name of the watershed	Type of PIA	Scheme of funding	Year of formation (batch)	Year of completion	Area of village (ha)	Watershed area (ha)	Total population[a]	% SC and ST	LL[b]	SMF[b]	LMF[b]	Total[b]
S. Rangapuram	NGO	IWDP	1995–1996 (I)	1998–1999	339	816	407 (47)	34	10 (5)	11 (7)	66 (42)	87 (54)
Utakallu	GO	DDP	1999–2000 (V)	2002–2003	1373	500	1199 (47)	14	37 (5)	140 (43)	143 (43)	320 (91)
Basinepalle	GO	DDP	1998–1999 (IV)	2003–2004	883	500	2130 (49)	29	175 (10)	139 (49)	111 (41)	425 (100)
Thaticherla	GO	DPAP	1998–2000 (V)	2004–2005	1903	500	2015 (48)	15	45 (10)	206 (85)	14 (06)	265 (101
Penchikallupadu	GO	APRLP	2002–2003	2007–2008	974	500	2102 (49)	10	22 (05)	87 (52)	05 (03)	114 (60)
Vendutla	GO	DPAP	1998–1999 (V)	2003–2004	2512	500	5794 (48)	24	47 (05)	55 (41)	19 (14)	121 (60)

PIA, project implantation agency; SC, scheduled cast; ST, scheduled tribe; LL, land less; SMF, small and marginal farmers; LMF, large and medium farmers; GO, government agency.
[a]Figures in parentheses indicate the proportion of female population as per 2001 Census.
[b]Figures in parentheses are the sample size.

The NGOs supporting the APFMGS provided detailed information, including technical and socioeconomic, pertaining to groundwater and surface water systems.

2.10 PROFILE OF SAMPLE SITES

In this section, the hydrogeological features of the selected locations and interventions through APFMGS as well as WSD initiatives have been highlighted. The focus is mainly on the surface water bodies, groundwater development, and water-harvesting structures.

2.10.1 Maruvavanka HUN

This HUN (Figure 2.4) lies between the northern latitudes of 15° 16′ 19.86″ N and 15° 06′ 52.64″ N and longitudes 77° 34′ 06.65″ E to 77° 40′ 06.97″ E with an area of 5025 ha. A major part of Maruvavanka is located in the Kurnool District while a minor portion is located in the Anantapur District. The unit covers 17 habitations with a total population of 15,203. The female to male ratio is 950 females per 1000 males, which is much lower than the district average. The literacy rate in the HUN is 37%, which is again much lower than the district average (56%). Backward Castes (BC) are the socially dominant community in this HUN.

The highest elevation in the Maruvavanka HUN is 440 m above mean sea level (amsl), and is located in the northern part of the HUN. The lowest elevation is 355 m amsl, and is located in the southern part of the HUN. The direction of the slope is from north to south. Generally, topography controls the course of the drainage and the general flow direction of the streams in this region. Maruvavanka originates in the northeastern hilly area of the Thuggali Reserve Forest, which includes the Gooty Range. A number of first- and second-order streams contribute to the flow in Maruvavanka and join the Pedda vanka, a tributary of the Penna River, to the southwest of Lachanapalli Village.

The distribution of rainfall indicates that 17% of the rain is received during the southwest monsoon (June to September) and 75% during the northeast monsoon (October to December). Rainfall records of the Peapully MRO show that the normal rainfall in this area is about 902 mm. It is evident from the data that 388 mm is the lowest rainfall recorded (from 2002 to 2003), while the highest rainfall recorded was 1499 mm (from 2000 to 2001). The average number of rainy days in a year is 42. It is interesting to note that from 1999 to 2000 and 2003 to 2004, in spite of having an excess number of rainy days, there was deficit rainfall. All other years correlate the amount of rainfall to the number of rainy days. The years 2002−2003 and 2003−2004 can be referred to as drought years, as they show negative deviation in the amount of rainfall received. On the whole, the average rainfall shows a declining trend, as far as the amount of rainfall is concerned. The decline seems to be steady from 1998 to 2002, after which it takes a sharp downward plunge. While there is an increasing occurrence of rainfall events from April until August, through May, June, and July, there is a declining trend from September to October. Daily distribution of rainfall is observed even during the deficit rainfall years, while it is skewed in the case of surplus rainfall years. The skewed distribution of daily rainfall also indicates the occurrence of storms during the surplus years, contributing heavily to the total annual rainfall.

Red soil accounts for about 72% of the area, followed by mixed soil (17%), sandy loam soil (6%), and black soil (4%). The nature and constitution of the soil in an area is generally controlled by the mineral and textural composition of the rock type. It is observed that of the geographical area, 57% is cultivated while the remaining is either fallow (24%) or wasteland (19%). Of the culti-vated land, about 80% of the area is dry land.

Agriculture is the main source of livelihood for 97% of the population, while the remaining 3% is engaged in employment. Of the 97% of agri-based families, 61% are small and marginal farmers and 13% are agriculture laborers.

Groundwater development in Maruvavanka HUN is mainly through bore wells, in the absence of any springs, natural or artificial. The density of bore wells is higher in the northwestern plain of the HUN, which uses 84% of the total groundwater consumed presently. The average depth of the bore wells in the HUN is 50−99 m. In general, bore well depths are observed to be increasing from medium to deep in the upper part of northwest plains (NWP), while in the lower plain of NWP and southwest tail, the depth of the bore wells varies from shallow to medium.

All the irrigation bore wells use submersible pumps, while drinking water bore wells are fitted with India mark II hand pumps. Groundwater recharge (64%) is done through the area underlain by fractured/cavernous limestone, which constitutes 36% of the geographical area. The present annual ground-water draft in Maruvavanka HUN is 162% of the annual groundwater recharge, categorizing it as "overexploited."

The Maruvavanka HUN consists of a total of six water bodies scattered over the central part of the HUN. Both the area of submergence as well as the ayacut of a tank is of importance from the groundwater recharge perspective, as they have the potential to augment the natural groundwater recharge. The recharge of tanks is 2% of the total groundwater recharge. In Maruvavanka HUN, 61 check dams were constructed by Panchayat Raj Department, DPAP, and the Vana Samrakshana Samithi (VSS). As many as 12 existing kuntas and 15 farm ponds are presently used as percolation tanks to enhance the groundwater recharge in this basin, while a few of these have been de-silted under the "Neeru-Meeru" program for storing more water, which resulted not only in the enhancement of the recharge of groundwater in this watershed area but also in increased storage capacity of the water bodies (tanks and kuntas). Agricultural use accounts for 87% of the groundwater demand in the HUN, while the remaining 10% is used for domestic purposes; hence, an efficient cropping system is the most important factor in the demand-side management of groundwater resource. The present source of domestic water supply is groundwater-based public water supply, while only four sprinkler systems and three drip irrigation systems are found to be in use in the entire HUN. Paddy uses 23% of the area under groundwater irrigation, followed by onion (20%); sunflower (11%); tomato (9%); and vegetables, maize, bajra, groundnut, castor, and horticultural crops (20%).

2.10.2 Vajralavanka HUN

This HUN (Figure 2.4) is very similar to the Maruvavanka HUN in many technical aspects. It lies between the northern latitudes 15° 18′ 32.94″ N and 15° 11′ 36.70″ N and eastern longitudes 77° 35′ 40.55″ E and 77° 44′ 39.85″ E, and forms the southeastern part of the Kurnool District. The area of the HUN is 10,567.81 ha spread over 14 habitations with a total population of 7882. The female to male ratio is 976 females per 1000 males, which is much lower than the district average. The literacy rate in the HUN is 71%, which is much higher than the district average (56%). BC is the dominant community accounting for 41% of the total households.

Red soil covers 66% of the area while black soil covers 14%, and the remaining area is covered with mixed soils. More than 50% of the cropped area is irrigated. The rainfall pattern, groundwater exploitation, and livelihoods pattern are similar to that of Maruvavanka.

All the irrigation bore wells use submersible pumps, while drinking water bore wells are fitted with India mark II hand pumps. Most of groundwater recharge (64%) in Vajralavanka HUN is through the area underlain by fractured/cavernous limestone, which constitutes 36% of the total area. The present annual groundwater draft is 162% of the annual, categorizing Vajralavanka HUN as "overexploited." Groundwater recharge is affected through 16 water bodies scattered over the central part of the HUN accounting for just 2% of the total groundwater recharge. A total of 58 check dams were constructed by the Panchayat Raj Department, DPAP, and VSS. Two existing kuntas and four farm ponds are presently used as percolation tanks to enhance the groundwater recharge in this basin, and a few of these have been de-silted under the Neeru-Meeru program for storing more water, which resulted not only in the enhancement of groundwater recharge in this watershed area but also in increased storage capacity of the water bodies (tanks and kuntas).

Agriculture uses about 87% of the groundwater in the HUN, while the remaining is used for domestic purpose. Paddy accounts for 23% of the area under well irrigation, while 20% of the area is under cotton crop, sunflower (11%), and sorghum (9%); other crops such as groundnut and vegetable account for the remaining area under groundwater irrigation.

2.10.3 Peethuruvagu Hydrological Unit

This HUN (Figure 2.5) is one among 29 HUNS in the Gundlakamma Basin aimed to be covered under the APFMGS project by Development Initiatives and Peoples Action Giddalur. Topographically, the highest point in this HUN is about 739 m located at the Ankalamma Bodu Reserve Forest, while the mouth of the basin at 200 m is located to the northeast of Besthavaripeta Village. The elevation difference of 539 m over a length of 21 km creates a rapid runoff from the HUN. Peethuruvagu flows through 14 habitations. The HUN is spread between two continuous hill ranges and occupies an area of 9498.29 ha. Most of the course of the drainage from the Velikonda Reserve Forest, Ankalamma Bodu, and Gogulla Konda has a slope of 20−30 degrees and is of dendritic type. The streams experience rapid runoff and converge near Penchikalapadu to form Peethuruvagu (a tributary to the Gundlakamma River).

The HUN falls under the scarce rainfall agro-climatic region (IV). The average maximum temperature ranges from 36 to 46°C while the average minimum temperature ranges from 23 to 28°C. The pattern of rainfall in the HUN can be studied from the data collected from five RG stations. The southwest monsoon starts from June and continues until September (average rainfall 374 mm). This is followed by the northeast monsoon from December to January (average rainfall 215 mm). Rainfall data from Komarolu RG station show that the average rainfall recorded is 739.9 mm and 692.1 mm in Kumbum, which is below the normal rainfall in the district. Further, the amount of rainfall received and the number of rainy days are observed to be erratic and uneven. This has adversely affected the filling of irrigation tanks and the performance of percolation tanks. Because of the situation, dependence on groundwater has considerably increased, leading to a heavy decline in the groundwater table.

During 1996 the number of rainy days recorded was 26, which is less than the normal number of rainy days (37 days), while the total rainfall recorded at Komarolu for that year (1533 mm) was more than the normal (750 mm). This resulted in flooding and damage to crops. There are three minor irrigation tanks in the HUN, while a major tank with a submergence area of 24 ha and an ayacut of 120 ha is located near Thaticherla habitation. In all, 13 tanks are scattered over the entire HUN.

From 1999 to 2004, precipitation was less over the HUN, causing a drought-like situation that resulted in the depletion of groundwater levels in all habitations. There was water scarcity in most of the habitations, both for agriculture and for drinking—the maximum fall in the water table of 100 m (300 ft) was recorded in Pusalapadu and Pandillapalle. The Groundwater

Estimation Committee, 1997, declared Pusalapadu as "overexploited." During the year 2000, the annual rainfall was 262.6 mm (district normal, 750 mm). The area is classified under the Hard Rock Province of Peninsular India, where groundwater occurs mostly under unconfined to semiconfined conditions.

Although wells are distributed evenly across the HUN, the density of wells is higher in the eastern part of the HUN, where land is suitable for cultivation. In the HUN, perennial groundwater sources include 663 bore wells and one spring. Additionally, 298 bore wells are seasonal (June to September). The pumping levels and depths of bore wells range from 200 ft (60 m) to 600 ft (182 m) and yields are very poor; the discharge ranges between 2575 and 7096 gallons per hour. These wells are connected to electric motors with 3−7.5 horsepower. There are about 53 groundwater sources (mainly bore wells) that have gone dry at the time of drilling, while 46 have become defunct over the last 10 years.

In this HUN, the upstream side of the basin consists of undulating hills connected to each other with an elevation difference of 539 m amsl. The rapid runoff causes heavy soil erosion and accumulation of silt in tanks, which adversely affect groundwater recharge. This also causes silt to accumulate in the mouth of the watershed area.

Water-harvesting structures (11 check dams) were constructed by Panchayat Raj Department and District Water Management Agency (DWMA) (watershed activity). The existing kuntas (four), farm ponds (16), and tanks (3) are presently used as percolation tanks to enhance groundwater recharge in this basin. Six tanks were de-silted under the Neeru-Meeru program for storing more water. This has helped to enhance the recharge of groundwater in this HUN. About 26 new ponds were constructed by DWMA under the watershed program in four habitations. However, as they were constructed in unsuitable places, the rate of recharge is very low. There are four RG stations (three by APFMGS and one by the Revenue Department) in the HUN for collecting daily rainfall data. Further, APFMGS has demarcated 56 observation bore wells in 14 habitations for detailed groundwater monitoring in the HUN.

The HUN constitutes forest and wastelands—44% of the HUN area is wasteland and 13% is forest. Most of the land area has become undulated due to soil erosion. Black soil accounts for about 60% and red soil for about 40% of the area in the HUN while 4% of the area is covered with problematic soils. These soils are poor in organic matter, have low water-holding capacity, and are poor in micronutrients. Farmers mostly grow rainfed crops (like cotton, sunflower, red gram, bajra, cotton, and vegetables) in black soil because of its water-holding capacity.

Most of the soil conservation activities were performed out under the watershed program. Contour bunding has considerably helped to check soil erosion in addition to reducing nutrient loss in the soils. Check dams were constructed in eroded gullies to restrict soil transportation.

REFERENCES

[1] Syme GJ, Ratna Reddy V, Pavelic P, Croke B, Ranjan R. Confronting scale in watershed development in India. Hydrogeol J 2012;20(5):985−93.

[2] Calder I, Garratt J, James P, Nash E. Models, Myths and Maps: Development of the Exploratory Climate Land Assessment and Impact Management (EXCLAIM) tool. Environ Model Software 2008;23:650−9.

[3] Sophocleous MA. Combining the soil water balance and water-level fluctuation methods to estimate natural ground water recharge: practical aspects. Jour Hydrol 1991;124:229−41.

[4] Moon SK, Woo NC, Leeb KS. Statistical analysis of hydrographs and water-table fluctuation to estimate groundwater recharge. Jour Hydrol 2004;292:198−209.

[5] Maréchal JC, Dewandel B, Ahmed S, Galeazzi L, Zaidi FK. Combined estimation of specific yield and natural recharge in a semi-arid groundwater basin with irrigated agriculture. Jour Hydrol 2006;329(1−2):281−93.

[6] Batelaan O, Smedt F. GIS-based recharge estimation by coupling surface−subsurface water balances. Jour Hydrol 2007;337(3/4):337−55.

[7] Sibanda T, Nonner JC, Uhlenbrook S. Comparison of groundwater recharge estimation methods for the semi-arid Nyamandhlovu area, Zimbabwe. Hydrogeol Jour 2009;17:1427−41.

[8] Sukhija BS, Rama Sc FA. Evaluation of groundwater recharge in semi-arid region of India using environmental tritium. "Indian Acad. Sci. LXXVII" Sec. 'A', vol. VI; 1973. 279−92.

[9] Chandra S, Ahmed S, Rangarajan R. Lithologically constrained rainfall (LCR) method for estimating spatio-temporal recharge distribution in crystalline rocks. Jour Hydrology 2012;402:250−60.

[10] Carney D. Sustainable Rural Livelihoods. London: DFID; 1998.

[11] Davies S. Adaptable Livelihoods. London: Macmillan; 1996.

[12] Rennie JK, Singh N. Participatory Research for Sustainable Livelihoods. Winnipeg: International Institute for Sustainable Development; 1996.

[13] Baumann P. Sustainable Livelihoods and Political Capital: Arguments and Evidence from Decentralisation and Natural Resource Management in India. London: Working Paper 136, Overseas Development Institute; 2000. 44pp.

[14] Frank Ellis. Livelihood Diversification and Sustainable Rural Livelihoods. In: Carney D, editor. Sustainable Rural Livelihoods. London, UK: DFID; 1998.

[15] Kelly (Letcher) RA, Jakeman AJ, Barreteau O, Borsuk ME, El Sawah S, Hamilton SH, et al. Selecting among five common modelling approaches for integrated environmental assessment and management. Environ Model Software 2013;47:159−81.

Part II

Hydro-geological and Bio-physical Aspects of the Watersheds

Chapter 3

Investigating Geophysical and Hydrogeological Variabilities and Their Impact on Water Resources in the Context of Meso-Watersheds

P.D. Sreedevi*, S. Sarah*, Fakhre Alam*,§, Shakeel Ahmed*,
S. Chandra* and Paul Pavelic¶

*CSIR-National Geophysical Research Institute, Hyderabad, India, §Presently at CGWB (MER), Patna, India, ¶International Water Management Institute, Vientiane, Lao PDR

Chapter Outline

Integrated Assessment of Scale Impacts of Watershed Intervention
http://dx.doi.org/10.1016/B978-0-12-800067-0.00003-7. Copyright © 2015 Elsevier Inc. All rights reserved.

3.1 INTRODUCTION

Groundwater is an important natural resource, and it plays the most vital role in supporting mankind. The presence of a safe and reliable source of water is an essential prerequisite for establishing a stable community as well as the socioeconomics of a country. Groundwater is the only major source for agriculture in the rainfed regions of India and it plays a key role in the socioeconomic development in its agrarian economy. Overexploitation in these regions has resulted in declining groundwater levels, and groundwater flow is now confined only to the deeper, weathered/fractured zones. This makes the complexity of management difficult, and because this resource is hidden and generally occurs in a complex system of rocks, its precise assessment is difficult. As a result, there is a large mismatch between the demand and availability of groundwater. In other words, the natural recharge, which is the main input to the system, is not enough to meet the demand. For these reasons and to meet the groundwater supply challenges in water-stressed areas, efficient groundwater management practices, such as change in cropping pattern or use of micro-irrigation systems, are being encouraged in many parts of India.

Groundwater users and managers practice watershed development (WSD), which includes the rigorous activity of arresting the water and allowing it to recharge the groundwater by various means, depending on the prevailing conditions. However, its quantification and selection of the optimum location for a managed aquifer recharge still remains a challenge due to the problems encountered in tackling the complexity of the host rock formations. Further, the widespread water-harvesting interventions through WSD in India, particularly in hard rock areas, failed to improve the groundwater situation because of the uncontrolled exploitation and the poor design of such interventions due to the absence of information on aquifer geometry and its characteristics.

While watershed interventions are expected to improve groundwater recharge through better soil and water conservation practice, the actual availability of groundwater for final use depends on the suitability of interventions to the aquifer system. In the context of watershed interventions, it is often presumed that groundwater recharge improves as one moves from upstream to downstream locations. However, these observations are not based on scientific information regarding aquifer geometry and drainage systems. Consequently, practices for WSD are not as effective as they should be.

Thus, hydrogeological investigation with geophysical methods provides a clear link to socioeconomics. A precise knowledge of the subsurface is helpful in two ways: ensuring efficacy of the WSD and suitability of the type of the WSD activity, as well as planning for judicial use of groundwater ensuring the sustainability of such a vital resource. Because of high variability of the parameters and the complex heterogeneous nature of the system, new and sophisticated techniques of geophysical logging, such as electrical resistivity imaging (ERI), have been deployed to characterize the system. The specialized methods (geological, hydrogeological, and geophysical) always prove helpful in demarcating the extent of aquifers in a given hydrological unit. These investigations are further used in this chapter to identify the groundwater potential zones and the sites that can be used for a managed aquifer recharge, which is an important part of WSD.

Most existing techniques applied to quantify the groundwater resource lack scientific rigor and rely on poor quality data, resulting in estimates that are approximate at best. The prevailing systems are so fragile that approximations may lead to erroneous decisions, which in turn badly affect the socioeconomics of the area. In this chapter, attempts have been made to utilize advanced techniques for developing clear and precise estimates to demonstrate the associated impact of WSD interventions that can support improved planning for management.

The role of hydrogeology in terms of water resource potential and its use in the context of watershed interventions is clearly demonstrated through the studies performed in two areas: Vajralavanka—Maruvavanka and Peethuruvagu meso-watersheds.

The objectives of the study are to

1. Delineate the aquifer geometry to understand the groundwater storage capacity of the study area
2. Estimate the amount of groundwater recharge under natural conditions
3. Estimate the changes in groundwater storage in the study area for a specified time and space
4. Educate the respective farmers and agency staff in the study area about the utilization and of the aquifer

3.2 STUDY AREAS

To meet the previous objectives, we selected two study areas located in different hydrogeological and meteorological setups.

3.2.1 Vajralavanka—Maruvavanka Watershed

The Vajralavanka—Maruvavanka watershed is situated in the drought-prone area of the Rayalaseema region in Andhra Pradesh, South India. The study area falls in the Survey of India (SOI) toposheet no. 57 E/11 and E/12, within latitudes 15.13° N and 15.30° N and longitudes 77.57° E and 77.73° E. The total area of the watershed is 149.27 km^2 (Figure 3.1a).

The prevalent climatic condition in the watershed is semi-arid, marked by a hot summer and mild winter. The temperature varies from 36 to 43°C during summer and 18 to 25°C during winter. The contribution of southwest monsoon

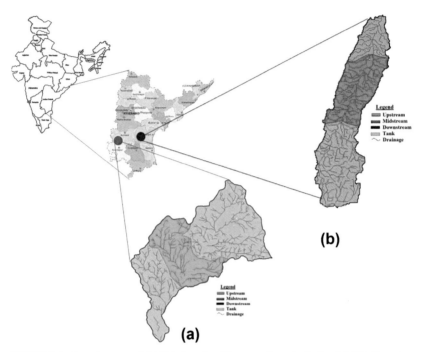

FIGURE 3.1 Location map of (a) Vajralavanka—Maruvavanka watershed and (b) Peethuruvagu watershed.

to annual rainfall is more than the contribution of northwest monsoon rainfall, and the normal annual rainfall is 597 mm. The average annual rainfall from 2000 to 2013 was 629.74 mm with significant seasonal variations. The annual rainfall distribution and its deviation from normal are depicted in Figure 3.2. Usually, the region receives its first rainfall from pre-monsoonal convectional showers in the month of May. Between January and May is the main dry season and the region receives little rain due to convection currents or winter cyclonic disturbances.

The drainage pattern of the watersheds ranges from dendritic to sub-dendritic at higher elevations and from parallel to subparallel at lower elevations. There are no major streams in the study area. Based on the drainage orders, the watershed is classified as a fourth-order basin that drains into the Gooty Cheruvu located at the southern boundary of the watershed.

The slope in the watershed generally exhibits an undulating topography and controls the momentum of runoff in the watershed. In the study area, the slope varies from 0 to 6° with a mean slope of 2° and standard deviation of 1.8°. A high degree of slope is observed in the northern and northeastern parts of the watershed (Figure 3.3a).

Most of the watershed is covered by red soil. The fertility of the soil decreases from north to south; whereas the northern part of the study area is

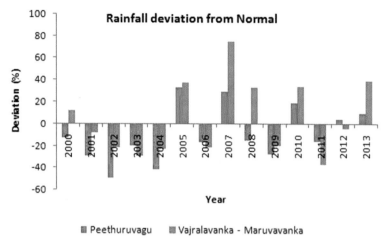

FIGURE 3.2   Rainfall deviation from normal.

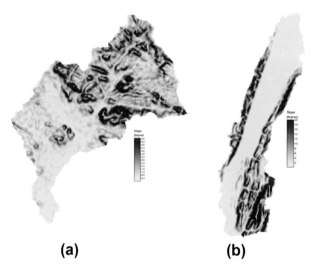

FIGURE 3.3   Slope map of (a) Vajralavanka–Maruvavanka and (b) Peethuruvagu watershed.

covered with fertile red soil, the southern part is covered with black loamy soils.

The terrain of the watershed is undulated with several denudational ridges, i.e., hills. The area exposes mainly rock types belonging to the Peninsular Gneissic Complex (PGC) of the Archaean age, granites, and other basic and acidic intrusions. The PGC is wider spread and is mainly represented by banded and streaky gneisses and granitoids. The gneisses comprise hornblende—biotite gneisses, hornblende gneisses, and biotite gneisses. The granitoids in the form of plutons, or dome-shaped bodies of varied dimensions, are seen amid the gneisses. These granitoids, which are massive and foliated, comprise granite and granodiorite [1,2]. The PGC is intruded by K-rich granites of lower Proterozoic age. These granite bodies, which are of varied dimensions, are either gray or pink, and the latter is younger. Quartz veins and dolerite/gabbro dykes are seen including all the previously mentioned litho units and show various trends. The general trend of foliation in the rocks of PGC and metamorphic rocks are found in the NNW–SSE with steep to subvertical dips. Joints are observed along NNW–SSE, S–E and N–S trends.

3.2.2 Peethuruvagu Watershed

The Peethuruvagu watershed is located toward the southwest of Prakasam District and has an area of about 98.81 km². It lies between latitudes 15.33° N and 15.55° N and longitudes 79.03° E and 79.12° E, and forms part of the SOI toposheet no. 57 M/2 and M/3 (Figure 3.1b).

The area is semi-arid with an average annual rainfall of 763 mm. The average annual rainfall from 2000 to 2013 was 693.08 mm with significant seasonal variations. The annual rainfall distribution and its deviation from normal are depicted in Figure 3.2. The intensity and amount of rainfall is unpredictable during the southwest monsoon period (June to September). The highest amount of rainfall occurs in the watersheds during the northeast monsoon period (October and November). The mean daily maximum temperature during summer is 40.7°C while the mean daily minimum temperature during winter is 19.7°C.

The drainage pattern of the watershed ranges from dendritic to subdendritic at higher elevations and from parallel to subparallel at lower elevations. There are no major streams in the study area. Based on the drainage orders, the watershed comes under the sixth-order basin, which drains into the Gundlakamma River at Besthavaripeta Village; it is located in the northern boundary of the watershed.

The slope in this region varies from 0 to 20° with a mean slope of 6° and standard deviation of 7.02°. A high degree of slope is observed in the southeastern part of the watershed (Figure 3.3b).

The important soil types in this area are red, black, and alkaline soils. Red soil covers 40% of the watershed followed by black soil and alkaline soil. Alkaline soil is saline and considered to be poor in terms of fertility.

Geologically, Peethuruvagu watershed is grouped under the geological formation of the Cuddapah basin. The area is covered with sedimentary rocks such as quartzite, shale, and phyllite. According to King's [3] classification, the rock formation of the Cuddapah basin is divided into lower and upper Cuddapah, comprising the Papagni, Cheyair, Nallamalai, and Krishna groups, respectively. Each of the major divisions of the Cuddapah system is marked by an unconformity, but the major one occurs at the base of the Nallamalai group. The rock formations in the study area come under the Nallamalai group of cumbum formation [3].

3.3 MATERIALS AND METHODS

Groundwater is a hidden but renewable resource and hence, knowledge of its availability and spatiotemporal variability has been the main issue. A number of geophysical and hydrogeological techniques as described in the following sections have been used to fulfill the objectives outlined in Section 3.1.

3.3.1 Geophysical Surveys

The main aim of the geophysical investigations in the present study is to understand the subsurface hydrogeological conditions accurately and adequately. Since the base of any geophysical method is the contrast between the physical properties of the target and the environment, the greater the contrast or anomaly, the clearer the observed geophysical response and hence, ease of identification. The efficacy of any geophysical technique lies in its ability to sense and resolve the hidden subsurface hydrogeological heterogeneities or variations. In the present study, the most advanced geophysical investigations, namely, electrical resistivity tomography (ERT) and electrical logging, were performed to determine the aquifer geometry based on the geophysical signature along with aquifer resistivity properties.

3.3.1.1 ERT

ERT has become an important and essential tool in groundwater assessment, addressing engineering problems and environmental site investigations. The 2D ERIs are created by inverting hundreds to thousands of individual resistivity measurements recorded within a short span of time to produce an approximate or true model of the subsurface resistivity variation. In most investigations, ERT data are collected along transects. The application of geophysics in shallow investigations such as environmental, geotechnical, and hydrogeological studies requires the development of fast, reliable, and high-resolution field equipment and interpretation techniques. The improvement of resistivity methods over the conventional DC resistivity method using multi-electrode arrays has led to an important development of electrical imaging for subsurface surveys [4−6]. Such surveys are usually performed using a large number of electrodes, 24 or more, connected to a multicore cable. A laptop microcomputer, together with an electronic switching unit, is used to select automatically the relevant four electrodes (i.e., pair of current and potential electrodes) for each resistivity measurement. Apparent resistivity measurements are recorded sequentially, sweeping any quadruple (combination of current and potential electrodes) within the multi-electrode array arrangements. As a result, high-definition pseudosections with dense sampling of apparent resistivity variation at shallow depths (0−100 m) are obtained in a short span of time with good precision, provided the acquired data are of good quality. It allows the detailed interpretation of 2D resistivity distribution in the ground below the surface [7,8]. The present field techniques and equipment to carry out 2D resistivity surveys are fairly well developed [9,10]; a resistivity meter called the SYSCAL Junior Switch, with 48 electrodes connected to the resistivity meter through a multicore reversible cable, has been used in the present study for carrying out the required work.

In the present study, ERT was performed covering the whole watershed along 11 profiles, using the Wenner—Schlumberger configuration. The total line length of the survey was 480 m, and 48 electrodes were used with 10 m interelectrode spacing. The inversion of the electrical measurements provided the distribution of the resistivity along the profiles, from the surface down to a depth of about 92 m.

3.3.1.2 Electrical Resistivity Logging

The most common electrode arrangement is normal or potential sonde in which one current electrode and two potential electrodes are located on the sonde. The other current electrode is kept on the surface. The curves obtained by potential or normal resistivity logs are symmetrical in form in which maximum indicates a layer with higher resistivity and minimum indicates a layer with lower resistivity.

As many as 10 resistivity loggings were performed in the study area in representative bore wells to understand the geologic sequences and to obtain lithological information as well as the groundwater flow system. Logging provides more continuous data on the vertical and lateral distribution of the well section and depends on the sensitivity of the sondes. Most of the resistivity logging surveys was performed at nearby ERT sites.

3.3.1.3 Interpretation of 2D ERI Sections and Resistivity Logging Data

The field ERT data were first processed for eliminating any noisy or bad data points using the PROSYS software. In most profiles, the quality of the data obtained was good as this was initially taken care of in the field during the data acquisition stage; this is apparent from the inverted 2D sections in terms of root mean square error between the observed and calculated apparent resistivity. The processed data were then inverted using the standard RES2DINV software [11], using the Finite Difference Numerical Approach, to reproduce the subsurface true resistivity variation. Thus, these true, resistivity subsurface models were finally interpreted.

The resulting 2D resistivity profiles were compared with the bore well logging data, which helped while interpreting the resistivity imaging data and in correlating the various thicknesses and resistivities of the geological formations.

3.3.2 Lithologically Constrained Rainfall Method

Recharge from rainfall is the most important parameter for groundwater availability. Although rainfall is the primary source of recharge, it also depends on the medium, i.e., the soil through which water moves. Thus, some parameters that affect the estimates of recharge are time invariant while some are time variant as well as nonlinear in nature. Hence, an appropriate methodology called lithologically constrained rainfall (LCR) [12] was used.

Spatial and temporal variation of natural recharge to groundwater is estimated using the LCR method in the study area [12]. In this method, lithological constraints are coupled with the rainfall in terms of soil resistivity (ρ_s) and vadose zone thickness (H) as follows:

$$R = k(\rho_s H)^{0.7} P^{0.5}. \tag{1}$$

Here, R is natural recharge; k is the constant that brings the effect of the other nonconsidered parameters into the relationship (the calibrated k value is 0.0006); ρ_s is soil resistivity in Ωm; H is vadose zone thickness, i.e., water levels below the ground level in meters; and P is rainfall in meters.

The method facilitates estimation of the distribution of natural recharge for a fairly long time series, incorporating both temporal and spatial heterogeneities. In the study area, natural recharge was estimated at 14 locations for understanding the spatiotemporal variation of recharge within the watershed with respect to rainfall and lithological constraints.

3.3.3 Change in Groundwater Storage (ΔS)

The change in groundwater storage ΔS describes the volumetric loss or gain of groundwater from the aquifer system between two time periods. This value is assessed by multiplying the difference in groundwater elevation between two monitoring periods with the specific yield of the formation and the area overlying the groundwater basin. Estimation of aquifer water storage variability is of high importance for the management of water resources.

The change in storage (ΔS) is computed as follows:

$$\Delta S = \Delta h * A * S_y. \tag{2}$$

Here, Δh is the change in water levels during the given time period in meters, A is an area influenced by the wells in m^2, and S_y is the specific yield value usually obtained from literature [13].

3.4 RESULTS AND FINDINGS

3.4.1 Top Layer Thickness and Resistivity

The soil types in the study areas include red loam, clayey loam, and sandy loam with variable thickness at the top layer. Soil type and thickness plays an extremely important role in controlling the movement of water over and through the watershed. Furthermore, soil thickness is the controlling factor for infiltration rates in the hydrological process [14]. Relatively thin soils are more prone to saturated overland flows compared to thicker soils, which have greater water storage potential [15]. These parameters are important for implementing a successful watershed enhancement or management activity.

In Vajralavanka−Maruvavanka watershed, the soil thickness varies from 0.70 to 2.40 m. The soil thickness is classified into low, moderate, and high thickness zones in the study area. The thickness of soils is high in the upstream region followed by downstream and midstream areas.

In Peethuruvagu watershed, the soil thickness varies from 0.40 to 2.60 m. At the starting point of the midstream areas, the soil thickness is high, followed by moderate and shallow thickness toward the end. In the downstream areas, the soil thickness is moderate at the starting point, followed by high thickness toward the end.

Soil composition, moisture content, and temperature control the soil resistivity. Soil is rarely homogenous and the resistivity of the soil varies geographically and at different depths. The apparent resistivity of the soil varies from 3 to 274 Ωm in the Vajralavanka−Maruvavanka watershed and from 2.60 to 41.00 Ωm in the Peethuruvagu watershed; the variations in resistivity are due to changes in soil moisture.

3.4.2 Delineation of Aquifer Geometry

The potentiality of the aquifer depends on the thickness of the saturated weathered zone and the number of fractures present in the zone. Electrical logging data, which provides the true resistivity of the geological formations in the bore wells, were correlated with the ERT data. Based on these studies the aquifer is classified as a two-tier coupled system—weathered and fractured/fissured layers—that exist almost over the entire area.

In the Vajralavanka−Maruvavanka watershed, the thickness of the weathered zone varies from 4.5 to 20 m with resistivity varying between 4 and 223 Ωm, followed by fractured/fissured zone, which extends from 7 to 71 m with resistivity varying between 115 and 1616 Ωm. This is followed by a basement depth that varies between 10 and 83 m.

In Peethuruvagu watershed, the thickness of the weathered zone varies from 4.10 to 13.80 m with resistivity varying between 3 and 131 m, followed by fractured/fissured zone, which extends from 7 to 31 m with resistivity varying between 252 and 1275 Ωm. This is followed by a basement depth that varies between 12 and 45 m.

3.4.3 Aquifer Characteristics

In the Vajralavanka−Maruvavanka watershed, the aquifer comprises crystalline rocks of igneous origin (granite, granodiorites, diorite, and gabbro dykes) and metamorphic origin (quartzite and gneiss). The groundwater in these formations occurs in the weathered and fractured zones under the water table and semiconfined conditions, respectively. In the study area, the weathered zone has been tapped extensively by the dug wells and dug-cum-bore wells, which invariably tap the fractures occurring below the weathered zone. Most of the dug wells and dug-cum-bore wells are located in the upstream and midstream areas of the watershed. The depth of open wells range from 7 to 15 m below ground level (bgl), and the depth of the water levels varies from 3.4 to 6.1 m bgl. These wells are located in the upstream of the watershed, i.e., S. Rangapuram and Utakallu villages in the midstream. The yield of the dug wells varies from 10 to 200 m^3/day and sustains a pumping period of 3–6 h/day [13]. Most of the wells are rectangular in shape and the depth of the wells is greater in the regional fractured directions.

Wells in the fractured zone generally yield low to moderate quantities of water depending on the number and interconnection of fractures tapped by the well. According to the Central Groundwater Board (CGWB) report [13], the existence of deep fractures of up to 200 m is also found in the study area but the potential fractures are encountered between depths of 40 and 100 m. The cumulative yield of the fractured zones varies from 0.4 to 15.7 liters per second (lps). However, the general yield of the bore wells was between 1 and 5 lps. The E−W, N−S, and NNW−SSE fractures are tensile fractures and the yield of the bore wells drilled close to these fractures is between 1 and 8 lps. On the other hand, the NE−SW and NNE−SSW fractures are shallow in nature and their yield ranges between 0.2 and 6 lps. The transmissivity of the fractured aquifer varies from 0.5 to 316 m^2/day and storativity values range from 7.4×10^{-5} to 9.0×10^{-3}. The specific yield of the unconfined aquifers varies from 0.01 to 0.055.

Hydrogeologically, the Peethuruvagu watershed comprises the Cuddapah aquifer system of sedimentary origin. The aquifer system in the study area consists of quartzites and consolidated shales. The occurrence and movement of groundwater in these rocks depends on the extent of weathering, degree of compaction, fracturing, and the occurrence of bedding planes. Generally, shales form poor aquifers due to their impermeability. The thickness of the weathering profile is also low compared with the other formations such as the weathered zone in granites. Response to rainfall is very quick due to its less permeable nature and low specific yield. Bedding planes and fractures mostly form the aquifers. The dug wells in this formation range in depths between 8 and 15 m bgl and the depth of the water level varies from 2 to 90 m bgl in the bore wells. The exploratory drilling of the CGWB to depths of 150 m bgl in this formation has resulted in a yield ranging from 172.80 to 587.52 m^3/day with transmissivity of the aquifer varying from 6.87 to 158.22 m^2/day. In shale formation, the discharge varies from 164.20 to 316.26 m^3/day and the transmissivity varies from 6.59 to 22.8 m^2/day [16]. Three tanks exist in the study area out of which the two tanks located in the downstream side do not receive sufficient flows and are mostly dry in all seasons despite normal rainfall. The first tank situated in the upstream area has a rate of infiltration within the tank beds that is very high, resulting in its drying up in a few days.

3.4.4 Depths to Water Levels

Groundwater levels were monitored from 2010 to 2013 to observe the groundwater fluctuation behavior and seasonal variations. The groundwater level data collected between 2005 and 2009 from the Bharati Integrated Rural Development Society [17] were used for understanding the long-term trend of water levels in the study areas. As part of this research, a network of observation wells was established in both of the watersheds for monitoring the water levels before and after monsoon for the years 2005—2013. In each year, the pre-monsoon water levels are represented by the levels in the month of May and the post-monsoon water levels are represented by the levels in the month of November. The rise and fall of the water table is a direct reflection of recharge and discharge conditions in the groundwater reservoir. Hydrographs for selected bore wells are shown in Figures 3.4 a—c and 3.5 a—c, respectively, for the two watersheds.

The depth of the water level during pre-monsoon (2013) was observed to range from 6.1 to 53.16 m bgl in the Vajralavanka—Maruvavanka watershed. Deeper water levels of greater than 10 m bgl were mostly observed in the watershed. In the Peethuruvagu watershed, water levels varied from 17.45 to 87.75 m bgl. Deeper water levels of >15 m bgl were observed in most of the midstream areas and the beginning of the downstream areas in the watershed.

The depth of the water level observed during post-monsoon (2013) ranged from 1.1 to 37.7 m bgl in the Vajralavanka—Maruvavanka watershed. While the shallow water levels were observed in the midstream and extreme downstream areas, the deep water levels were observed in most parts of the upstream and small isolated patches of the northwest part of the downstream areas of the watershed. In Peethuruvagu watershed, the water levels ranged from 5.1 to 67.18 m bgl; the deep water levels were observed in most parts of the midstream and downstream areas of the watershed.

High water level fluctuation zones were observed in the midstream and upstream areas of the Vajralavanka—Maruvavanka watershed and in the midstream and downstream areas of the Peethuruvagu watershed. The aquifer in the upstream areas was observed to be shallow, since after rainfall the groundwater recharge first reaches the shallow aquifer [18], and only then recharges the unconfined aquifer followed by the deeper aquifers. Most of the shallow wells in the study areas dry up in summer; most upstream bore wells in Vajralavanka—Maruvavanka watershed and midstream bore wells in Peethuruvagu dry up in summer due to the shallow nature of the aquifer.

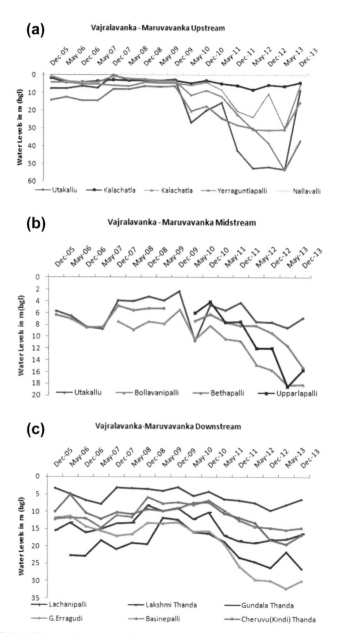

FIGURE 3.4   Hydrographs for selected bore wells in (a) upstream; (b) midstream; (c) downstream areas of Vajralavanka−Maruvavanka watershed.

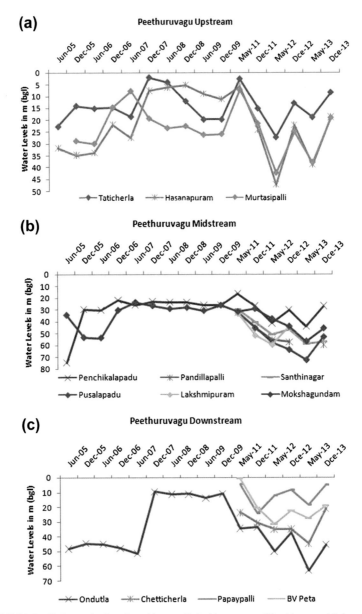

FIGURE 3.5   Hydrographs for selected bore wells in (a) upstream; (b) midstream; (c) downstream areas of Peethuruvagu watershed.

3.4.5 Estimation of Natural Recharge using LCR Method

Quantitative estimates of recharge to aquifer and changes in groundwater storage are important to manage the development of groundwater resources and assess the amount of groundwater that can be withdrawn without exceeding recharge. In hard rock areas, the most common methods for recharge estimation are groundwater balance, water table fluctuation, soil water balance, and chloride mass balance [19−23]. These methods require analysis of huge volumes of hydrological data such as precipitation, surface runoff, evaporation, and change in groundwater storage accumulated over a considerable time span, which is unavailable or unreliable in many areas [24]. Therefore, the lithologically constrained rainfall (LCR) method was adopted to estimate the natural recharge in the study area [12]. This method needs three input parameters—soil resistivity, vadose zone thickness, and precipitation. The advantages of this method include a reasonably good estimate of recharge with input parameters that can be obtained easily in the field with good accuracy, thus allowing cost-effective estimates in a shorter time compared with other methods. Rainfall data were collected from adjacent rain gauge stations in and around the watersheds. Soil resistivity was measured using geophysical methods and water levels were measured seasonally throughout the study area. Natural recharge was estimated for 8 years, from 2005 to 2013.

The natural recharge value was observed to vary throughout the watersheds depending on the heterogeneity and lithological characteristics of the area; these values are shown in Table 3.1 and Table 3.2. It is observed that these characteristics change very sharply within the close distances in the hard rocks. Since the lithological alterations occur very slowly in the geological timescale, it can be considered to be almost constant (say for ±50 years). Thus, rainfall is the only parameter varying with time for the study period.

3.4.6 Estimation of Changes in Groundwater Storage (ΔS)

To estimate the value of ΔS, the water levels are observed through a network of observation wells spread over the area during pre- and post-monsoon seasons. During the monsoon season, the recharge is greater than the extraction; therefore, the difference in the value of ΔS between the beginning and end of the monsoon season indicates the total volume of water added to the groundwater reservoir [25].

In the study areas the value of ΔS was estimated for the following nine hydrological years: June 2005−May 2006, June 2006−May 2007, June 2007−May 2008, June 2008−May 2009, June 2009−May 2010, June 2010−May 2011, June 2011−May 2012, June 2012−May 2013, and June 2013−Dec 2013.

Figures 3.6 a−c and 3.7 a−c show that the value of ΔS continuously declined from 2009 to 2012 in both watersheds. This indicates that

TABLE 3.1 Natural Recharge from rainfall in Vajralavanka–Maruvavanka watershed

No.	Village	Rainfall (in mm)									Natural recharge (in mm)								
		2005	2006	2007	2008	2009	2010	2011	2012	2013	2005	2006	2007	2008	2009	2010	2011	2012	2013
1	Lachanapalli	822	473	1046	794	480	800	376	571	832	16.68	17.56	16.84	14.03	10.22	17.84	16.77	23.37	28.17
2	Lakshmi Tanda	692	364	774	654	906	688	354	602	537	42.74	37.44	60.73	52.96	58.10	43.14	42.64	56.03	52.93
3	Gundala Tanda	692	364	774	654	906	688	354	602	537	52.51	38.75	52.01	28.88	33.45	37.64	38.78	55.34	53.24
4	G. Erragudi	822	473	1046	794	480	800	376	571	832	52.28	37.39	59.35	41.88	31.96	49.38	42.27	60.48	84.64
5	Cheruvu Tanda	822	473	1046	794	480	800	376	571	832	37.63	27.14	39.00	22.23	18.76	26.92	26.36	39.81	52.66
6	Basineapalli	822	473	1046	794	480	800	376	571	832	33.69	26.56	36.78	22.40	17.51	25.01	23.47	33.78	47.30
7	Bethapalli	822	473	1046	794	480	800	376	571	832	17.81	19.65	19.79	20.19	0.00	22.11	19.36	27.74	38.19
8	Utakallu	822	473	1046	794	480	800	376	571	832	46.72	37.72	35.69	24.82	17.65	70.59	75.88	110.18	51.85
9	Bollavanipalli	692	364	774	654	906	688	354	602	537	25.11	21.95	15.49	13.45	13.25	15.80	14.43	24.57	26.61
10	Upparlapalli	692	364	774	654	906	688	354	602	537	0.00	67.72	76.27	62.17	71.91	20.36	20.73	31.65	43.15
11	S. Rangapuram	880	609	935	642	505	710	596	604	604	27.98	45.59	42.67	37.97	34.20	39.58	56.09	54.34	44.03
12	Kalachatla	880	609	935	642	505	710	596	604	604	31.68	61.06	37.09	40.47	72.38	84.29	137.68	165.62	134.06
13	Yerraguntapalli	880	609	935	642	505	710	596	604	604	137.53	108.84	83.68	63.42	57.91	140.74	184.70	206.92	213.87
14	Nallavalli	880	609	935	642	505	710	596	604	604	75.33	64.83	51.40	37.72	34.23	52.52	108.18	109.07	67.24

TABLE 3.2 Natural Recharge from rainfall in Peethuruvagu watershed

No.	Village	Rainfall (in mm)									Natural Recharge (in mm)								
		2005	2006	2007	2008	2009	2010	2011	2012	2013	2005	2006	2007	2008	2009	2010	2011	2012	2013
1	Thaticherla	901	611	1038	851	667	1248	686	753	1093.5	33.98	27.60	19.74	26.19	29.50	37.00	36.74	33.13	34.35
2	Hasnapuram	901	611	1038	851	667	1248	686	753	1093.5	87.77	52.39	32.87	22.84	34.44	38.86	57.86	56.64	67.17
3	Muttarajupalli	901	611	1038	851	667	1248	686	753	1093.5	63.79	32.97	51.88	52.29	51.05	54.53	46.29	53.18	59.7
4	Penchikalapadu	880	609	935	642	505	710	596	604	811.4	16.95	11.38	14.63	12.33	11.62	12.98	12.59	13.92	15.18
5	Mokshagundam	880	609	935	642	505	710	596	604	811.4	36.38	20.43	23.05	19.89	16.79	21.08	20.49	25.27	27.36
6	Ondutla	1018	638	985	650	555	911	588	799	839.6	59.55	49.40	19.81	17.92	16.21	41.45	43.17	61.04	61.9

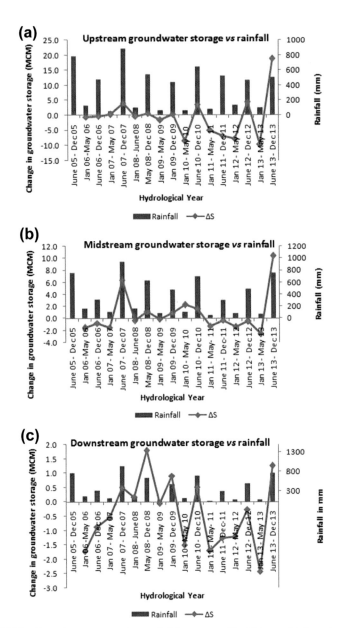

FIGURE 3.6 Groundwater storage changes in (a) upstream; (b) midstream; (c) downstream areas of Vajralavanka–Maruvavanka watershed.

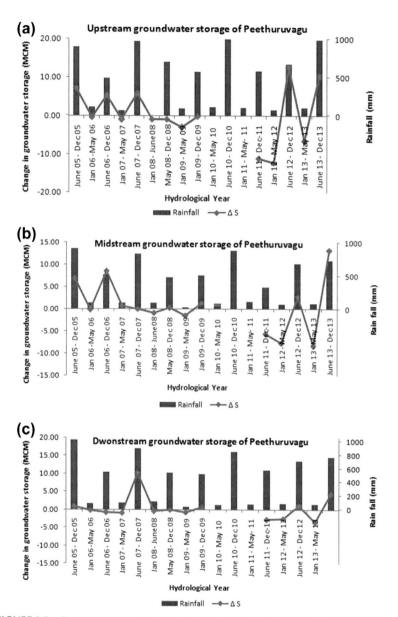

FIGURE 3.7   Groundwater storage changes in (a) upstream; (b) midstream; (c) downstream areas of Peethuruvagu watershed.

groundwater extraction is greater than the annual groundwater recharge, which may be due to low rainfall compared to the annual average rainfall in both watersheds. It is further observed that the value of ΔS is positive during the hydrological cycle June 2012–December 2012. This may be due to the transition in agricultural land use—from irrigated crops to horticulture—in the upstream areas of the Vajralavanka–Maruvavanka watershed and the entire Peethuruvagu watershed.

3.5 DISCUSSION

In the study areas, the soil type includes clayey loam, red loam, and sandy loam with variable thickness, which forms the top layer. Soil thickness varies from 0.7 to 2.4 m and from 0.4 to 2.6 m and the apparent resistivity varies from 3 to 274 Ωm and from 3 to 41 Ωm, respectively, in the Vajralavanka–Maruvavanka and Peethuruvagu watersheds. Soil type and thickness play an extremely important role in controlling the water movement over and through a watershed. The thickness of the soils are high in the upstream areas, followed by downstream areas in the Vajralavanka–Maruvavanka watershed, while the lower portion of the upstream areas, the beginning of the midstream areas, and the lower portion of the downstream areas have high soil thickness in the Peethuruvagu watershed. These areas are the controlling locations for infiltration rates in the hydrological process in the study areas. Similarly, an area with soil cover followed by hard rock is not suitable for any intervention. These conditions prevail in the upstream and midstream areas of the Vajralavanka–Maruvavanka watershed and the lower portion of the upstream areas, the beginning to the center of the midstream areas, and the lower portion of the midstream areas in the Peethuruvagu watershed.

Based on the geophysical surveys, the aquifer is divided into two layers, i.e., weathered and fractured zones, in both the study areas. The thickness and resistivity of the weathered zone varies as follows: in regions where thickness varies from 4.5 to 20 m the resistivity varies from 4 to 223 Ωm, and in regions where thickness varies from 4.1 to 13.8 m the resistivity varies from 3 to 131 Ωm. Similarly, the thickness and resistivity of the fractured/fissured zone varies as follows: in regions where the thickness varies from 7 to 71 m the resistivity varies from 115 to 1616 Ωm, and in regions where the thickness varies from 7 to 32 m the resistivity varies from 252 to 1275 Ωm. Below these regions exists a basement that is encountered between 10 and 83 m and 12 and 45 m, respectively, in the Vajralavanka–Maruvavanka and Peethuruvagu watersheds.

The thickness of the weathered and fractured zones is classified as low, moderate, and high in the study areas. The Vajralavanka–Maruvavanka watershed is observed to comprise predominantly moderate weathered zone thickness, followed by shallow weathering in the upstream areas, NW of the midstream areas, and the lower part of the downstream areas, while a deep weathered zone is observed in the central part of the downstream areas.

However, a fractured zone is also observed in most parts of the upstream and SW portion of the downstream areas—the zone has moderate thickness in most of the midstream and central part of the downstream areas, while the deep fractured zone is observed in the edges of the SE portion of the midstream and downstream areas, as well as an isolated patch in the midstream areas. The depth of the basement map shows a crescent shape in the Vajralavanka—Maruvavanka watershed—the upper part of the crescent is the shallow basement, followed by moderate and deep thickness.

In the upstream areas of the Peethuruvagu watershed, the weathered zone thickness is observed to be high at the starting point, which continues up to Hasnapuram Village, followed by moderate and low zones. In the midstream areas, the weathered zone thickness varies from low to moderate, while near Laxmipuram Village an isolated high zone is observed. Further, in the downstream areas starting near Chetticherla Village moderate weathered thickness is observed, which is followed by high weathered zone thickness. A moderately thick fractured zone is observed up to Muttarajupalli Village, followed by shallow thickness in the upstream areas. The study shows that the fractured zone thickness increases from near Santhinagar Village and then decreases from moderate to shallow up to the end of the midstream areas. The thickness of the fractured zone is observed to increase gradually from the beginning to the end of the downstream areas. The depth to the basement varies from 12 to 45 m in the Peethuruvagu watershed. A shallow to deep basement is observed in the center of the midstream areas, which then gradually decreases toward the end of the midstream areas. The depth of the basement is observed to have an increasing trend in the downstream areas of the watershed.

These studies reveal that weathering thickness of more than 5 m is feasible for water recharge by way of water spreading methods such as check dams, percolation tanks, and farm ponds in both the study areas. The areas tapping groundwater from deep fractured zones are ideal for artificial recharge through injection wells in both the study areas.

The soil type and thickness are important for implementation of a successful watershed enhancement or management activity in the study areas. The low and moderate thickness of weathered zones after rainfall infiltration contributes much to the shallow subsurface runoff, and this will appear on the soil cover or nearby streambeds. These conditions were observed near Penchikalapadu tank in the Peethuruvagu watershed—within a few days after rainfall, the tank becomes dry.

It is observed that high weathered zone thickness facilitates the maximum amount of infiltration to saturate both the weathered and fractured aquifers in the study area. Similarly, low and moderate thicknesses of fractured zone areas are indicated by less interconnectivity resulting in moderate to low yields. On the other hand, highly fractured zones will possibly result in good interconnectivity and good yield of water in the study area.

Thus, a detailed knowledge of the subsurface aquifer geometry and properties are important at the watershed scale for implementing the watershed management decisions and programs, particularly in the hard rock areas. Furthermore, adequate surface water and aquifers of suitable nature are a prerequisite for interventions, and aquifer suitability in terms of storage space and permeability is also required. Even very high permeability results in loss of recharge water due to subsurface runoff, while low permeability reduces the rate, as observed in the subsurface runoff near the Penchikalapadu tank in the Peethuruvagu watershed.

Physical characteristics including permeability and water level gradient or slope are important for understanding flow direction and identification of overexploited zones. In both study areas, water levels were observed from 2005 to 2013, and were found to be declining in both the seasons. This indicates that the farmers are exploiting the static groundwater resource in the study areas.

High water level fluctuation zones are observed in the midstream and upstream areas of the Vajralavanka−Maruvavanka watershed and in the midstream and downstream areas of the Peethuruvagu watershed. The interventions required for a watershed tapping only the weathered zone are different from those tapping the fractured zone.

It is observed that groundwater is tapped from deep aquifers in both watersheds. In such areas, the injection method will produce the desirable results for recharging the aquifer. In this method injection of rainwater through injection bore wells or shallow tube wells may be facilitated.

Similarly, in the upstream areas of the Vajralavanka−Maruvavanka watershed and in Karadikonda, the aquifer is shallow. In such areas, groundwater recharge is affected through water-spreading methods such as check dams, percolation tanks, and form ponds. Trends of water level fluctuations are the key to assessing the groundwater recharge, which directly depicts the changes in the groundwater storage in a given area.

We observe that natural groundwater recharge varies throughout the study area. This indicates that the hard rocks are heterogeneous and the lithological characteristics change sharply within short distances. Hence, estimation of rainfall recharge (R) is very useful for the development of groundwater allocation policies. Annual groundwater use must be significantly less than the groundwater recharge to ensure that wells do not go dry. Therefore, sustainable groundwater allocation policies should ensure that allocation is less than the recharge (A) and that the actual use is less than the allocation (U), i.e., $R > A > U$. It is observed that the groundwater levels in the study areas have been continuously declining from 2005 to 2012. This is because the allocation and the actual use of groundwater in these watersheds is greater than the rainwater recharge.

A qualitative estimation of the change in groundwater storage is important to manage the development of groundwater resources and to assess the amount of groundwater that can be withdrawn without exceeding recharge. This is particularly important in regions with large demands for groundwater supplies, where such resources are the key to economic development. The values of ΔS are observed to be continuously declining from 2009 to 2012, which indicates that the groundwater extraction is more than the annual groundwater recharge in the study areas. This is an alarming situation. However, the values of ΔS are found to be positive during the June 2012—December 2013 hydrological cycle, which may be due to a shift in the cropping pattern from irrigated crops to horticulture in the upstream areas of the Vajralavanka—Maruvavanka watershed and the entire Peethuruvagu watershed.

3.6 ZONES SUITABLE FOR DIFFERENT TYPES OF ARTIFICIAL RECHARGE INTERVENTIONS

In both study areas, dependence on groundwater resources has increased tremendously in recent years due to the vagaries of monsoon and scarcity of surface water. Further, groundwater extraction is more than the natural recharge by rainfall. By visualizing the groundwater basin as a large natural underground reservoir, it is clear that overdevelopment of groundwater resources in one portion of a basin directly affects water supplies throughout the remainder of the basin. This has led to basin-wide planning and development of groundwater. To maintain sustainability, a hydrologic equilibrium must exist between all waters entering and leaving the basin [26]. The imbalance between the rapid and excessive discharge and slow and deficient recharge of the groundwater must be reduced through scientifically designed artificial groundwater recharge systems.

The essential requirements of artificial recharge systems are availability of noncommitted runoff, suitable site selection, and site-specific design of recharge structures. Recharging will be effective only if the shallow subsurface and underground formation has enough space to hold water recharged into it. Further, porosity and the water holding and releasing capacity of formations and geological structures vary with the presence of lineaments, fractures, and folds. Similarly, the surface soil properties such as infiltration capacity, terrain slope, and drainage density determine the infiltration opportunity time and thereby the recharge rate to the aquifer material. Hence, selection of a suitable recharge site is an important step in the artificial recharge planning.

Based on the aquifer geometry, the study areas can be categorized into three zones:

1. Moderate to deep weathering and fracturing zone, which is suitable for artificial recharge measures using water-spreading methods
2. Areas with deep fractures, which are suitable for artificial recharge methods such as injection methods
3. Areas with a very shallow basement, which are not suitable for any intervention (Figure 3.8 a and b)

Based on the drainage order, mini-percolation and percolation tanks can be proposed on the first- to third-order stream areas belonging to the first

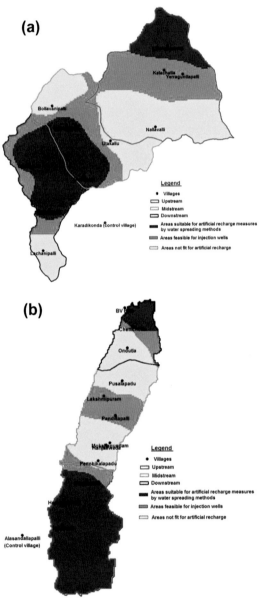

FIGURE 3.8 Zones suitable for different types of artificial recharge interventions in (a) Vajralavanka—Maruvavanka; (b) Peethuruvagu watershed.

zone as identified in the map, while check dams can be proposed when the area gets into plane topography. Injection wells may be proposed preferably in the streambeds of the areas belonging to the second zone. Furthermore, interventions such as farm ponds, gully plugging, and control trenching can also be considered not only for groundwater recharge but also for maintaining the required soil moisture during dry seasons. The areas belonging to the third zone cannot be considered for any artificial recharge intervention, as any amount of inbounding may lead to surface outflow directly from the bottom of the soil cover.

3.7 CONCLUSION

The areas under study depend extensively on groundwater for all utilizations and almost entirely for drinking water purposes. The groundwater system gets recharged from the rainfall, which is the main source. However, rainfall occurs for a very limited period during the year. Climate change and climate variability have shown that the number of rainy days is reducing, even if the total yearly rainfall remains unchanged.

In the past, the demand for groundwater has been much less compared with the recharge because of the rainfall. The system has been working very well without any deficit of groundwater and the renewability of the system was ensured. However, the demand in the form of withdrawal of groundwater has increased severalfold during the last two decades. Because of these factors, the groundwater draft has increased over the years while the natural groundwater recharge remains the same or has even reduced due to the change in the variability of the rainfall. As a result of this imbalance, problems such as overexploitation, progressive water level decline, and deterioration of groundwater quality have cropped up and have started to adversely affect the drinking water sources in the study areas. Drinking water scarcity is aggressive during summer in S. Rangapuram and the Utakallu villages in the Vajralavanka—Maruvavanka watershed as well as most parts of the Peethuruvagu watershed, including Alasandapalli (control village), due to overexploitation and the shallow nature of the aquifer. Thus, in most cases, the system has not been completely renewable and is certainly not sustainable.

The overexploitation and the progressive groundwater depletion situation can be remediated in two ways: by reducing our demand to that of the rainfall recharge or by enhancing the recharge artificially to match the demand. However, for both actions, complete knowledge of the system is a must. Hence, for a sustainable watershed program, planners need to understand the aquifer geometry, water level trends, groundwater recharge, and change in groundwater storage. Such programs are known as interventions or more precisely watershed development (WSD). WSD initially has been purely on an ad hoc basis, which clearly could not solve the problem in an optimal way.

The present study has utilized the state-of-the art techniques of geophysical investigations and has tried to prepare a realistic geometry of the existing aquifer system by making use of punctual information such as ERI, borehole logging, etc. The recharge from rainfall has been estimated through a newly developed approach using nonlinear relationships of the most relevant parameters. Thus, a groundwater balance prepared on the basis of advanced techniques has been more realistic and has provided the information on groundwater availability at various space and timescales.

This knowledge is needed to implement area-specific WSD programs with good soil cover, weathering thickness, and so on. The areas tapping the first fracture can be treated with water-spreading methods, while the areas tapping deep fractures should have injection wells. Thus, a complete knowledge of the system with details on the varying weathered thickness and presence of fractures as well as the groundwater availability helps in judiciously planning the WSD and in appropriately and optimally planning the interventions.

ACKNOWLEDGMENTS

We are grateful to the Director of the National Geographic Research Institute for permission to publish this work. We also thank Dr. SAR Hashimi, Ret. Deputy Director A.P. Groundwater Department, Hyderabad, for useful discussion and support during the preparation of this manuscript. This work was performed under an Indo-Australian project funded by the Australian Center for International Agricultural Research.

REFERENCES

[1] GSI. Geological Quadrangle map 57 F. Madras: Printed at Info maps; 1995.

[2] GSI. Geological Quadrangle map 57 E. Hyderabad: Printed the map printing division; 2004.

[3] King W. Kadapah and Karnul formations in Madras Presidency. Mem Geol Surv India8 1872:1−293.

[4] de Franco R, Biella G, Tosi L, Teatini P, Lozej A, Chiozzotto B, Giada M, Rizzetto F, Claude C, Mayer A, Bassan V, Gasparetto Stori G. Monitoring the saltwater intrusion by time lapse electrical resistivity tomography: the chioggia test site (Venice Lagoon, Italy). Jour Appl Geophys 2009;69(3−4):117−30.

[5] Brunet P, Rémi C, Christopher B. Monitoring soil water Content and deficit using electrical resistivity tomography (ERT) − a case study in the Cevennes area, France J Hydrol 2010;380(1−2):146−53.

[6] Zaidi FK, Osama MK, Kassem. Use of electrical resistivity tomography in delineating zones of groundwater potential in arid regions: a case study from Diriyah region of Saudi Arabia. Arab J Geosci 2012;5:327−33.

[7] Loke MH, Barker RD. Rapid least-squares inversion of apparent resistivity pseudosections using a quasi-Newton method. Geophys Prospect44 1996:131−52.

[8] Store H, Storz W, Jacobs F. Electrical resistivity tomography to investigate geological structures of earth's upper crust. Geophys Prospect48 2000:455−71.

[9]   Loke MH. Electrical imaging surveys for environmental and engineering studies: A practical guide to 2-D and 3-D surveys; 1997. 61 pp.

[10] Loke, M.H. (1997b). "Software: RES2DINV". 2D interpretation for DC resistivity and IP for windows 95. Copyright by M.H. Loke. 5, Cangkat Minden Lorong 6, Minden Heights, 11700 Penang, Malaysia.

[11] Loke MH. "Software: RES 2D INV", Ver. 3.50, Rapid 2-D resistivity and IP inversion using the least square method; 2002.

[12] Chandra S, Ahmed S, Rangarajan R. Lithologically constrained rainfall (LCR) method for estimating spatio-temporal recharge distribution in crystalline rocks. Jour Hydrology402 2012:250−60.

[13] CGWB. Groundwater information Anantapur District, A.P.. Southern region Hyderabad: Central Groundwater Board, Ministry of water resources, Govt. of India; 2007.

[14] Woolhiser DA, Fedors RW, Smith RE, Stothoff SA. Estimating infiltration in the upper split watershed, Yucca Moutain, Nevada. Jour Hydrol Eng 2006;11:123−33.

[15] Pelletier JD, Rasmussen C. Geomorphically based predictive mapping of soil thickness in upland watersheds; 2009. Water Resources Research45, WO9417,doi:10.1029/2008 WR007319.

[16] CGWB. Groundwater information Prakasam District, A.P.. Southern region Hyderabad: Central Groundwater Board, Ministry of water resources, Govt. of India; 2007.

[17] Bharati Integrated Rural Development Society (NGO). Nandyal, Kurnool (Dist).

[18] White PA, Hong YS, Murry D, Scott DM, Thorpe HR. Evaluation of regional models of rainfall recharge to groundwater by comparison with lysimeter measurements, Canterbury, New Zealand. Jour Hydrol (NZ) 2003;42(1):39−64.

[19] Sophocleous MA. Combining the soil water 664 balance and water-level fluctuation methods to estimate natural ground water recharge: practical aspects. Jour Hydrol 1991;124:229−41.

[20] Moon SK, Woo NC, Leeb KS. Statistical analysis of hydrographs and water-table fluctuation to estimate groundwater recharge. Jour Hydrol 2004;292:198−209.

[21] Maréchal JC, Dewandel B, Ahmed S, Galeazzi L, Zaidi FK. Combined estimation of specific yield and natural recharge in a semi-arid groundwater basin with irrigated agriculture. Jour. Hydrol 2006;329(1−2):281−93.

[22] Batelaan O, Smedt F. GIS-based recharge estimation by coupling surface− subsurface water balances. Jour Hydrol 2007;337(3/4):337−55.

[23] Sibanda T, Nonner JC, Uhlenbrook S. Comparison of groundwater recharge estimation methods for the semi-arid Nyamandhlovu area, Zimbabwe. Hydrogeol Jour 2009;17: 1427−41.

[24] Sukhija BS, Rama FASc. Evaluation of groundwater recharge in semi-arid region of India using environmental tritium. In: Indian Acad. Sci. LXXVII; 1973. Sec. 'A', Vol. VI, pp. 279−292.

[25] Kumar CP. Assessment of groundwater potential. Int J Eng Sci 2012;1(1):64−79.

[26] Sooraj KPV, Mathew EK. GIS and remote sensing for artificial recharge study in a degraded western ghat terrain; 2013. http://www.csre.iitb.ac.in/%7Ecsre/conf/wp-content/uploads/fullpapers/OS4/OS4_14.pdf.

Chapter 4

Application of a Simple Integrated Surface Water and Groundwater Model to Assess Mesoscale Watershed Development

Paul Pavelic*, Jian Xie *, §, P.D. Sreedevi ¶, Shakeel Ahmed ¶ and Daniel Bernet #

*International Water Management Institute, Vientiane, Lao PDR, §Beijing Normal University, College of Water Sciences, Beijing, PR China, ¶CSIR-National Geophysical Research Institute, Hyderabad, India, #Institute of Geography & Oeschger Centre for Climate Change Research, University of Bern, Bern, Switzerland

Chapter Outline

4.1 INTRODUCTION

Watershed development (WSD) programs have achieved significant positive biophysical and socioeconomic benefits in India since their introduction in the 1980s, and are still actively promoted and implemented throughout the drier, drought-prone regions of the country [1–3]. The regions and the specific types of benefits accrued are described in Chapter 7. Watersheds, from an Indian rainfed production system's context, are complex landscapes that entail

Integrated Assessment of Scale Impacts of Watershed Intervention
http://dx.doi.org/10.1016/B978-0-12-800067-0.00004-9. Copyright © 2015 Elsevier Inc. All rights reserved.

socio-political-ecological elements that play a critical role in shaping the livelihood support base for the food, social, and economic security of rural communities.

In simple terms, WSD promotes livelihood improvement by increasing agricultural productivity brought about by the increased availability of local surface water, soil moisture, and groundwater. Specifically, *ex situ* construction of structures that impound water along drainage lines (check dams), along with *in situ* water management practices (contour bunds and pit excavations that retain water in fields) improve infiltration capacity and water-holding capacity of the soil, result in higher crop water availability, replenish groundwater reserves, and reduce erosion and land degradation [4−6].

When looked at from the narrow, yet vitally important hydrological perspective, WSD is a form of conjunctive water management as it draws upon both surface water and groundwater supplies and takes advantage of their differing characteristics, particularly in terms of temporal and spatial variability and availability [7,8]. Conjunctive water management recognizes the natural hydrologic connection between surface water and groundwater, and attempts to develop and manage water resources in an integrated manner to improve efficiency and equity. However, goals such as this are made difficult since WSD targets areas where water scarcity is combined with high water demand. Therefore, issues related to trade-offs associated with shifting water in space and time and inequalities can emerge among communities in different watersheds or even within an individual watershed.

Assessing the impacts of WSD must be preconditioned on an understanding of the upstream/downstream relationships that are naturally present within watersheds. Even though these can be quite complex as well as site and scale specific [6], some generalizations can be made. At the individual watershed level, it is the upstream areas that are typically most water limited as water yields and storage capacities are restricted; on the other hand, the middle or lower parts of the watershed are relatively more water abundant. As the scale considered increases, and/or as a greater number of interventions are established in the upstream areas to capture and use water, the natural endowment of the downstream areas is diminished and these areas are the most impacted in relative terms.

Evidence has shown that the natural differences that exist between upstream and downstream farmers can be enhanced by WSD [1,9,10]. The potential negative impacts on communities downstream have been recognized from the WSD activities, and include reduced inflows to downstream reservoirs [11] or out of watersheds [1]. On the other hand, the positive net benefits of WSD that emerge at the local scale do not necessarily carry over to the larger scale when externalities are fully accounted for [5]. Kumar et al. [7], for example, demonstrated the case where the benefits derived from WSD in upper watersheds of the Krishna River Basin are generally outweighed by the higher opportunity costs if the water was used for higher valued purposes downstream. Thus, WSD leads to increased agricultural development and intensification as well as significantly reduced downstream runoff to major surface storages and greater overexploitation of groundwater [1]. With the depletion of groundwater resources in downstream areas, more intensive and deeper drilling of wells is needed, which affect the poor most, leading to inequitable access to those resources. Thus, WSD constrains the downstream movement of surface water runoff and potentially the flow of groundwater, depending on the degree of interception through pumping.

Various approaches have been developed for evaluating the positive and negative impacts of WSD. In the Indian context, there has been a tendency to employ simple yet conceptually sound approaches based on water-accounting principles. For example, Batchelor et al. [12] performed a water balance for the Chinnahagari watershed in Karnataka and found that WSD had a significant impact on the patterns of water use that had the potential to exacerbate inequity issues. Glendenning and Vervoort [13,14] developed a conceptual water balance model to understand the watershed scale impacts of WSD in the Arvari River Catchment, Rajasthan, India. The model revealed that WSD increased the overall availability of water for irrigated agriculture.

More complex, process-based models of WSD systems have traditionally been developed with a focus on surface water or groundwater resources separately, regardless of the degree or complexity of the interactions between them. For instance, Garg et al. [15] applied SWAT, the surface hydrologic model, to the Kothapally watershed in Andhra Pradesh; while in terms of groundwater modeling, MODFLOW is most commonly used [16]. On the other hand, coupled hydrologic models are needed if the conjunctive nature of WSD is to be taken into account meaningfully. Such models are useful for analyzing complex water resource problems because they consider linkages and feedback processes affecting the dynamics of evapotranspiration, surface runoff, moisture movement through the unsaturated zone, and groundwater interactions [17]. To our knowledge, the models developed for coupled hydrologic modeling have not been applied to any significant degree in a WSD context in India.

Hence, modeling, be it simple or complex, is recognized as a vital and effective tool to aid the planning and implementation of WSD programs. However, there appears to be a large divide between the information and the tools available for scientific works such as those described earlier, and the operational practices and needs of the decision makers and implementing agencies. WSD planning is usually made with very rudimentary information about the hydrological regime and the associated effects and externalities that new projects may impose.

The aim of this chapter is to describe a new distributed model based on simple mass balance principles that can couple surface water and groundwater flow systems and can be applied in rainfed areas at any representative scale to assess the impacts of watershed interventions, land use, and the effects related to climate change on water availability. The model is applied to the mesoscale study site at Prakasam District in Andhra Pradesh.

4.2 MODEL DEVELOPMENT

4.2.1 Model Principles

Representing the key processes within the hydrological cycle with sufficient accuracy is a basic necessity of the model. While the simpler models describe these processes in a more limited way that makes them disadvantaged in terms of accuracy, they provide advantages in terms of reduced data requirements and potential for application by non-modeling specialists. Simplicity and utility are of paramount importance for maximizing the usefulness of the model for policy makers, planners, and practitioners. Employing the principle of parsimony, the simplest possible equations and algorithms were selected and a number of simplifying assumptions were made during the development of the model.

The model is based on the conservation of (water) mass principles, whereby all water inputs are matched by the sum of outputs and storage changes within the system. The study area is defined by hydrological boundaries based on Digital Elevation Model (DEM) data in raster format. The spatially distributed model domain is represented by gridded cells that distinguish four land coverage types: (1) impervious cells, (2) drainage cells, (3) surface water storage cells, and (4) vegetated cells.

4.2.2 Model Components

The following section, which describes approaches used in various water balance components, is derived largely from the more detailed descriptions reported by Bernet [18].

Evapotranspiration of crops and open water surface is estimated using the standardized Penman–Monteith equation, which is dependent upon climatic parameters such as radiation, air temperature, air humidity, wind speed, and crop coefficients.

Surface runoff is determined using the Soil Conservation Service Curve Number method [19], which assumes that direct runoff is characterized by a curve number that is dependent on watershed characteristics and accounts for antecedent moisture conditions.

Soil moisture storage is determined from the standard soil water balance equation, which accounts for the effective precipitation and irrigation, evapotranspiration, groundwater recharge, and resultant change in soil moisture. Recharge is assumed to occur only when the field capacity of the soil is exceeded.

Groundwater flow is described by the explicit finite difference method, which was chosen because it is relatively easy to implement. The aquifer is assumed to be unconfined with homogenous and isotropic characteristics.

Water can be used by tapping either open surface water or groundwater. The redistribution of water by pumping is handled in a simplified manner in which water extracted from one grid cell is used locally. Therefore, only vertical transfer of water is considered. In the case of pumping wells, water is used close to the well and is not transported over large distances. Although transfer of water for irrigation from open water bodies is not accounted for, this is not considered a major constraint because the major source of water for irrigation in the semi-arid rainfed areas where WSD is mostly undertaken is from groundwater [20]. Domestic, livestock, and industrial abstractions consume water fully without return flow, whereas irrigation may generate return flows.

4.2.3 Model Coupling

The interactions between surface water and groundwater are typically very complex, but can be simplified when the groundwater table is assumed to be well below the streambed and the root zone. This is generally the case in the drought-prone areas targeted for WSD where the intensity of groundwater use is high and the groundwater levels are deep. Therefore, any change in groundwater storage due to processes other than irrigation and consumptive abstractions are not accounted for.

Groundwater and surface water interactions are not fully coupled, and are limited to downward-directed recharge or upward transfer from pumping.

4.2.4 Model Implementation and Data Requirements

The current configuration of the model is scripted in MATLAB. Its structure comprises 11 functions plus three scripts responsible for declaration, pre-calculation, and initialization [18]. Input data files are provided in both ASCII and Excel spreadsheet formats.

The suite of data needed to run the model is listed in Table 4.1. Much of the input data can easily be acquired from open online sources, while some can be approximated from relevant literature and/or applying the "rule of thumb." Surface or groundwater hydrograph data, if available, enable the fine-tuning of the model.

4.3 SITE DESCRIPTION AND MODEL PARAMETERIZATION

The model was applied to the Peethuruvagu watershed in Andhra Pradesh (78.99−79.16°E, 15.33−15.60°N; Figure 4.1). The watershed, nested between two parallel hill ranges, covers an area of about 95 km^2 and consists of 14 villages (note that only selected villages are shown in Figure 4.1). Hydrologically, the watershed lies within the Gundlakamma River Basin. The topographic elevation varies by >500 m over a longitudinal distance of just

TABLE 4.1 Overview of data requirements of the model

Group	Variable	Typical source
Climate	Air temperature (max/min)	Climate station
	Relative humidity	
	Radiation	
	Wind speed	
	Rainfall	
Topography	DEM	Online (e.g., http://asterweb.jpl.nasa.gov/data_products.asp)
Soil properties	Soil depth	Soil maps or literature
	Water content	
Land use	Land cover type	Remote sensing images (e.g., http://www.landsat.org/ortho)
	Cropping pattern	
Crop characteristics	Crop coefficient	Literature for various crop types
	Rooting depth (max)	
	Irrigated volume	
Hydrogeology	Aquifer depth	Hydrogeological reports
	Hydraulic conductivity	Literature/rule of thumb and calibration
	Storativity	
	Initial groundwater level	
WSD	Storage capacities of surface water structures	Water table contour maps
	Surface runoff	Literature or estimated
Data for model refinement	Groundwater levels	Field measurements or literature
		Observation well data

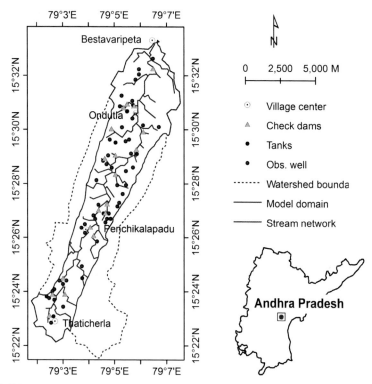

FIGURE 4.1 Location of the Peethuruvagu watershed indicating the watershed area, modeled domain, watershed development interventions, and observation wells.

over 20 km. This long, narrow watershed drains toward the north. The soils include shallower red soils and deeper black soils with depths of up to 90 cm.

The average rainfall recorded at Penchikalapadu rain gauge station (79.06°E, 15.44°N, 206 m asl[1]) from 2004 to 2010 was 628 mm/year. Records from this period show that 56% of the rainfall occurs during southwest monsoon period (June to September) and 33% during northeast monsoon period (October to December). The average maximum and minimum temperatures are approximately 41 and 25°C, respectively. Climate warming is significant (0.16°C/year; $R^2 = 0.78$) according to the derived minimum annual temperatures in Peethuruvagu from 1974 to 2010.

WSD was established in this area in 2004, and the surface water intervention structures include check dams, rock-filled dams, farm ponds, and kuntas, along with the tanks that have been historically present. In the watershed, only the 13 check dams and 8 tanks that represent the major

---

1. Meters above sea level.

structures were considered; the other structures were not considered for the sake of simplicity and lack of relevant information. The maximum capacities of the tanks and check dams were derived from five structures with available information. Leakage (percolation) takes place from the bed of the structure when there is excess water to percolate according to the modeled water balance.

Hydrogeologically, the area features two major aquifers: the unconfined aquifer consisting of highly fractured cumbum shales and the confined aquifer consisting of Proterozoic quartzite. The deeper nonweathered shale acts as a confining layer. Only the unconfined aquifer is represented explicitly in the model with a thickness of $\sim$45 m. The assigned hydraulic conductivity of the weathered cumbum shales is 5×10^{-5} m/s and the specific yield is 0.115. The depth to water table ranges from 2 m in the stream valleys to 40 m in the upland areas (October 2005 data).

The study area was discretized into a total of 14,703 grid cells (87 columns and 169 rows), each having dimensions: 120 m $\times$ 120 m. The model domain extends approximately $\sim$60 km^2 and accounts for $\sim$63% of the watershed area. Areas elevated above 327 m asl where cultivation is not practiced were not explicitly included in the model domain. However, the accumulated surface runoff was taken into account from DEM data and imported into the modeled area.

The groundwater boundary conditions assigned on the basis of water table maps include no-flow boundaries over most of the boundary perimeter, constant head boundary at the outlet, and specified flux upstream. The base of the aquifer is set as a low permeability boundary, allowing vertical exchange of water with the underlying confined aquifer system.

The dominant crops grown in the watershed include paddy, groundnut, red gram, bajra, jowar, and cotton. A simplified cropping calendar was established with the aid of Landsat Remote Sensing Images from the years 2000, 2001, and 2006. The modeled crop schedule was assumed to be characterized by cotton during the wet season (June to November) and vegetables during the dry season (December to May). Variable rooting depths, constrained by soil thickness and crop factors, were adopted to represent the plant growth over time in the vegetated cells of the model. Evapotranspiration is the combination of evaporation from bare soil and transpiration of the crops for vegetated cells.

All water used for irrigation is assumed to be pumped from wells, with no assumed pumping from tanks or check dams, which is close to reality. The assigned groundwater pumping rates, extrapolated from well discharge data collected by the Andhra Pradesh Farmer Managed Groundwater Systems (APFAMGS) project, was 35 mm/month for June to November, 28 mm/month for December to February, and 20 mm/month for March to May. The smaller areas under irrigation during the dry seasons appear to outweigh the relatively higher irrigation demand in these periods.

For convenience, the seasonal irrigation patterns are kept constant over the years, irrespective of the variable water demands driven largely by rainfall variability.

4.4 RESULTS

4.4.1 Model Performance

The behavior of the model is illustrated in terms of groundwater levels, because these can be easily compared against the measured data. In addition, they offer a good general indication of model performance since they reflect the storage changes in the aquifer as a function of the surface runoff, evapotranspiration, recharge, and pumping.

Figure 4.2 shows the comparison between the simulated and measured hydrographs taken for selected observation wells over a 5 year period (2005–2009). It can be seen that some of the general trends are captured by the model, but the specific characteristics of individual wells are not reflected very well. This is believed to be due to the simplifications associated with the model, particularly in terms of the fixed annual irrigation patterns. Dynamic irrigation demand requires reliable simulation of crop conditions and is beyond the scope of the model.

The model and observation data consistently show that between 2005 and 2009 groundwater levels became progressively shallower. This trend is thought to be due to increased recharge associated with the WSD structures, in conjunction with the changed cropping patterns and the more judicious use of groundwater brought about by the implementation of the APFAMGS project

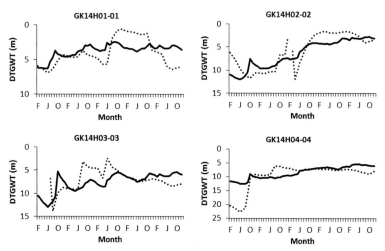

FIGURE 4.2   Comparison between simulated (solid line) and observed (dotted line) depths to groundwater table (DTGWT) from February 2005 to December 2009.

focusing on community-driven groundwater management [21]. The groundwater flow field across the watershed is also reproduced to a very high degree by the model with discrepancies largely restrained to ±2 m (data not shown).

4.4.2 Hydrological Processes

Figure 4.3 presents the simulated time series outputs for the various water balance components. Runoff events (Figure 4.3, top) are infrequent and short lived, with only two major events occurring over the 5 years. Despite the unavailability of surface hydrological observations, the simulated runoff (as well as evapotranspiration) shows a reasonable response to the rainfall pattern during the simulation period. The runoff coefficient ranges from <0.01 to 0.7, with an average of 0.1.

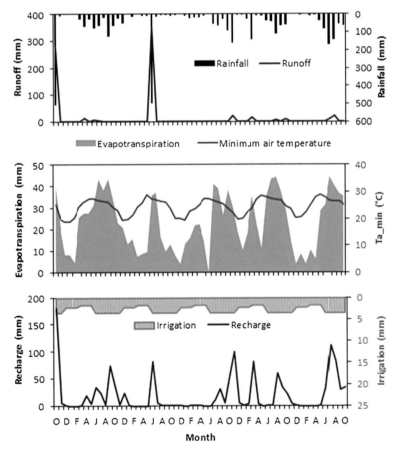

FIGURE 4.3 Simulated hydrologic components under the "with watershed development and irrigation" scenario. Note: The fluxes shown are averages across the model domain from October 2005 to October 2010.

Evapotranspiration (Figure 4.3, middle) is generally elevated during the cropping season. This corresponds reasonably well with minimum air temperatures (Ta_min) as well as rainfall, indicating the large degree of control that these two factors play (Figure 4.3, middle and bottom). Despite the consistently high evapotranspiration with warmer temperatures and intensive irrigation, moderate rainfall events rather than major storms are responsible for much of the evapotranspiration.

Groundwater recharge (Figure 4.3, bottom) is biased toward the months of June through November when the southwest monsoon rainfall prevails and evapotranspiration rates are highest. Soil infiltration rate test in several of the major recharge structures in the watershed (K. Brandes, personal communication) reveals high infiltration potential (4−49 mm/h) suggesting that the low intensity rainfall events rather than soil conditions constrain groundwater recharge.

4.4.3 Scenario Modeling

The potential to apply the model to assess alternative management options was tested for a limited selection of scenarios, which are of a relatively generic nature. The effect of those scenarios on the hydrologic processes and specifically in terms of water availability were examined in terms of the major water balance components. The baseline case was without WSD and irrigation, while the three options considered were with irrigation but no WSD, with WSD and irrigation, and with WSD and irrigation, together with a reduced rainfall by 10% (Table 4.2).

TABLE 4.2 Soil water balance components for various scenarios

Water balance components (mm)	Scenario[a]			
	No WSD or IRR	IRR only	IRR and WSD	IRR and WSD (low RF)
Rainfall[b]	547	547	547	493
Irrigation	0	56	56	56
Evapotranspiration	417	441	441	419
Surface outflow	39	39	31	24
Recharge	84	94	118	97
Δ Soil moisture storage	7	6	13	9

[a]WSD: watershed development; IRR: irrigation; RF: rainfall.
[b]Average for climatic year October 2005 to October 2010.

The results show that water and land management activities have a profound effect on the water balance components in the watershed. Before the introduction of the WSD program, approximately 76% of the rainfall was evapotranspired, 15% was recharged to the aquifer, and 7% flowed out of the watershed. With the WSD program in place, evapotranspiration, available soil moisture, and groundwater recharge all increased, while surface runoff decreased—runoff discharging out of the watershed was reduced from 39 to 31 mm/year. If dry climate was to occur in the future, a generic 10% decrease in annual rainfall would result in a 23% decrease in the amount of surface runoff and an 18% decline in the amount of recharge.

4.5 DISCUSSION

In this chapter, we have shown the application of the model at the mesoscale. Previously, Bernet [18] applied the model at the microscale at Kothapally watershed in Andhra Pradesh where it was compared against calibrated SWAT simulation results, both in an uncalibrated state where best estimates were used as well as in a refined state where key parameters were adjusted. Further, the model was able to provide information on groundwater dynamics that SWAT could not achieve.

Currently the model is written in MATLAB, which makes it accessible only to modeling specialists. However, it is being improved such that it is more user friendly for nonspecialists. Input from key stakeholders from government and nongovernment agencies in India are being sought to ensure that the model meets the needs of the users as much as possible. Other general improvements are being added to the model and processes that may be important in particular settings, such as evaporation from shallow water tables, are being incorporated.

4.6 CONCLUSIONS

We have presented a simple, integrated surface water—groundwater modeling simulation tool that provides assessments of the availability of surface water and groundwater resources on a monthly basis for a range of watershed interventions, land use, and climate-related scenarios.

In light of the scale- and tradeoff-related issues in WSD, and given the limitations with the existing suites of tools to support evaluating and planning WSD programs, there would seem to be scope for a pragmatic broad scale approach for developing more robust and equitable WSD programs. Being "simple" in formulation, the model aims to offer a tool that is as generic as possible and requires limited amounts of data for climate, topography, soils, land use, hydrogeology, and watershed interventions that can usually be met from secondary sources.

The model was tested at a mesoscale watershed in Andhra Pradesh, India, building on earlier work where it had been tested at a microscale watershed also in Andhra Pradesh. The performance of the model was found to be reasonable in terms of the groundwater components, which can be most easily verified. The surface runoff and evapotranspiration can be well simulated.

Overall, the model can provide a convenient tool for evaluating the potential impact of the watershed development, hence, shedding light on designing and implementing improved watershed development strategies. Current efforts are underway to make the tool more user friendly for nonspecialists so that it can be taken up by relevant government and nongovernment agencies to support planning and decision making. Being relatively generic in nature, its application is not necessarily limited to WSD-related issues in India; it could potentially be used to address other agricultural water management problems.

ACKNOWLEDGMENTS

We acknowledge the financial support of the Australian Centre for International Agricultural Research through the project LWR/2006/072. This work also contributes to the CGIAR Research Programs on Water, Land, and Ecosystems (WLE) and on Climate Change, Agriculture, and Food Security (CCAFS). We would like to thank Dr. Surinaidu Lagudu and Sreedhar Acharya for providing access to data and for their constructive suggestions. Two locally based non-governmental organizations, DIPA and BIRDS, are kindly acknowledged for providing much of the data used in the model. Constructive comments on this manuscript were provided by Dr. B.T. Yen from the Soils and Fertilizers Research Institute, Hanoi, Vietnam.

REFERENCES

[1] Calder I, Gosain A, Rao MSRM, Batchelor C, Snehalatha M, Bishop E. Watershed development in India. 1. Biophysical and societal impacts. Environ Dev Sustainability 2008a;10: 537—57.

[2] Government of India. Common guidelines for watershed development projects; 2008. http://www.eSocialSciences.com/data/articles/Document1812200890.6061823.pdf [Last accessed 10.03.11].

[3] Wani SP, Joshi PK, Raju KV, Sreedevi TK, Wilson MJ, Shah A, Diwakar PG, Palanisami K, Marimuthu S, Ramakrishna YS, Sundaram SSM, D'Souza M. Community watershed as growth engine for development of dryland areas — executive summary: a comprehensive assessment of watershed programs in India. Patancheru 502324, Andhra Pradesh, India: International Crops Research Institute for the Semi-Arid Tropics; 2008. p. 36.

[4] Reddy VR, Gopinath Reddy M, Galab S, Soussan J, Springate-Baginski O. Participatory watershed development in India: Can it sustain rural livelihoods? Dev Change 2004;35: 297—326.

[5] Kerr J, Milne G, Chhotray V, Baumann P, James AJ. Managing watershed externalities in India: theory and practice. Environ Dev Sustainability 2007;9(3):263—81.

[6] Syme GJ, Reddy VR, Pavelic P, Croke B, Ranjan R. Confronting scale in watershed development in India. Hydrogeol J 2012;20(5):985−93.

[7] Kumar S, Surinaidu L, Pavelic P, Davidson B. Integrating cost and benefit considerations with supply- and demand-based strategies for basin-scale groundwater management in South-West India. Water Int 2012;37(4):460−77.

[8] Marques GF, Lund JR, Howitt RE. Modeling conjunctive use operations and farm decisions with two stage stochastic quadratic programming. J Water Res Plann Manag 2010;136(3):386−94.

[9] Batchelor CH, Rama Mohan Rao MS, Manohare Rao S. Watershed development: A solution to water shortages in semi-arid India or part of the problem? Land Use and Water Resources Research 2003;3:1−10. Available from, http://www.luwrr.com/uploads/paper03-03.pdf [last accessed 8.9.10].

[10] Calder I, Gosain A, Rao MSRM, Batchelor C, Snehalatha M, Bishop E. Watershed development in India. 2. New approaches for managing externalities and meeting sustainability requirements. Environ Dev Sustainability 2008b;10:427−40.

[11] Sakthivadivel R, Scott CA. Upstream-downstream complementarities and tradeoffs: opportunities and constraints in watershed development in water scarce regions. In: Sharma BR, Samra JS, Scott CA, Wani SP, editors. Watershed Management Challenges: Improving Productivity, Resources and Livelihoods. International Water Management Institute; 2005. pp. 173−85.

[12] Batchelor CH, Singh A, Rama Mohan Rao MS, Butterworth J. Mitigating the potential unintended impacts of water harvesting. In: IWRA International Regional Symposium 'Water for Human Survival', Vol. 26.; 2002. p. 29.

[13] Glendenning C, Vervoort RW. Quantifying the impacts of rainwater harvesting in a case study catchment: The Arvari River, Rajasthan, India. Aust J Water Res 2008;12(3):12.

[14] Glendenning CJ, Vervoort RW. Hydrological impacts of rainwater harvesting (RWH) in a case study catchment: The Arvari River, Rajasthan, India: Part 2. Catchment-scale impacts. Agric Water Manag 2011;98(4):715−30.

[15] Garg KK, Karlberg L, Barron J, Wani SP, Rockstrom J. Assessing impacts of agricultural water interventions in the Kothapally watershed, Southern India. Hydrol Processes 2012;26(3):387−404.

[16] Surinaidu L, Bacon CGD, Pavelic P. Agricultural groundwater management in the Upper Bhima Basin: current status and future scenarios. Hydrol Earth Syst Sci 2013;17:507−17.

[17] Werner AD, Gallagher MR, Weeks SW. Regional-scale, fully coupled modelling of stream aquifer interaction in a tropical catchment. J Hydrol 2006;328:497−510.

[18] Bernet D. Simple surface and groundwater balance model to assess watershed interventions at different scales—model development and application at a micro-scale watershed in Andhra Pradesh, India. MSc. Thesis. Zürich, Switzerland: Swiss Federal Institute of Technology (ETH); 2011.

[19] SCS (Soil Conservation Service). National Engineering Handbook, Section 4: Hydrology. Washington: USDA; 1972.

[20] Shah T. Taming the anarchy: groundwater governance in South Asia. Resources for the Future Publishers; 2009. p. 311.

[21] Das, S.V.G., and Burke, J. (2013). Smallholders and sustainable wells. Retrospect: Participatory groundwater management in Andhra Pradesh (India). FAO Report.

Chapter 5

Modeling the Impact of Watershed Development on Water Resources in India

Barry Croke *, Peter Cornish § and Adlul Islam ¶

*Australian National University, Canberra, Australia, §University of Western Sydney, Hawkesbury Campus, Australia, ¶Natural Resources Management Division (ICAR), New Delhi, India

Chapter Outline

Integrated Assessment of Scale Impacts of Watershed Intervention
http://dx.doi.org/10.1016/B978-0-12-800067-0.00005-0. Copyright © 2015 Elsevier Inc. All rights reserved.

99

5.1 INTRODUCTION

Watershed development (WSD), which originated in India to facilitate soil conservation and water resource management, has now also become a cornerstone program to improve the livelihoods of farmers, particularly those with no access to a major irrigation infrastructure [1]. These people are mostly marginal and small landholders, and belong to the poorest and least food-secure families in India. While WSD is a major strategy for livelihood improvement, often the strategy is not underpinned by sound hydrology in terms of both the local as well as downstream impacts of water-harvesting structures [2,3]. This chapter describes the development of a hydrological model based on the observations made at study sites in West Bengal as part of an Australian Centre for International Agricultural Research (ACIAR)-funded project (LWR/2002/100), which were subsequently adapted to a very different agro-ecological area in southern Andhra Pradesh as part of another ACIAR-funded project (LWR/2006/072). The study sites for both projects are shown in Figure 5.1, along with the state boundaries.

5.1.1 West Bengal

The two study sites (Pogro and Amagara) for project LWR/2002/100 (marked in red; Figure 5.1) are in the Purulia District in far western West Bengal, within the East India Plateau (EIP). The mean annual rainfall at these sites is approximately 1300 mm/year, which is mostly confined to 3 months, resulting in high runoff and drainage during the wet season, but a shortage of readily available water at the end of the dry season, because the ponds and shallow wells become dry—shallow, annually recharged groundwater from open wells is important for domestic use along with surface water (ponds).

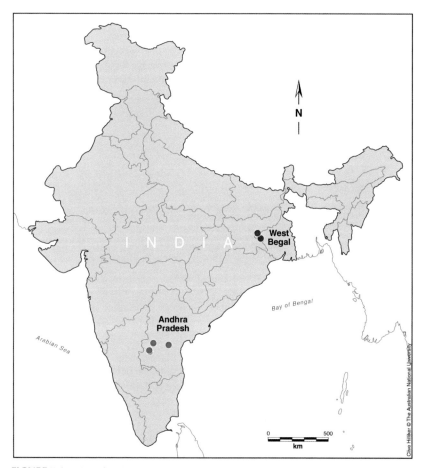

FIGURE 5.1   Map of India showing state boundaries and study site locations in Andhra Pradesh (blue circles) and West Bengal (red circles).

Agriculture in these areas is primarily monocropped rice in a rice-fallow system on terraced and bunded fields. There is little irrigation infrastructure, without which agriculture is said to be "drought prone" despite the high rainfall. WSD is expected to address this climate risk and raise the overall agricultural productivity.

5.1.2 Andhra Pradesh

The study sites for project LWR/2006/072 (marked in blue, along with the nearest stream gauge; Figure 5.1) are in southern Andhra Pradesh, where the climate is semi-arid, with a mean annual rainfall of ∼500−600 mm/year, confined to the wet season, although exceeding potential evaporation. Farmers here access deep groundwater resources for both irrigation and domestic use in Andhra Pradesh.

Agriculture is rice-based here, but because of irrigation development cropping is more diverse and intensive compared to the EIP. Terracing is hydrologically less important in the Andhra Pradesh study sites, due to the focus on the impact of larger scale structures (e.g., check dams, see Chapter 3). Since water levels in regional aquifers are falling, there is focus on WSD to manage this problem.

5.2 EFFECTIVE AND SUSTAINABLE WSD: HYDROLOGICAL CONSIDERATIONS

Water can be stored in different parts of the landscape: surface water, soil water, and shallow and deep groundwater. These storages are linked as there is a significant flux of water between them, and this must be considered while designing WSD interventions (for example, extracting water from the shallow aquifer will potentially reduce recharge to a deeper aquifer, and reduce surface water). A catchment can be divided into two zones: recharge zone in the uplands and the medium uplands, and discharge sites in the lowlands and medium lowlands. Structures that can be used in WSD include structures that increase the surface water storage (e.g., ponds, bunds, pits) in the recharge areas, which can lead to an increase in recharge. Other structures (e.g., seepage pits, wells) in the discharge areas can permit access to enhanced groundwater resources due to the WSD work in the upland areas. Sometimes, a recharge pit can act as a temporary seepage pit.

The possible structures include:

- *30 × 40 plots*: Incorporate bunded 30 × 40 m plots with pits located at the lowest point of each plot.
- *5% pits*: Five percent of a paddy field is converted into a pit to store surface water in recharge areas, but as the medium uplands can have shallow water up to November, these pits can also provide seepage water for irrigation early in the rabi season.
- *Ponds*: Broad shallow depressions that collect surface flows and store this for domestic use, and irrigation. Such ponds can also contribute to groundwater recharge.
- *Recharge pits*: Very large pits dug deep into the ground that collect surface flows and store this for irrigation purpose; these can also lead to enhanced groundwater recharge.
- *Seepage pits*: Variable-sized pits ranging from 10 to 200 m^2 with depths depending on the depth of the shallow groundwater table (up to about 10 m deep in drier areas, or about 2 m in wetter parts of the catchment).
- *Wells*: Deep pits with narrow openings that tap into shallow and sometimes deep groundwater resources.

Further, in-stream structures such as check dams can be used on a larger scale (contributing areas of several kilometers squared or more) to create surface water bodies that can be used for irrigation and domestic uses, as well as having the potential to increase recharge to aquifers (accessed using tube wells and bores), making them more resilient.

Understanding the changes in the hydrological behavior of a catchment because of WSD requires models representing the various types of water-harvesting structures that might be used and information regarding how these interact to produce a catchment-scale response. This requires monitoring of the water storages and fluxes in the catchment to understand the catchment response and to develop the necessary models.

5.3 STUDY SITES IN WEST BENGAL

The Pogro study site lies ∼25 Km to the south of Bokaro, and 80 km to the east of Ranchi, and is an ∼2 km^2 headwater catchment of the Chapai Nula, which ultimately drains into the Damodar River just upstream of the Panchet Reservoir. The study site lies in the Chhota Nagpur or EIP, which covers 65,000 km^2 including much of the state of Jharkhand and parts of adjoining West Bengal, Chhattisgarh, Bihar, and Odisha. Although rich in natural resources, it is one of the poorest regions in India, with high population density and mostly subsistence agriculture on small landholdings. Rainfall is high (1100−1600 mm, 80% of which falls between June and September) with high runoff and soil erosion, and frequent dry spells during the monsoon. Rural livelihoods are based largely on monocropped kharif rice. Cropping during the post-rainy season is limited by a paucity of irrigation resources and uncontrolled cattle grazing. The uplands are degraded and make little contribution to the overall productivity. Although rice has been grown on lowlands for generations, population growth has created pressure to crop more marginal lands, leading to terracing of mid-slopes and uplands creating" medium uplands" (Figure 5.2) that now comprise the major area for rice production.

Agricultural development in the region lags behind the rest of India partly because it lacks the irrigation infrastructure that fueled the "green revolution." This area only recently became a priority target for development by the government, so rural electrification (that elsewhere has driven groundwater exploitation) and other infrastructure development is lagging. It also has a significant tribal population that is relatively new to agriculture, which presents a particular challenge to development as the generations of experience that underpinned development in other parts of India is lacking.

Despite high rainfall, the region is characterized by low cropping intensity and diversity as well as low water productivity. With little irrigation capacity, the single rice crop per year is said to be drought prone, although Cornish et al. [4] argued that drought is a perception based on the need for permanent water with transplanted rice, and alternative kharif (monsoon) crops could be grown

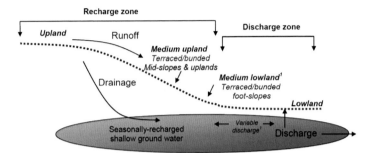

FIGURE 5.2 Micro-watershed landscape schematic (catchment area approximately <10 km², with relief ~50 m) Medium lowlands (upslope from lowlands) are a significant discharge area during wetter years.[1]

risk free. Furthermore, the area of irrigated rabi (winter season) crop is small and little use is made of the residual soil water following rice cultivation.

The study site has a well-defined discharge point at a culvert under the railway line that forms the eastern boundary of the study site. There is also a second culvert, which is the outlet from a smaller subcatchment to the northeast of the study site, which is used as a reference. The site is called the Pogro study site after one of the villages located in the subcatchment. The second study site (Amagara; lower red circle in Figure 5.1) is an ~1 km² subcatchment, although there is no defined outflow from this site. Due to the nature of the sites, the Pogro catchment has been used for the hydrological study of the impact of WSD works, while the Amagara site demonstrated insights into the behavior of established WSD works.

Figure 5.3 shows the digital elevation model (DEM) derived from the differential geographical positioning systems (GPS) data collected during an electromagnetic survey of soil and shallow groundwater resources, as well as the catchment boundaries draining into the two culverts (blue) and the lowland areas in the main subcatchment (green). The interpolation of the data points was constrained using streamlines also mapped as part of the data collection process. Due to the patchy nature of the data, the use of differential GPS data and the highly modified structure of the landscape, there is considerable uncertainty in the derived DEM: areas with a higher density of points have a smaller uncertainty than the more sparsely covered regions, although the influence of extensive bunding in the area introduces error in the interpolation, as this assumes a smoothly varying land scale and does not take into account the presence of terraces.

5.4 DATA COLLECTION

Monitoring of the water fluxes within the Pogro study site has been done at the point scale as well as by mapping the variation in conductivity across the catchment. In addition to water fluxes, measurements of the soil properties

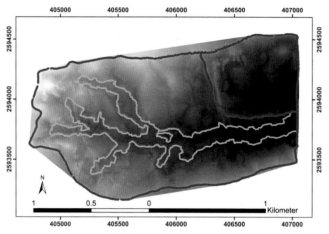

FIGURE 5.3 DEM of study site derived from differential GPS data, showing the catchment boundary (blue) for both subcatchments (blue), and the lowland areas (green).

(both hydrological and agronomic) have also been made. The goal of the monitoring work was not only to gain an understanding of the hydrology of the site, but also to generate the data needed for the development of the hydrological models.

In separate agronomic studies, as many as 60 fields in each watershed were monitored for rice yield and related management factors, and the farmers were asked to rate each year for its suitability for rice production along with reasons. This provided a range of data for indirectly verifying the predictions of the hydrological model. Extensive soil sampling established values for saturation and the lower limit of water extraction by rice; these were used in a water balance model that also provided independent verification of some hydrological model predictions.

5.4.1 Weather Monitoring and Recording

Two weather stations were installed in the study site: a Stevenson screen in the lowlands near the Pogro Village for recording temperature data, and a manual rain gauge in the village (on the roof of the house of the main data collector). An automatic weather station (AWS) was co-located with the manual rain gauge for recording minute values of rainfall, temperature, relative humidity, wind speed, and solar radiation. In 2010, standalone rainfall and temperature/relative humidity sensors were added to the station in the village to provide backup for the AWS. The rainfall data was event based, with each tip of the bucket recorded. This gives a good measure of the rainfall intensity. Due to the limited memory of the standalone temperature/relative humidity sensor, data were recorded every 15 minutes. In addition, manual measurements of temperature were collected twice daily along with the manual rainfall recorded after events.

5.4.2 Shallow Groundwater Assessment via Wells and Piezometers

Water fluxes within the Pogro site were monitored at the point scale to understand the site hydrology and generate the data needed to develop models. The monitoring network is shown in Figure 5.4, in which the hydrological boundaries are marked in black and the lowlands are marked in gray.

The western boundary is not fixed as there is a large pond just outside the boundary and while the main flow path is south, at high water levels, this pond can also spill into the study catchment. The study site is bounded in the east by a railway embankment with two culverts. Each culvert drains a separate subcatchment. The smaller subcatchment (B, northern culvert in the Figure 5.4) was used as a reference, with no project-driven WSD work performed within the life of the project. While not a classic "paired catchment" study, subcatchment B has been used to test the ability of the models and capture the impact of climate variability, thereby establishing the detection of the impact of WSD within subcatchment A.

5.4.3 Monitoring Network for the Pogro Site

The monitoring network consisted of 14 wells, 4 ponds, 15 piezometers, and 2 weather stations. Most of the open wells in the study site were monitored— selection was based on giving as much coverage of the study site as possible.

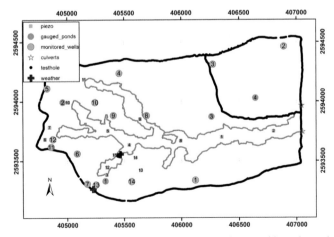

FIGURE 5.4 Monitoring of water resources in the Pogro study area with catchment boundaries (black) and lowlands areas (gray).

A subset of four ponds was selected for monitoring, comprising two upland ponds and two ponds located closer to the catchment outlets. The piezometer network was focused mainly on the lowlands, with the aim of developing an understanding of the subsurface flux through this region. A transect from the upland (piezometer 13), through the medium upland (piezometer 14), to the lowland (piezometer 4) was also included to investigate the relative behavior of these regions. The network was mostly installed during November 2005 and completed in early 2006. Manual measurements were made by trained villagers, and data were recorded at differing frequencies depending on the expected rate of change—mostly daily during the wet season and fortnightly during the dry season.

5.4.4 Weather Monitoring and Recording

This analysis puts rainfall in the project years into the longer term context. Figure 5.5 shows the comparison between long-term rainfall exceedance curves for the area around Pogro, based on the 0.5 × 0.5 degree gridded rainfall data from the Indian Meteorology Department (Rajeevan and Bhate, 2008 [12]), and the data collected in the Pogro catchment throughout the project.

The data show a similar variation in rainfall, suggesting that the annual rainfall in Pogro over the duration of the project is representative of the long-term frequency distribution of annual rainfall averaged across a 0.5 × 0.5 degree area.

On the other hand, the mean monthly rainfall derived from the same datasets (Figure 5.6) suggests slightly more pre-monsoon rain (May) in Pogro, less in July and August, and an increase in September. With only 6 years of data, no comment can be made regarding long-term change in the monsoon pattern, although reduced rainfall in July and August can have a significant impact on paddy rice, especially for transplanting; while greater pre-monsoon rain may create new cropping opportunities.

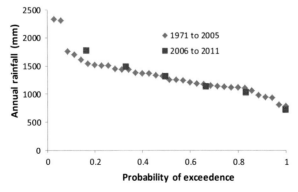

FIGURE 5.5 Annual rainfall exceedance curves for the Pogro study site. Long-term data are from the Indian Meteorology Department gridded dataset.

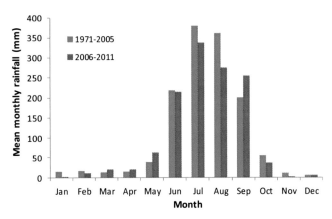

FIGURE 5.6 Mean monthly rainfall for Pogro (2006–2011) and the local area (gridded data).

5.4.5 Farmer Assessments for Each Year

Farmer assessments indicate that during the high rainfall years, 2007 and 2011, there was enough water to provide the long duration of ponding required by rice, but during 2006 some fields in the medium uplands ran short of water suggesting a marginal shortage of rainfall for rice. The monsoon ended prematurely (for rice) during the years 2008 and 2009, while the year 2010 was so dry that rice could not be transplanted in any field, implying little runoff as well as seepage (Table 5.1).

TABLE 5.1 Amount of rainfall and farmer assessment of rice in medium uplands

	Rainfall (mm): total/June to September	Assessment
2006	1303/1140	Bad on some medium uplands: delayed transplanting plus early draining
2007	1774/1518	Good
2008	1139/1004	Moderately bad: early monsoon and transplanting but early end to monsoon
2009	1029/944	Very bad: transplanting delayed, plus early end to monsoon
2010	723/603	Bad: no transplanting of rice including lowlands
2011	1429/1225	Good

5.4.6 Monitoring of Water Levels

5.4.6.1 Ponds

In addition to the culverts, water levels in the four ponds in the Pogro study site were manually recorded by one of the trained villagers on a daily to fortnightly basis. Three of these ponds lie inside the main study subcatchment (catchment A), while the other is in the reference subcatchment (catchment B).The observed water levels for the 2006−2007 to 2009−2010 hydrological years are shown in Figure 5.7. We see that only during the very wet monsoon in 2007 (July to September rainfall of 1518 mm) did all ponds fill and presumably over top, while during the monsoon of 2008 (1004 mm) and 2009 (944 mm) there was insufficient runoff to fill any of the ponds. The average July to September rainfall at Pogro was 1072 mm for the six observed years (2006−2011). These observations suggest that despite high average annual rainfall, there might not be enough runoff during some years to be captured by WSD and to address the agronomic problems identified by the farmers in Table 5.1.

We also observe from Figure 5.7 that generally, all ponds showed a decrease in water level of about 10 mm/day during February, of which 2.5 mm/day can be accounted for evaporation. However, Pond 2 appears to have a more rapid decline in water level from November to January, although there are little data for this pond due to damage to the staff gauge in 2006. The decrease in water level was not due to consumptive use, as the ponds were not significantly used for irrigation. Therefore, the loss of water is primarily due percolation into the shallow aquifer.

5.4.6.2 Shallow Groundwater

The dug wells in the catchment are typically 8 m deep, with an annual variation in water level of 4−7 m. The water levels in some of the wells (4, 8, and 14, Figure 5.4) rise to within 1 m of the surface during the wet season, indicating a strong interaction between shallow groundwater and soil water in some locations. On the other hand, wells 4 and 14 are near ponds, and are likely to be directly fed by recharge from the pond. Interestingly, the villagers reported that

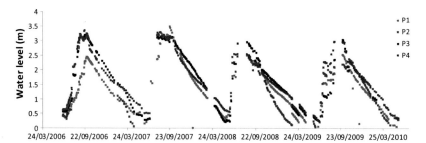

FIGURE 5.7   Observed water levels in the four monitored ponds (P1−P4).

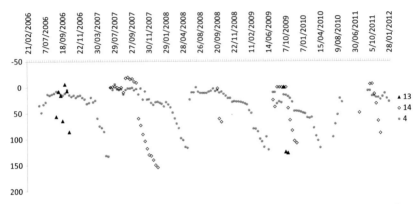

FIGURE 5.8 Depth of groundwater along transect from upland (piezometer 13), through medium upland (piezometer 14) to lowland (piezometer 4).

in the north of the catchment there are some tube wells that had been dug to a depth of about 80 m without finding a significant source of water, while elsewhere the deeper groundwater (depth >20 m, accessed using tube wells) gave a secure source of water for domestic use. These data give information on the behavior of the deeper groundwater systems, which is important when understanding the hydrology of the study site, even though the focus of the interventions is on the shallow (annually recharged) groundwater systems.

The piezometer network showed that during every year from 2006 to 2012, except for 2010, water was at least briefly at the surface in the medium uplands (Figure 5.8).The year 2010 was the driest during the project (total rainfall of 723 mm), resulting in water levels considerably below the surface (although still present). Even in the lowlands (piezometer 4, Figure 5.4), the shallow groundwater did not rise to the surface in 2010, explaining why the farmers reported that rice failed in the lowlands (Table 5.1) that normally benefit from both runoff as well as seepage.

Figure 5.8 shows the difference in residence time for water in the shallow aquifer along a transect from the uplands to lowlands (piezometers 4, 13, 14). In the uplands (piezometer 13), while water was occasionally observed near the surface, this did not last long, draining down through the medium uplands to the lowlands. On the other hand, the residence time of water in the medium uplands (piezometer 14) was considerably longer, with water within a meter of the surface through mid-November. Hence, it is sometimes recommended that 5% of the area of each medium upland field be set aside for a "5% pit" to capture local runoff for use in "rescue irrigation" for rice [5], as well as possibly increased aquifer "recharge." The fact that groundwater may remain close to the surface until November indicates the potential for such pits to also act as "seepage" pits that might provide irrigation for a short period after the monsoon to supplement residual soil water for a rabi crop [4]. Needless to say, if a small watershed was heavily populated with small pits to capture runoff (and later supply shallow groundwater) and the potential to impact on the hydrology is significant, then this is an important question for model application.

The observations for all years show that constructing ponds in upland drainage lines immediately transforms the land below the pond into lowlands. The ponds store runoff water during the monsoon, which continues to "leak" into the drainage line after the monsoon recedes. This was particularly evident in 2010 in the transect running down the lowland draining from Pond 1 (piezometers 3, 4, and 12), which showed a strong influence of the pond that had received some runoff from hard surfaces in the Pogro Village earlier during the monsoon. Pond construction, whether large ponds or 5% pits in medium uplands [5], is the main mechanism for converting "quick flow" to "slow flow" and potentially to transpiration.

The piezometer data also indicate a flow constriction downstream of piezometer 12, where water was observed at the surface well into the dry season during all years (late March). A pump test conducted on a trial seepage pit (located between piezometers 4 and 12) during February 2009 showed a strong recovery, with a recession time constant of 5.5 days, indicating that the subsurface flow constriction was downhill from this location.

Figure 5.9 shows the observed groundwater levels for the 14 monitored wells, with very similar rates of decline observed in most wells through the dry season, while some wells filled to just below the surface (wells 4 and 14, in particular) during the wet season. Figure 5.10 shows modeled groundwater storage assuming a time constant of 150 days, and recharge equated to infiltration from the surface store of the study site-scale hydrological model—water level data have been linearly scaled to the modeled groundwater storage. As a consequence of not including the impact of the soil store on the recharge, the modeled water level is observed to be too high during the dry season and to

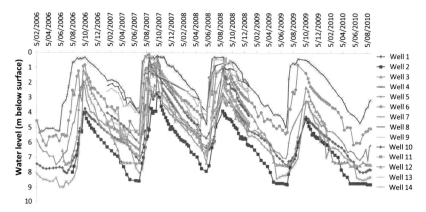

FIGURE 5.9   Observed water levels for the monitored wells.

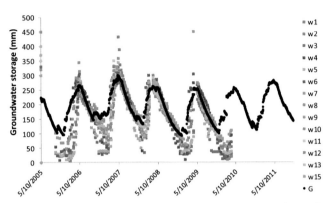

FIGURE 5.10 Modeled groundwater (black circles) and scaled water level for monitored wells.

respond too quickly during the start of the wet season. Similarly, physical interpretation of the time constant for groundwater storage is difficult as it is influenced by the transmissivity, storativity (or effective porosity), and the length of the aquifer. Furthermore, the amplitude of the variation in the groundwater levels depends on both the recharge rate and the storativity, and therefore, without additional information, only a crude estimate of the groundwater resource is possible at this stage.

Generally, lowlands have access to groundwater resources for a longer time (levels are maintained by subsurface flux from the upper parts of the catchment), making it "safe" for rice. On the other hand, there is limited access to groundwater in the upper parts of the catchment outside the wet season, although even a short period of access may be important for agricultural development. Furthermore, upland sites may be influenced by local effects, including the location of ponds that are a persistent source of recharge, and flow constrictions that limit the downhill flow of groundwater, increasing the residence time.

Overall, the data show that

1. Ponds are a significant source of groundwater recharge (recession rates of ponds during February are as high as 20 mm/day compared with a Penman evaporation rate of ~2.5 mm/day). This is also reflected in the height of the shallow groundwater table monitored below the pond, as well as in the distribution of the lowlands.
2. There is significant spatial variation in the transmissivity for the shallow groundwater, resulting in variations in the water-holding capability across the catchment, which will affect the effectiveness of WSD structures.
3. Medium uplands have limited capacity to hold free water (up to 2 months for piezometer 14, Figure 5.8), although this may be useful for irrigation during short periods after the monsoon, while the uplands are unable to hold water at least to the depth monitored by the piezometers (about 1.5 m).

4. Variability in the response of the shallow aquifers suggests that an experimental approach should be adopted while planning WSD interventions, using test holes to explore the potential water resource before implementation of the WSD works. This recommendation is included in a set of "guidelines" for WSD developed within the project.

5.4.7 Gauging Runoff

There are two component discharge volumes: water level (also known as stage height) and a rating curve relating water level to discharge volume. The rating curve is derived from flow velocity measurements and the flow cross section. Staff gauges were installed in both culverts in 2005, where the approximate water level was manually recorded daily by trained villagers through the wet season. For continuous recording of the flow data at both culverts, weir structures were constructed and water level loggers (pressure sensors whose atmospheric variations are removed using a reference sensor located in the Pogro Village) were installed in both culverts in 2008 (Figure 5.11). As the culverts are a thoroughfare during the dry season, the weirs were designed to have minimal impact on all forms of traffic. A slight increase in height toward the walls of the culvert was added to increase sensitivity to low flows. A small channel was included in the center for installing a STARFLOW instrument so that a rating curve could be obtained. This was not very successful; however, and the flow velocities were measured manually using a current meter during the 2011 monsoon.

Rainfall through the 2011 wet season and the resulting discharge through each culvert are shown in Figure 5.12. This shows the similarities (e.g., very similar event profiles) and differences (e.g., magnitude of response to the event in late October) between the flows through each culvert, due to a combination of the land use and degree of WSD, as well as possible spatial variations in rainfall. Some remaining problems with the flow data are still to be resolved (e.g., the flow peak through culvert A in late September exceeds the value expected if 100% of the rainfall is converted to discharge).

FIGURE 5.11 Culvert B (control subcatchment) weir designed for adequate gauging of flow with minimum impact on users.

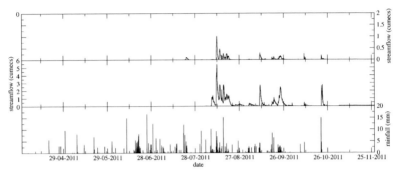

FIGURE 5.12   Rainfall plus observed flow at both culverts (B, top; A, middle), 2011 wet season.

5.5 MODELING HYDROLOGICAL RESPONSE

5.5.1 Water Balance Modeling

A water balance model was used to better understand the risks and opportunities of growing transplanted rice in medium uplands [4], which are by far the largest areas of rice, and comprise about 70% of the total watershed area in most micro-watersheds. What happens in the medium uplands is thus hydrologically significant. The model provided daily estimates of plant-available water (to a depth of 90 cm), surface runoff, and drainage below the root-zone from 2006 to 2012, and thus provided a more complete representation of the study period compared to any of the hydrological measurements. Model assumptions are discussed by Cornish et al. [4]. As this is a point-scale model, runoff and drainage estimates cannot be scaled up to a watershed, although they do provide estimates of possible runoff and drainage to shallow aquifers. These estimates provide another opinion on runoff and drainage along with later hydrological estimates.

The prediction of high runoff in 2007 (Table 5.2) is consistent with the pond-level data for 2006–2009 (Figure 5.7) showing that the ponds were filled only in 2007. Similarly, drainage predictions are found to be consistent with the much-diminished groundwater rise observed in 2010 (Figure 5.8). The inter-annual variation in drainage is observed to be much less than that for runoff. This is because soils always saturate during the monsoon, mostly for extended periods during which drainage occurs. Shallow groundwater therefore appears to be a more reliable potential source for irrigation compared to surface storage.

Water balance modeling shows that transplanted rice suffered from insufficient ponding for many years, and accounts for the low rice yields as well as crop failures in the region [4]. One way to address this would be to grow direct-seeded rice that does not require puddling and any mandatory ponding ("aerobic" rice). This may be crucial for rice-based food security,

TABLE 5.2 Hydrological and agronomic assessments for the rice-fallow in medium uplands at Pogro site

	2006	2007	2008	2009	2010	2011	Mean	CV
Annual ET (mm)	797	846	663	576	641	768	715	0.15
Rice ET (mm)	488	482	465	453	433	441	460	0.05
Predicted runoff (mm)	122	568	167	131	0	290	213	0.93
Predicted drainage (mm)	360	366	327	316	69	401	307	0.39
Soil water to 90 cm at September 30 (mm)	281	266	96	193	102	219	193	0.41

because in some years there is insufficient runoff to fill ponds that might be used to irrigate rice. Hence, the technology for aerobic rice is being developed. Without wet tillage for puddling, it is likely that the soil structure will improve over time and soil drainage rates will increase.

Thus, water balance modeling shows that the partitioning between drainage and runoff is highly sensitive to drainage rate. Therefore, if there was a shift from transplanted rice to aerobic rice over a significant area there would be profound hydrological effects and important implications for designing WSD. These are important questions to be explored through future model applications.

5.5.2 Models that Capture the Function of Water-harvesting Structures

Using the data collected in the Amagara and Pogro study areas, simple spreadsheet models representing the behavior of selected water-harvesting structures have been created. These models are a key input to the development of the guidelines for WSD design. These models include the

1. Influence of runoff controls (bunds, pits, 30 × 40 plots, and ponds) in increasing local soil moisture as well as recharge to shallow and deep groundwater systems, and subsurface flow to downhill areas
2. Local and downhill impact of paddy scale interventions in recharge areas (e.g., 5% pits)
3. Performance of seepage pits in discharge areas.

5.5.3 Hydrological Model for the Pogro Study Site

The primary model used in this study is IHACRES [6] rainfall-stream flow model, using the catchment moisture deficit (CMD) version of the nonlinear loss module [7,8]. The model has been modified for use in this study in two ways:

1. Inclusion of a surface store to account for the impact of ponds, bunds, pits, etc., on the infiltration and runoff. Five land surface types are considered in the model: upland areas draining into ponds, ponds, uplands, medium uplands, and lowlands.
2. The CMD module has been modified to partition water between the shallow and deep aquifers, and only models subsurface fluxes (surface fluxes are handled by the surface store module).

The conceptual diagram of the model is shown in Figure 5.13. Overland flow from the uplands is assumed to drain either into ponds or to uplands that have water-control structures (bunds, pits, etc.). Overflow from the ponds and upland structures then contribute to the lowland storage, and any overflow from the lowland storage appears as stream flow at the culvert. This structure ignores some of the finer details of the Pogro catchment (e.g., the forestry area in the southeast of the catchment, part of which drains directly into the culvert, and the fact that while most of the ponds are located in the upper parts of the catchment, some are located lower in the catchment), but does capture most of the characteristics of the study site.

The input data needed by the model are

- Area of catchment, and proportion of each land class
- Infiltration rate (K_{sat}) for each land class
- Storage capacity of each land class

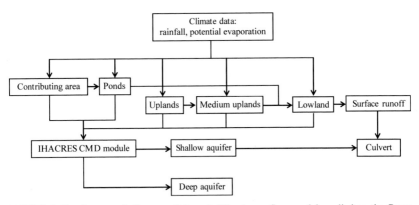

FIGURE 5.13 Conceptual diagram of the rainfall–stream flow model applied to the Pogro study site.

- Evaporation/infiltration threshold (currently used for the medium uplands and lowland land classes only)
- CMD module parameters (d, h, n, and f)
  - d is the first flow threshold
  - h is the second flow threshold
  - n is the fraction of effective rainfall partitioned to the second flow component under saturated conditions
  - f is the plant stress threshold parameter (stress threshold is fd)
- Rainfall and potential evaporation time series

The surface module produces estimates of overland flow, evaporative loss, storage, and infiltration. The overland flow is convolved with a transfer function to generate the contribution to stream flow at the catchment outlet, using two identical stores in series (Nash cascade) to reproduce the delay in peak as well as the overall shape of the peak. The infiltration is passed to the CMD module to provide estimates of the inputs to the shallow and deep aquifers. The shallow aquifer is assumed to contribute to the stream flow within the study catchment through a single exponentially decaying store, while the deep aquifer produces a subsurface flow that contributes somewhere downstream of the gauge.

The model can operate at any temporal scale, although this will impact the parameter values, particularly the infiltration rates. Initially, the model was applied at a 10 minute resolution, so that the impacts from loss of information regarding rainfall intensity could be avoided. For application to the EIP, the model will be applied at a daily timescale due to the resolution of the available rainfall data. This means that the infiltration rates used in the model will need to be reduced significantly to adequately capture the runoff.

5.5.3.1 Surface Store

The surface store module is a simple mass balance calculation that takes into account direct rainfall on the surface, runoff from the uphill contributing area, evaporation loss, and infiltration into the subsurface system. For the lowland land class, the evaporation and infiltration decrease when the CMD decreases (i.e., catchment becomes wetter) below a set threshold (150 mm). This is because under wet conditions, the surface storage in the lowlands is maintained by the water from the shallow aquifer system (as seen in the piezometer data where the groundwater level was above the surface for many of the piezometers in the lowlands).

5.5.3.2 CMD Module

The modifications to the CMD module involved rewriting the module to produce two outputs: U (contribution to the shallow aquifer) and R (contribution to the deep aquifer) rather than just the effective rainfall, although without addition of extra parameters (over the two-segment form in [7]). The evaporative loss from the moisture store uses the original functional form adopted by Croke and Jakeman [7,8].

The revised drainage equation was derived using the same approach as the original module, with the assumption that at a particular soil moisture level, there is a set fraction of rainfall that goes to U and a different set fraction that goes to R.

This is represented as follows:

$$\frac{\Delta U}{\Delta P} = n(1 - \mathbf{f}_U(M)), \quad \frac{\Delta R}{\Delta P} = (1 - n)(1 - \mathbf{f}_R(M))$$

$$\Delta M = -\Delta P + \Delta U + \Delta R$$

$$\frac{\Delta M}{\Delta P} = n\mathbf{f}_U(M) + (1 - n)\mathbf{f}_R(M)$$

Here, n is the fraction of drainage that goes to the shallow store under saturated conditions, $\mathbf{f}_U(M)$ and $\mathbf{f}_R(M)$ are functions that determine how the flux to both aquifers varies with catchment moisture deficit M, and P is the input to the moisture store (in this application, this is the infiltration from the four land classes).

Taking the limit as Δ tends to zero gives the differential equation:

$$\frac{dP}{dM} = \frac{1}{n\mathbf{f}_U(M) + (1 - n)\mathbf{f}_R(M)},$$

which can be expressed as follows:

$$P_k = \int_{M_i}^{M_f} \frac{dM}{n\mathbf{f}_U(M) + (1 - n)\mathbf{f}_R(M)}. \tag{1}$$

The conditions on $\mathbf{f}_U(M)$ and $\mathbf{f}_R(M)$ are

1. They lie between 0 and 1
2. $\mathbf{f}_U(0) = 0$ and $\mathbf{f}_R(0) = 0$
3. They are nondecreasing functions (derivative never negative)
4. $\mathbf{f}_U(x) \rightarrow 1$ and $\mathbf{f}_R(x) \rightarrow 1$ as $x \rightarrow \infty$
5. $\displaystyle\int_0^{x_1} \frac{dx}{\mathbf{f}_U(x)} = \infty, \quad \int_0^{x_1} \frac{dx}{\mathbf{f}_R(x)} = \infty, \quad x_1 > 0.$

The last condition states that an infinite amount of rainfall is needed to reach a completely saturated condition. An additional condition is that the above integral (Equation 1) can be solved analytically for M_f, as well as for U and R.

The simplest functional form that meets all five conditions above is the linear form, where both $\mathbf{f}_U(M)$ and $\mathbf{f}_R(M)$ are given by

$$\mathbf{f}_U(M) = \min\left(1, \frac{M}{d}\right) \text{ and } \mathbf{f}_R(M) = \min\left(1, \frac{M}{h}\right),$$

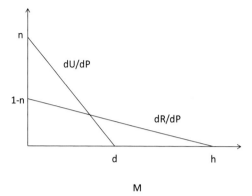

FIGURE 5.14 Fraction of rainfall that becomes U and R as a function of catchment moisture deficit M.

as shown in Figure 5.14. The solution for this set of equations is as follows:

$$
M_f = \begin{cases}
M_f = M_i - P & \begin{aligned} U = R = 0 \end{aligned} & M > h \\[2ex]
M_f = M_i e^{-P(1-n)/h} - \dfrac{nh}{1-n}\left(e^{P(1-n)/h} - 1\right) & \begin{aligned} U &= 0 \\ R &= P - M_i + M_f \end{aligned} & d < M \le h. \\[3ex]
M_f = M_i e^{-P/\Gamma}, \quad \Gamma = \dfrac{dh}{nh + (1-n)d} & \begin{aligned} U &= n\left(P + \dfrac{M_f - M_i}{\Gamma d}\right) \\ R &= P - M_i + M_f - U \end{aligned} & M \le d
\end{cases}
$$

5.5.4 Application

The model has been applied to both gauged culverts and the observations of pond water level. The area of each catchment was determined from the GPS survey of the study site, coupled with analysis of the Hydro1K 7.5 arcsecond DEM (Verdin and Greenlee, 1996) [9]. Initial K_{sat} values were set using the infiltration measurements carried out through the project, with the final values allowed to vary slightly during the calibration of the model. The initial values were: contributing area for ponds (1 mm/hr — mostly compacted areas around villages), pond (0.4 mm/hr), uplands (30 mm/hr), medium uplands (5 mm/hr) and lowland (0.1 mm/hr). Note that after puddling, the infiltration rate in the paddy fields decreases significantly, and must be less than 3 mm/d for fields to retain water throughout the rice growing period (So and Kirchoff, 2000 [10]). Finally, the time constant for the shallow aquifer is set to 5.5 d based on the 2009 pump test. Having defined these values, the calibration is a two-stage process:

1. The upland and pond storage modules are calibrated using the observed pond water levels. The observed levels are assumed to be representative of

all the ponds in the study site (the model does not include individual ponds, just a single representative pond; so if the pond in the model overflows, it is assumed that all ponds overflow at the same time).

2. The remainder of the model is calibrated to the observed stream flow through the culverts.

5.5.4.1 Pogro Model

The model was applied to the Pogro catchment at three temporal resolutions: 10 minute, hourly, and daily. The model runs a value of the coefficient of determination for non-linear models (also known as the Nash–Sutcliffe efficiency; R^2_{NS}), which was calculated using all points with the available stream flow data (manual as well as automatically recorded). The available data suggest that the catchment area for culvert A needs to be increased from 1.98 km^2 by at least a factor of 2 to generate the estimated stream flow.

There are three possible causes:

- Underestimation of the rainfall (a common problem, but usually of the order of 5%, not 50% [10])
- Overestimation of the stream flow
- Catchment area under estimated

To the west of the catchment, there is a dam that can overflow into the study site. Assuming that the entire overflow from the dam comes into the study site, there would be an increase of up to 3.79 km^2 in the catchment area. The match between this and the required increase suggests that this may be a valid solution. This implies that during the late wet season, the runoff from the catchment is about 100% of the rainfall for most years, although this is not unexpected due to the very high rainfall during the wet season. This is further confirmed by water balance modeling, where all the rainfall runs off once the soil is saturated, except for the drainage into the shallow aquifer and evapotranspiration (ET) losses, which are 5–6 mm/day combined.

5.5.4.2 10 Minute Resolution

Manual calibration of the model was carried out for both culverts. The adopted parameter values are given in Table 5.3. The R^2_{NS} for culvert A was 0.65 and that for culvert B was 0.19. While the result for culvert A is reasonable, the low performance for culvert B suggests deficiencies in the input data. This could be because either the data used (stream flow data used to calibrate the model or the input climate data) or the model structure is not suitable for the catchment. The catchment draining into culvert B includes a large pond near the culvert, which intercepts most of the runoff from the catchment. This leads to a large impact on the flows in the culvert, resulting in a potentially complex behavior that is not being captured by the model. Alternately the flow data could have significant errors, which hinder model performance.

TABLE 5.3 Parameter values for the Pogro model

Parameter	Culvert A	Culvert B
Pond contributing area		
Maximum storage (mm)	3	3
Saturated infiltration rate K_{sat}(mm/h)	1	1
Ponds		
Maximum storage (mm)	3000	3000
K_{sat} (mm/h)	0.4	0.4
Uplands		
Maximum storage (mm)	3	3
K_{sat} (mm/h)	30	30
Medium uplands		
Maximum storage (mm)	100	100
K_{sat} (mm/h)	0.5	0.5
Infiltration threshold (mm)	130	130
Lowlands		
Maximum storage (mm)	100	100
K_{sat} (mm/h)	0.05	0.05
Infiltration threshold (mm)	180	180
CMD module		
Infiltration threshold (mm)	150	150
Flow threshold (mm)	200	200
Recharge threshold (mm)	150	150
Stress threshold (mm)	150	150
Fraction of recharge when saturated	0.1%	0.1%
Unit hydrograph module		
Quick flow time constant	0.2	0.08
Number of stores	2	2
Shallow aquifer time constant	5.5	5.5
Number of stores	1	1
Deep aquifer time constant	30	30
Number of stores	1	1

Continued

TABLE 5.3 Parameter values for the Pogro model—cont'd

Parameter	Culvert A	Culvert B
Land use		
Pond fraction	2.5%	3.5%
Medium upland fraction	52.8%	34%
Lowland fraction	22%	40.2%
Pond contributing area	12.5%	17.5%

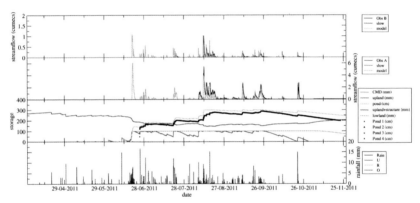

FIGURE 5.15 Model results for both culverts for the 2011 wet season.

Note that the parameter values were fixed over the entire period, meaning that the impact of the WSD work done during the project (mostly during the 2009–2010 dry season) is not taken into account. The result is a tendency for the modeled flows to start before the recorded flows during 2010 and 2011 (Figure 5.15 shows the results for the 2011 wet season). Hence, it should be noted that the large flow event predicted by the model was not observed as the water level sensors were installed after this date (sensors were removed when flows stopped for security, and reinstalled each year after the first significant rainfall event). While this could be taken as an indicator for the impact of the WSD work on the flows through culvert A, there is a similar effect on the flows through culvert B, which indicates that further work is needed on the model to adequately capture the climate-driven impacts on the generation of stream flow through the two culverts.

To reproduce the observed flows during the late wet season, a threshold for infiltration was included in the model leading to no infiltration from the lowlands or the medium uplands when the catchment moisture deficit (determined in the CMD module) is below its respective thresholds. Evaporation was also switched off and transferred to the CMD module (simulating the replenishing of the surface water in areas where the shallow groundwater level was above the surface). Initially, it was perceived that the CMD module would generate the flow through the shallow aquifer. However, the model (given the current calibration) does not produce a suitable input to the intermediate flow storage to represent this (the slow component is not visible in the plots shown, and the recession of the flow peaks are not reproduced).

5.5.4.3 Daily Resolution

The model has also been applied to both culverts at a daily resolution. If rainfall is assumed to be constant across a time step (i.e., 1 day), then this results in a significant decrease in the intensity and an increase in the duration of rainfall, either of which will result in a need to modify the model to use information on rainfall intensity—either by making some assumption about what the intensity was or modifying the K_{sat} values for the areas with little surface storage (i.e., not the ponds or bunded paddy fields). Given that the model will be applied on a much larger scale using gridded daily rainfall and temperature data, the latter option (modifying K_{sat}) was adopted. For both culverts A and B, the K_{sat} value for the contributing area for ponds was decreased from 1 to 0.7 mm/h, based on the modeled storage in the pond surface store; all other parameters were fixed at the values shown in Table 5.3.

The daily modeled results over the entire data period (Figure 5.16) capture the large inter-annual variation in stream flow that was also predicted by the water balance modeling (Table 5.2). This shows the impact of low rainfall in 2010 on the stream flow and, consequently, on the storage in the

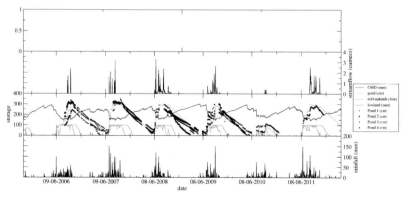

FIGURE 5.16 Modeled results for culvert A for the study period. Top shows the observed discharge, middle shows the modeled storages (line) and observed pond levels (note the different units shown in the legend), and the bottom shows the daily rainfall depth.

ponds. It also demonstrates the small surface storage in medium uplands in 2010, which means a lack of water security for transplanted rice in the medium uplands. The impact of weather patterns for each year on the duration of inundation in the medium uplands and lowlands can also be seen in Figure 5.16, center.

A comparison of the modeled pond storages and the observed water levels shows that "the model captures the variation in the recorded pond storage reasonably well, indicating that the combination of the contributing area draining to the ponds coupled with the storage and K_{sat} values for the ponds and their contributing area adequately captures the fluxes into and out of the ponds. Large inter-annual variations in pond storage, and the minimal storage in 2010, highlight the limitation of WSD work during dry years, when there may be little or no runoff captured to provide "rescue irrigation" for either rice or any other kharif crop, let alone for the following rabi. This stresses the need for other options to ensure food security, which are discussed by Cornish et al. [4].

While 2010 was a dry year, there was over 600 mm of rainfall from July through September, with the estimated rice evapotranspiration (ET) only slightly less than the average over 7 years (Table 5.2). This demonstrates that rice failure in 2010 was not a direct result of a shortage of water, but the absence of ponding that is necessary for transplanting and good rice production.

Using the gridded data from 1971 to 2005, a long-term prediction for the runoff coefficient (discharge divided by rainfall) through culvert A is shown in Figure 5.17, along with a fit shown in the figure, and shown in Equation 2 (formulated to asymptotically approach a runoff coefficient of 100%). Note that the scatter in the plot is due to the influence of the distribution of rainfall within the year, and reflects the natural variability. The formula for runoff coefficient can be used to predict the volume of discharge based on the annual rainfall and hence the upper limit for water harvesting. For example, an average annual rainfall of 1200 mm has a runoff coefficient of 0.37, corresponding to a total discharge through the culvert of approximately 440 mm. Of this, approximately 30% is generated in the lowlands (22% of the catchment area), and 55% in the medium uplands (53% of the catchment area), with 11% generated in the uplands (10% of the catchment area), and a small contribution (4%) from the ponds and their contributing area (15% of the catchment area). This highlights the importance of a distributed system of water storage (e.g., 5% pits), and the difficulty in optimizing the storage for a variable climate. Future work will need to consider rainfall variability (both in time and amount) and land area required for storage (resulting in a decrease of productive land), as well as defining the primary goal of WSD work (providing security during the wet season, access to water in the early to mid-dry season, or a combination).

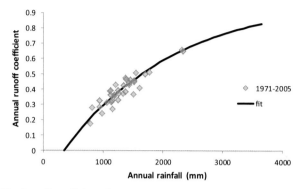

FIGURE 5.17 Runoff coefficient for 1971–2005 predicted by catchment-scale model for Pogro site.

$$r = 1 - e^{-(P-350)/1850} \qquad (2)$$

5.5.4.4 Developing and Applying Models to Evaluate the Potential of Out-of-catchment Impacts as WSD is Scaled up Over Larger Areas of the EIP

One of the objectives of this project was to simulate the impact of watershed management interventions through hydrological modeling for planning and management of WSD activities. The model developed for the catchment-scale modeling of the Pogro study site is being used to explore the impact of WSD across the EIP, using long-term gridded climate data from the Indian Meteorology Department. These modeled flows are being compared with gauged flows at a collection of sites, including:

- Dams in the vicinity of Pogro and Amagara (Parga, Shaharajore, and Kumari, having catchment areas of 18, 43, and 95 km^2, respectively).
- Stream gauges near Hazaribagh, although the data are of poor quality (Hurdag, Nagwan, Olidih, and Banikdih and Usri, with areas of 23, 92, 34, 64, and 731 km^2, respectively).
- Stream gauges in the Brahmani basin with good quality data: Tilga (2987 km^2), Jaraikela (11,641 km^2), Gomlai (21,644 km^2), and Jenapur (36,667 km^2) see Croke et al. 2011 [12].

Application of the Pogro model to these catchments indicated a reasonable reproduction of the overall volume of discharge, although there were deficiencies in the temporal distribution of the stream flow. This is predominantly due to errors in the stream flow data, which have very poor correlation with rainfall, indicating significant time errors. Water balance modeling for medium uplands using climate data from other parts of the region, with no runoff predicted when annual rainfall is less than 1000 mm, is shown in Figure 5.18.

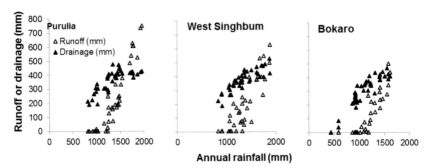

FIGURE 5.18   Water balance modeling of runoff and drainage for three sites in the EIP.

5.5.5 Guidelines for Designing WSD Intervention Plans Using Models, Monitoring, and a Dose of Common Sense

The goal of WSD interventions is to hold water in the landscape in a form that can be accessed during the early to mid-dry season, as well as increasing security for the wet-season crops. In case of the uplands where agro-forestry and perennial horticulture options are used, the aim is to increase the water retention time to enable more access to water for plants, thus increasing the crop yields.

To facilitate effective planning of WSD work, guidelines have been created for the design of WSD interventions/structures based on intervention scale models (see earlier) that represent the behavior of the different structures, including the interactions between structures, coupled with a basic under-standing of hydrological response. The guidelines are designed to assist the planner in deciding the type and arrangement of structures to be installed on a site, ranging from a hill slope to a small catchment. This is achieved by un-derstanding how the structures operate and what factors might limit their effectiveness, as well as the likely limits to the volume of harvestable water.

With all the water-harvesting structures, the trapping efficiency (the frac-tion of retained water that can be successfully extracted) needs to be consid-ered, taking into consideration the fact that some water will flow through, around, or under the structures depending on the local terrain, geology, etc. Such flow past the structure becomes available for the downhill/downstream users. Similarly, surface water bodies (e.g., ponds) have losses through infil-tration to groundwater or evaporation. Hence, the guidelines adopt a 20% trapping efficiency (although this may be higher, maybe 50%, for surface water bodies, depending on when the water would be used). The estimated value of 20% is a conservative value based on the likely recovery from the shallow aquifer storage, and is intended to ensure security in terms of access to water. This is a reasonable estimate of the limit of the efficiency of the WSD structures, with significant improvement possible only at a considerable cost. Thus, if 150 mm of water has been retained by the WSD structures, the available water for irrigation is estimated to be 30 mm. This limits the amount of land that can be irrigated and reduces the risk of running out of water, hence, increasing security for the farmers.

These are the general guidelines:

1. Consider the contributing area (for runoff/groundwater recharge) as well as the residual water (rainfall−evaporation−runoff) from the wet season. This includes the structures to be installed in the contributing area (i.e., ensures consideration of the hydrological connectivity of the landscape).
2. Determine key hydrological characteristics: rainfall/evaporation rates, slope, infiltration rate, and indications of depth to groundwater (auger holes, as well as local knowledge).
3. In recharge areas (mostly medium uplands), consult the farmers regarding sites with surface water, and observe catchments in November (early dry season) to look for evidence of subsurface water (presence of green vegetation or by using an auger). Revisit in February to determine the depth of groundwater (again, using an auger) to assess the potential duration for which shallow groundwater will be available. If water is available for only 2 months or so after the monsoon, then a 5% pit may be used not only for rescue irrigation of rice [5] but also to establish a rabi crop that will mature on residual soil water left by rice [4].
4. In areas where shallow groundwater persists longer, a larger seepage pit may permit fully irrigated crops such as vegetables. In discharge areas, as a general rule, the volume of the seepage pit should be approximately 50% of the required volume of water per irrigation, although this will vary depending on the local conditions, and can be larger if fish rearing is planned. Pit design for irrigation should be based on the fill rate as well as the depth of the water table (i.e., use an adaptive approach to pit design—stop at the point when it is not possible to remove seepage water from the pit if this is reached before the designed depth is achieved).
5. Structures in the upper parts of a catchment area that increase water retention also increase recharge, leading to an increase in shallow groundwater storage. This is an advantage to downhill/downstream users by increasing the water available for irrigation, but it can also be detrimental if there is a significant increase in the inundation period. If it is necessary to ensure the availability of water in the upper parts of the catchment, "leaking" structures may be sealed to limit infiltration loss, which can be evaluated following observations of the performance of the structure over 1−2 years.

5.6 APPLICATION TO ANDHRA PRADESH

Andhra Pradesh has a semi-arid climate, with an average yearly rainfall of 450−600 mm, which occurs mostly between July and October. The main study sites are the Gooty catchment (Anantapur/Kurnool districts) and Vendutla catchment (Prakasam District). As these do not have stream flow gauges, a gauged site close to Gooty (Lakshmipuram) was used to test and modify the model developed using the data collected for the Pogro catchment in West Bengal. The area of the gauged catchment is ~ 2750 km^2, which is an order of magnitude larger than the study sites, and is located to the immediate north of the study site, with the stream flowing northeast toward Kurnool (Figure 5.19). In comparison, the Gooty catchment flows southeast, into the Penneru River, while the Vendutla catchment flows north, joining the Gundlakamma River.

5.6.1 Applying the Model to Lakshmipuram Catchment

This section describes the development of a hydrological model sensitive to the WSD impacts on the gauged catchment in Andhra Pradesh. The starting point for the model is the model developed for the Pogro study site in West Bengal (described earlier in this chapter). Due to the differences in the spatial

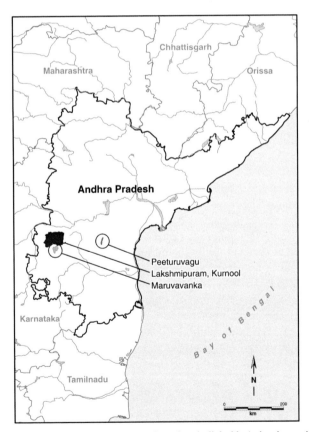

FIGURE 5.19 Map of Andhra Pradesh (state boundary in light blue) showing main rivers and water bodies as well as the study sites and gauged catchment area (Lakshmipuram).

scale, pattern of land use, and climate, it is likely that the model structure for the Pogro study site will not necessarily be suitable for use in the study sites in Andhra Pradesh. Hence, the gauged Lakshmipuram catchment is used to explore differences in the hydrological response between the Pogro study site and southern Andhra Pradesh. Changes to the model are based on the differences in the pattern of land use and climate.

5.6.1.1 Available Time Series Data

Data on the daily average stream flow are available for the Lakshmipuram catchment from June 1, 1988 to May 31, 1996. Two sources of daily rainfall data are available: 0.5 × 0.5 degree gridded data [13], spanning from 1971 to 2005, and data for 10 rain gauges spanning from 1971 to 2007, although there are missing data that need to be handled. This was done by varying the number of gauges used for estimating the rainfall for each day based on the availability of data.

As measured values of evapotranspiration are not available for this area, the maximum daily temperature (obtained from the 1 × 1 degree Indian Meteorology Department gridded dataset [14]) is used as a surrogate, using potential evaporation (PE) as a coefficient times the maximum daily temperature. For example, in Australia, a common factor calibrated on several datasets is 0.166 [15]. Using the calculated PE data for the Pogro catchment, a value for the coefficient is 0.1446; this value was also applied to the Andhra Pradesh catchment. While such an estimate for PE is very poor on a daily timescale, the influence of the error in the estimated PE value is small at that temporal scale because of the difference in magnitude between the soil water storage and the daily PE value during the wet season—errors in the PE value become significant in a hydrological model typically at a monthly scale. This approach limits the use of the model for studying climate change impacts, however, as the relationship between temperature and PE can vary.

5.6.1.2 Autocorrelation and Cross-correlation Analysis

A cross-correlation analysis can be useful for gaining insight into the relationship between rainfall and stream flow as well as for testing the dataset. The approach is to calculate the autocorrelation of the driver (in this case rainfall), and the cross-correlation of the output (in this case stream flow) with the driver. Figures 5.20–5.22 show the results of the analysis for the Lakshmipuram catchment.

Figure 5.20 shows the correlation functions using the gridded rainfall data to estimate the catchment area's rainfall, while Figure 5.21 shows the correlation functions using the individual rain gauge data. The higher value of ~0.5 in the cross-correlation function at lag = 0 in Figure 5.21 suggests that the individual gauges produce a better estimate of the areal rainfall compared to the gridded data.

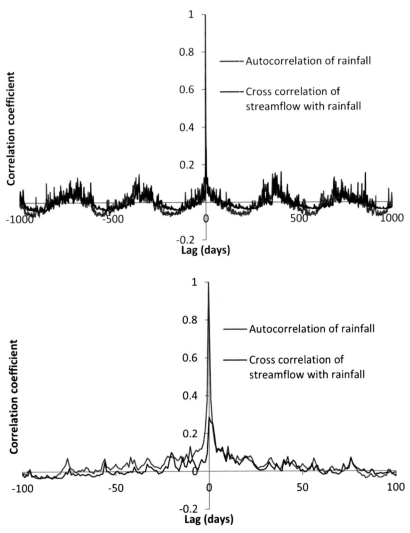

FIGURE 5.20 Cross-correlation analysis for the Lakshmipuram catchment using the gridded rainfall data.

The secondary peak visible at lag = 3 in Figure 5.21 is almost entirely due to a single event on October 10, 1994. When this event is removed from the stream flow record (leaving a 6 day gap in the streamflow data), the secondary peak almost totally disappears (Figure 5.22), and the peak of the cross-correlation function increases slightly. This indicates a timing error in the rainfall and/or the stream flow for that event.

While the gauged rainfall dataset give a better correlation with observed stream flow at a daily scale, there is no strong relationship for either rainfall

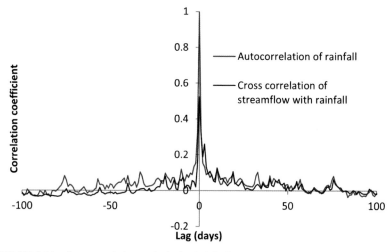

FIGURE 5.21 Cross-correlation analysis for Lakshmipuram catchment using the interpolated individual rain gauge data.

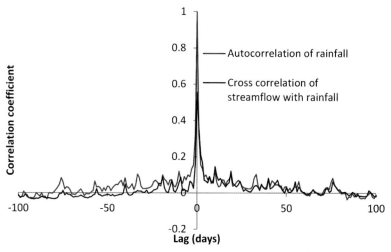

FIGURE 5.22 Cross-correlation analysis for Lakshmipuram catchment using the interpolated individual rain gauge data after removing the peak on October 10, 1994.

dataset with observed stream flow. This is mostly due to the high observed stream flow in 1994, while both rainfall datasets show this to be the driest year over the data period. Removing this year results in a poor relationship between annual gauged rainfall and observed streamflow ($r = 0.27$), with a much more significant relationship between annual gridded rainfall and observed stream flow ($r = 0.68$).

Thus, while the gauged data provide a better representation of the temporal pattern of rainfall at a daily time scale, the long-term rainfall depth is poorly represented due to the limited number of gauges available. This implies that any model using the available data will have difficulty capturing the observed stream flow signal, with the source of the problem being the data used rather than the model structure.

5.7 MODIFICATION OF THE ORIGINAL MODEL

There are a number of differences between the study site in West Bengal and the gauged catchment in Andhra Pradesh. These include (but are not limited to) spatial scale, climate, and pattern of land use. These differences potentially lead to a change in the hydrological response, as well as to a need to modify the model structure. The revised model structure is shown in Figure 5.23.

Due to the larger spatial scale of the Lakshmipuram catchment, there are several large in-stream dams on the river; the largest is a dam on the Handri River, which has a surface area of approximately 10 km². Hence, there was a need to modify the model to include the impacts of such dams. Large in-stream dams are distinct from the ponds used in the model developed for the Pogro study site as the ponds are located in the upper parts of the catchment,

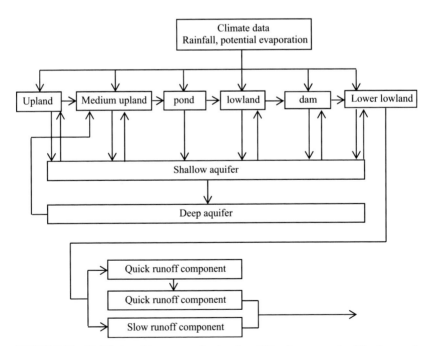

FIGURE 5.23 Modified model structure, showing the additional components of the deep aquifers, dams, and lower lowlands.

often only receiving water from relatively small upland contributing areas (of the order of 1 ha). In comparison, the dams are located in the lower parts of the lowlands, resulting in the development of a lower lowlands class, which are downstream areas of large dams, and can receive irrigation water from the dam. The lakes produced by the dam walls are much deeper and generally larger than the ponds; they are assumed to have very low infiltration rates because of a rock/nonpermeable bed. Exfiltration into the lakes formed by the dam walls can occur due to sub-surface inflow to the lakes from upstream land units (the exfiltration occurs at the sides of the lake or at the upstream of the lake).

Furthermore, the overflow of the medium upland class is perceived to contribute in two different ways in the catchments of Andhra Pradesh. One way is the overflow into the ponds (which include check dams), as with the Pogro case study discussed above. However, since the in-stream dam is located towards the middle of the gauged catchment, the medium upland class therefore causes overflow directly into the lower lowlands. The fraction of the medium uplands producing overflow in the ponds is estimated using a visual interpretation of satellite imagery of the area, with an adopted fraction of 25% of the medium upland land class causing direct overflow into the lower lowlands.

A deep aquifer has also been added to the model (Figure 5.23) as the climate of Andhra Pradesh in comparison with that of West Bengal is much drier, with significantly less rainfall. Hence, the shallow aquifer is much drier and therefore the inhabitants pump water up from the deep aquifers to irrigate their crops. The pumped water is added to the medium upland storage in the model because it is the driest cropping land class, which will need the water first. The total amount of pumped water from the deep aquifer to the medium upland storage is derived from a survey among villagers in Andhra Pradesh conducted within this project about cropping and water availability issues (see Chapters 2 (section 2.9) and 7 (section 7.2)). Adding the deep aquifers entails adding the fluxes between the shallow aquifers and the deep aquifers as well as the storage, or deficit, of the deep aquifers. Water percolates from the shallow aquifers to the deep aquifers as well as from ponds, which can be a major source of recharge to the deeper aquifers, if suitably sited. Because of the dry conditions in the catchment, it is assumed that the dominant flux from the deep aquifers is lateral flow, and as a result seepage from the deep aquifers to the shallow aquifers will not occur and, hence, this is not included in the model.

5.7.1 Evapotranspiration

The large spatial scale of the Lakshmipuram catchment means that there is a much more significant variation in the spatial distribution of evapotranspiration with time than in the Pogro study site. Hence, to model the behavior of the system, it may be necessary to include a weighting term for the

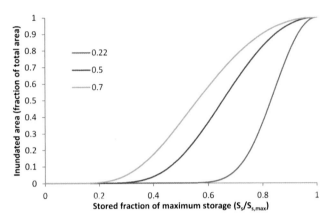

FIGURE 5.24 Cumulative distribution area fraction used in estimating the evapotranspiration.

evapotranspiration that is sensitive to the soil moisture storage. Because of the lack of information on the distribution of evapotranspiration, it is assumed the area fraction can be represented by a Gaussian function, as follows:

$$f_{s,k} = e^{-\left(\frac{x_2}{\mu}\right)^2} = e^{-\left(\frac{\ln(x_1)}{\mu}\right)^2}, \tag{3}$$

where $x_2 = \log(x_1)$, $x_1 = \frac{S_{s,k}}{S_{s,max}}$, with $S_{s,k}$ the surface storage at time k, and $S_{s,max}$ the maximum possible surface storage, and μ is a parameter that indicates the shape of the cumulative distribution (Figure 5.24). Every land class has its own characteristic function dependent on a certain μ value. The μ values give a relative indication of the slope between the areas; an area with a steep slope has a bigger μ value than an area with a lower slope value. Therefore, the μ values are set to 0.22, 0.5, 0.7 and 0.7 for upland, medium upland, lowland, and lower lowland land classes, respectively. The ponds in the area are constructed and assumed to have a flat bottom. Hence, this calculation of area fraction does not influence the evaporation of the ponds (this is a μ value of 1, giving an inundated area fraction of 1).

The evaporation of the surface layer ($E_{s,k}$) can now be calculated as follows:

$$E_{s,k} = f_{s,k} \max\left(S_{s,k}, {}^{P}E_k - E_{sa,k}\right),$$

where ${}^{P}E_k$ = maximum daily temperature × 0.1446 (Section 5.6.1.1). The evapotranspiration of the shallow aquifer ($E_{sa,k}$) is estimated by a nonlinear loss module developed by Croke and Jakeman [7], which represents the moisture state of the catchment as a catchment moisture deficit (CMD), which is represented here as either $M_{sa,k}$ (moisture deficit of the shallow aquifer) or $M_{da,k}$ (moisture deficit of the deep aquifer).

The relationship between the moisture deficit and the evapotranspiration from the shallow aquifer is defined as follows:

$$E_{sa,k} = \begin{cases} {}^{p}E_k & \text{for } M_{sa,k} < h \\ {}^{p}E_k e^{2\left(1 - \frac{M_{sa,k}}{h}\right)} & \text{for } M_{sa,k} \geq h \end{cases}$$

where h is a threshold that indicates when the vegetation is beginning to become stressed. When $M_{sa,k} < h$, the actual shallow aquifer evapotranspiration is assumed to decrease exponentially with increasing $M_{sa,k}$ [7].

5.7.2 Exfiltration

Because of the dry climate and deep groundwater tables in Andhra Pradesh, exfiltration to the surface is approximately zero in the uphill land classes (uplands, medium uplands, and dams). The calculation of the exfiltration is, hence, not based on the shallow aquifer flow but on the moisture deficit of the shallow aquifers.

The exfiltration is therefore calculated as a parameter (maximum exfiltration) multiplied by a fraction

$$X_k = \begin{cases} 0 & \text{for } M_{sa,k} > g \\ \dfrac{(g - M_{sa,k}) G^*}{g} & \text{for } M_{sa,k} \leq g. \end{cases}$$

When the moisture deficit decreases below the threshold (g), the exfiltration increases linearly with the decreasing moisture deficit of the shallow aquifer; G^* is the maximum exfiltration rate. The exfiltration first occurs on the lower lowlands and the lowlands, and proceeds uphill.

The lakes created by the dams have very low infiltration rates (water flowing down into the ground) and are assumed to have circular surface areas. The exfiltration into the lakes from the surrounding lowland areas occurs at the sides of the lake, and depends on the permeability value of the lowlands. The area over which the exfiltration into the lakes occurs is the perimeter of the lake multiplied by the depth of the shallow aquifer.

5.7.3 Percolation

Percolation from the shallow aquifers to the deep aquifers is also expressed as a fraction multiplied by a parameter that indicates the maximum percolation (F^*) during a time step:

$$F_k = \begin{cases} \dfrac{(p - M_{sa,k}) F^*}{p} & \text{for } M_{sa,k} < p \\ 0 & \text{for } M_{sa,k} \geq p. \end{cases}$$

The parameter p is a threshold indicating the value that determines when the percolation process will shut off.

For the percolation and exfiltration calculations, these simple approaches were chosen because of the lack of information about the processes. No data are available to verify any more complex approaches for these calculations.

5.7.4 Runoff

The modeled runoff is generated through a quick ($Q_{q,k}$) and a slow ($Q_{s,k}$) flow pathway, with effective rainfall ($U_{k-\delta}$) partitioned between these using a quick flow fraction (γ). The quick overflow from the lowlands is determined using unit hydrographs of two identical linear stores in series (i.e., a Nash cascade) producing runoff, while the slow overflow component is assumed to be caused by the exfiltration from the shallow aquifers back in the surface storages—this overflow component is translated into runoff using a single linear store. Both overflow components have a time constant τ and a possible delay δ, using the following formulas:

$$Q_{q,k} = -2\alpha_q Q_{q,k-1} - \alpha_q^2 Q_{q,k-2} + \left(\gamma\beta_q\right)^2 U_{k-\delta}$$
$$Q_{s,k} = -\alpha_s Q_{s,k-1} + (1-\gamma)\beta_s U_{k-\delta},$$

where, $\alpha_q = -e^{-1/\tau_q}$ and $\beta_q = 1 + \alpha_q$ (similar for the slow flow component).

5.7.5 Storages

The surface storage is calculated using a mass balance approach, including the previously defined processes that influence the surface storage. The shallow aquifer storage is defined as the maximum storage minus the deficit at that time step. The maximum storage is calculated as the depth of the shallow aquifer multiplied by the porosity of the saturated soil, while the deep aquifer is only expressed as a deficit with no exfiltration or percolation (see Figures 5.25 and 5.26).

Summarized, the formulas to calculate the storages are as follows:

Surface storage of each land class:

$$S_{s,k} = S_{s,k-1} + P_k + I_k + X_k - O_k - E_k,$$

where:

$$I_k = \min\left(S_{s,k}, K_{sat}A_{i,k}\right)$$
$$O_k = \max\left(S_{s,k} - S_{s,max}, 0\right),$$

and I_k is the infiltration frp, the surface storage to the shallow aquifer, and O_k is the overflow from the surface storage.

Shallow aquifer deficit mass balance is given as follows:

$$M_{sa,k} = M_{sa,k-1} + X_k + E_{sa,k} + F_k.$$

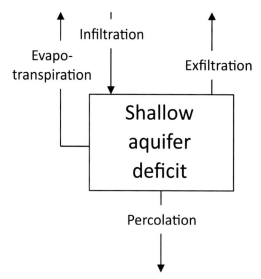

FIGURE 5.25 Shallow aquifer module.

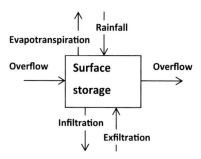

FIGURE 5.26 Surface module.

When the shallow aquifer deficit exceeds the maximum available storage in the shallow aquifer, the exfiltration and the percolation will be zero.

The deep aquifer deficit mass balance is given as follows:

$$M_{da,k} = M_{da,k-1} - F_k.$$

5.8 CALIBRATION AND VALIDATION MODEL ON THE LAKSHMIPURAM CATCHMENT

The model is calibrated using R_{NS}^2 for non-linear models and, in addition, the relative volume error (RVE) is used to compare the results. The calibration is done using the "least square nonlinear" (lsqnonlin) function in MATLAB. This function estimates the model parameters by minimizing the least-squared error in the runoff (i.e., maximizing R^2_{NS}). The model is calibrated as described by Klemes [16]: first the parameters are calibrated on the first 70% of the dataset and validated on the final 30%. Afterward, the first 30% is used for validation and the final 70% for validation.

R^2_{NS} was used as the performance indicator in the calibration, coupled with a visual interpretation of the modeled stream flow. The advantage of doing a visualization is "Details can be observed in the results which would have remained hidden in a quantitative evaluation, or which can help to direct the tools used for quantitative evaluation. Visualisation takes advantage of the strong human capacity for pattern detection and may allow model acceptance or rejection without determining strict formal criteria in advance" [17].

The area characteristics are estimated before the calibration and are displayed in Table 5.4. The characteristics are derived from a study performed by Cornish et al. [18]. These characteristics will not be changed during the calibration, except for the maximum storage values, which can easily be changed by human intervention.

The parameters that have to be calibrated are

- Quick overflow component (τ_q)
- Slow overflow component (τ_s)
- Proportion of quick overflow component (y)
- Maximum percolation rate (F^*)
- Maximum exfiltration rate (G^*)
- Plant stress threshold (shallow aquifers) (h)
- Exfiltration parameter (g)
- Percolation parameter (f)

TABLE 5.4 Area characteristics

Parameter values	Uplands	Medium uplands	Ponds	Lowlands	Dams	Lower lowlands
Maximum storage (S_{max}, mm)	3	50	3000	100	10000	100
Hydraulic conductivity (K, mm/h)	30	0.5	0.4	0.05	0.005 infiltration 0.05 exfiltration	0.05
Area proportion (A_c)	0.25	0.385	0.005	0.25	0.01	0.10

- Depth of shallow aquifer (D)
- Porosity of saturated shallow aquifer (ρ)
- Maximum surface storages ($S_{s,max,\ land\ class}$)

In addition to these parameters, the initial storages of the aquifers and any large dams also need to be set. The quick and slow delay components (δ) have not been calibrated, because data analysis showed that these parameters are zero.

The model was calibrated using two calibration periods: the first 70% and the last 70% of the dataset (ignoring the data for 1988, which is used to warm up the model). For each calibration period, the remaining data was used for validation. Optimising the parameter values using just the R^2_{NS} value gave a typical R^2_{NS} for most periods of between 0.52 and 0.58, though with a significantly high RVE value (13−15%). The exception was the first 30% of the dataset, when the R^2_{NS} value was 0.2, and RVE of 10%. This is influenced mostly by the observed response in 1990, where the observed flow response in late September/early October are larger than for neighbouring years given the rainfall depth. This indicates a general consistency in the hydrological response over the data period, though there is either a change in the catchment response or an error in the data towards the start of the data period.

Calibrating using a combination of the R^2_{NS} and the RVE gave much better RVE values (between 0.6 and 8%), with only a slight reduction in the R^2_{NS} value (typically 0.02). For this reason the optimal parameter values estimated using both performance measures was adopted. The calibrated parameter values are displayed in Table 5.5 and the estimated characteristics for each land use class shown in Table 5.6.

5.8.1 Modeled Runoff

While the model was able to reproduce some of the flow events satisfactorily, there were several events for which the model consistently performed poorly. The cumulative flow departure (CFD) is the cumulative sum of the flow minus mean flow over the period, and is a useful tool for exploring long-term model behaviour. Wet periods produce a positive slope in the CFD, while dry periods result in a negative slope. The observed and modelled CFD are shown in Figure 5.27. While there is generally a good match throughout most of the data, there is a slight overestimation of the flows in the wet season of 1990, an underestimation in 1993, and a significant underestimation of the flows in 1994. These errors combine to account for most of the difference between the modelled and observed CFDs at the end of the data period.

In general, two different possible causes could be identified for explaining these modeling differences. The first possibility is spatial variability in rainfall distribution, or errors in the rainfall dataset. The spatial difference of the rainfall can affect the model in overestimating as well as underestimating

TABLE 5.5 Calibration results

Parameter	Value
Quick overflow	0.049
Slow overflow	12.6
Proportion of quick overflow	54.5
Threshold percolation	1725
Threshold exfiltration	1725
Threshold evapotranspiration	910
Exfiltration parameter	4
Percolation parameter	15
Depth shallow aquifer	10000
Porosity saturated shallow aquifer	0.4
Initial value shallow aquifer deficit	1775
Initial value dam storage	9500

TABLE 5.6 Calibrated area characteristics

Parameter values	Uplands	Medium uplands	Ponds	Lowlands	Dams	Lower lowlands
Maximum storage (S_{max}, mm)	3	11	3000	126	10000	51
Hydraulic conductivity (K, mm/h)	30	0.5	0.4	0.05	0.005 infiltration 0.05 exfiltration	0.05
Area proportion (A_c)	0.25	0.385	0.005	0.25	0.01	0.10

events. The second possible cause is a timing error in one of the datasets, though timing errors are not visible in a cumulative graph.

The underestimation by the model near September 1 and 21, 1993, explain most of the overall model bias for 1993 as can be seen in Figure 5.28. These underestimations are probably caused by the spatial distribution of the rainfall.

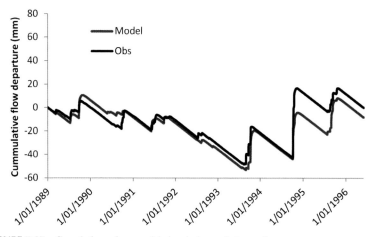

FIGURE 5.27   Cumulative values modeled and observed stream flow.

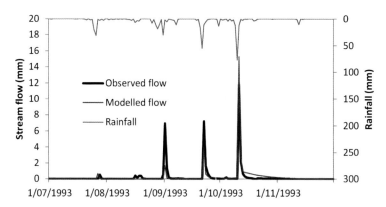

FIGURE 5.28   Output from the model for the wet season of 1993.

The rainfall dataset showed that the rainfall events during this period mostly occurred in the northeastern part of the catchment, near the outlet, while a significant lower amount of rainfall occurred in the rest of the area. As the model is spatially aggregated, the influence of this rainfall distribution will not be captured by the model. There is also a possibility that there is a significant error in the rainfall depth due to the limited number of rain gauges available.

The overestimation in 1995 (Figure 5.29) is probably caused by the input rainfall data as well. This is likely, because of a significant standard error in the mean (20%) of the averaged value of the rainfall on this particular day (August 29, 1995). This statistical error is calculated as the standard deviation of the rainfall gauge data for a particular day, divided by the square root of the number of gauges with the data for that day.

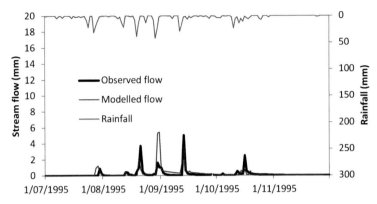

FIGURE 5.29   Model output in 1995.

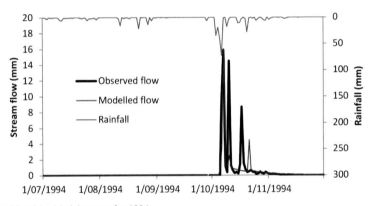

FIGURE 5.30   Model output for 1994.

This error is caused by excessive high values of rainfall in small number of rainfall gauges. These values could be the real measured values at those locations (probably because of small storms causing these amounts of rainfall in a small area), or erroneous data values. However, if these values are real, the calculated average rainfall may be relying too much on these (small) individual events leading to an over-estimation of the rainfall.

The second possible cause, as mentioned previously, could be a timing error in one of the datasets, as can be seen in the third flow peak in Figure 5.30. Such errors can easily occur during data collection or during transformation of the handwritten data into digital databases.

Originally there was an extra peak in the observed stream flow in Figure 5.30 on October 9, 1994. This peak was removed from the dataset (set to negative values and therefore not used in the calculation of performance indicators) during the calibration process, but included for the generation of the CFD. This peak was identified in Section 5.6.1.2 as being the main cause of the second peak in the cross correlation of stream flow with rainfall. This secondary peak is due to the high rainfall a few days earlier, and actually indicates a significant under-estimation of the rainfall for this event. The impact of this on the performance indicators is to significantly decrease the Nash-Sutcliffe efficiency as well as introduce a significant bias. Thus, these two events are the dominant contributors for the error observed in Figure 5.27 for 1994.

5.9 APPLYING THE MODEL ON AN UNGAUGED STUDY SITE IN ANDHRA PRADESH

After the original model was modified to perform at its best on the gauged catchment in Andhra Pradesh, the model was applied to the ungauged Mar-uvavanka study site, is located to the south of, and adjacent to, the gauged catchment—the Maruvavanka study site is approximately 205 km^2. To check the model performance on an ungauged study site, a general approach (the calcu-lation of the rainfall–runoff coefficient, C_{rr}) has been taken to compare the gauged catchment with the ungauged study site. The rainfall–runoff coefficient is the total modelled runoff over the data period divided by the total rainfall.

For the gauged catchment area in Andhra Pradesh the modelled flow over the data period is 166 mm, corresponding to a rainfall-runoff coefficient of 0.052, corresponding to a total discharge of 202 mm, and a total rainfall of 3905 mm. This means that 5.2% of the rainfall is transferred by the model into runoff (under the assumption that there are no other significant fluxes into or out of this catchment).

The rainfall-runoff coefficient of the ungauged study site is calculated using the same input temperature data source as the gauged catchment. For the input rainfall data, the individual rainfall gauges in or near the study site are used. Spatial information about the study site (e.g. distribution of land use classes) is obtained from satellite data and field observations. Because of the significant smaller magnitude of the study site and its spatial interpretation, the dam land class and the lower lowland land class are removed from the model structure (displayed in Figure 5.31).

The calibrated parameters are used to derive C_{rr} for the study site, which is adjacent to the catchment and therefore assumed to have the same behavior in hydrology (Table 5.7). The rainfall-runoff coefficient of the Maruvavanka study site as described in this section is 0.68, corresponding to a total streamflow over the period of 271 mm, and a total rainfall of 3967 mm.

The difference with the gauged catchment can be explained through the presence of the large dam in the Lakshmipuram catchment, resulting in signif-icantly lower annual flows. This influence is further enhanced by the smaller size of the Maruvavanka catchment, resulting in a tendency for more rainfall on fewer days. The change in land use in the ungauged study site (removal of the dams and the lower lowlands land classes), has a significant influence on the modeled runoff, with the discharge decreasing from 377 mm using the model structure developed for the Lakshmipuram catchment to 271 mm using

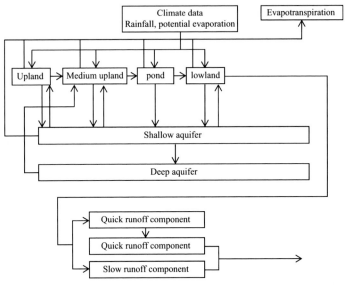

FIGURE 5.31 Model structure for the Maruvavanka study site.

TABLE 5.7 Rainfall for the gauged and ungauged catchments

Area	Rainfall (mm)	Total rainfall events	Average rainfall amount (mm/event)
Ungauged Gooty study site	3420 (~570 mm/year)	364	9.4
Gauged catchment Andhra Pradesh	3353 (~560 mm/year)	605	5.5

the model shown in Figure 5.31). With the dams and lower lowlands removed from the model structure, the land class producing overflow has a higher maximum storage (from 51 mm for the lower lowlands to 126 mm for the lowlands), causing less runoff.

5.10 DISCUSSION

5.10.1 Data Analysis

The analysis of the rainfall and stream flow data for the gauged catchment highlights a number of issues with data quality. This means that it becomes difficult to test model structures, and as a result, the model structure was based partially on the signal in the data, and partially on a general understanding of the hydrological characteristics of the region. As a result, the accuracy of the results depends on how well that understanding applies to the Lakshmipuram catchment.

5.10.2 Modeling the Gauged Lakshmipuram Catchment

A spatial interpretation of the area resulted in the change of the model structure developed for the study in West Bengal. Also some model processes were changed because these processes were either not included in the original model or were causing problems in the generated output by the model.

The assumptions used here have greatly influenced the results generated by the model. The simple linear-calibrated percolation and exfiltration processes are found to influence the modeled runoff significantly. A small change in parameter values has a significant impact on the runoff being generated by the model. Unfortunately, no additional information is available to implement and underpin a more complex approach. The parameters encapsulating these processes are calibrated during the research, and are the major drivers of uncertainty in the defined model processes.

Nevertheless, the simple structure and processes of the model, based on the visual interpretations of the catchment and study site using satellite imagery, along with a simple approach of unknown model processes, gave a better representation of the catchment's hydrology compared to the original model. The R^2_{NS} and RVE of the generated runoff of the model over the whole dataset are found to be 0.53 and -3.9%, respectively (using the calibrated parameters of the final 70% of the dataset and the first 30% for validation). A major limitation to model performance is data quality, particularly the estimated areal rainfall. While the temporal pattern of rainfall seems to agree with the observed streamflow, there is a problem with the estimated annual rainfall which does not match the annual streamflow data. This means that there is a significant data problem, which therefore impacts on the model performance. At an event scale, the modeled runoff events mostly match the observed flow peaks reasonably well. However, some measured runoff events could not be captured by the model. The majority of these mismatches could be related to errors in rainfall depth, mostly an under-estimation of the rainfall. This is likely to be due to the small number of available gauges not adequately capturing the spatial variability in the rainfall.

5.10.3 Modeling the Ungauged Maruvavanka Study Site

Due to a lack of stream flow data for the Maruvavanka study site, the model was not able to be adequately tested. Instead, only a comparison between the rainfall-runoff coefficients for the study site and the gauged catchment was investigated. The climate data used for the ungauged study site was extracted from rainfall gauges in this area and the gridded temperature data from the Indian Meteorological Department. These data are the best available for this study site.

The visual interpretation of the land use showed that the model structure needed to be changed to suit the study site, and hence the dam and lower lowlands land classes were removed from the model structure. This resulted in a decreased modeled runoff (and a lower rainfall−runoff coefficient).

Sensitivity analysis showed that the calibrated values of the gauged catchment mainly influence the exfiltration, infiltration, and percolation of the area. Without the dam and lower lowland land classes, these processes would not change significantly and the calibrated parameters should therefore be representative for this study site.

The small difference (increase) in the rainfall−runoff coefficients of the gauged catchment and the nongauged study site could mostly be related to impact of the large in-stream dam in the gauged catchment, and to a lesser extent, the difference in rainfall data between the catchments. The input data (rainfall) for the study site has more intense rainfall events, increasing the modeled runoff.

5.11 CONCLUSION

Good WSD design needs an understanding of the hydrological characteristics of the area treated, and the upstream/downstream linkages. This means developing a model that not only represents the hydrological response of the catchment, but which is also sensitive to the various water-harvesting structures that can be used in WSD. Further, improvement in access to water resources is just one aspect of the problem; agronomic limits to crop yield, crop choices, and farming systems in general also need to be considered to make the best use of the water resource. This chapter focuses on understanding the hydrological characteristics of the two contrasting study areas.

The model structure developed for the Pogro study site in West Bengal was applied to the gauged catchment in Andhra Pradesh after making modifications to account for the different land use pattern, as well as greater reliance on deeper groundwater resources. While the model's performance was generally not good, this was mostly because of data quality issues: while there is gridded daily rainfall data available for India, it needs to be used with care, even at a scale of 2750 km^2. Ideally, downscaling methods should be used to estimate the rainfall for a catchment, but this will require generation of a set of possible rainfall time series to which the model should be calibrated. While it may be possible to use the stream flow data as an input into the downscaling procedure, this will produce a cyclic argument as it would require the use of a rainfall—stream flow model to condition the generation of rainfall.

Data availability is a significant problem for mesoscale studies in Andhra Pradesh. The existing stream flow stations are focused on a larger scale (several thousand square kilometers or bigger). The lack of data means that the uncertainty in model predictions is considerably higher, and this needs to be taken into consideration while using the model outputs for designing WSD work and policies influencing WSD.

Similarly, while applying WSD on a small scale (i.e., headwater catchments of a few square kilometers), the focus needs to be predominantly at the plot scale, considering what structures might be used to trap water in a single field, or at most in ponds with a contributing area of a few hectares. This includes the use of bunds, pits, etc., to retain water for later use. This could be in the form of surface water storage, or infiltration into a shallow aquifer—in both cases, storage efficiency needs to be considered. Evaporation and infiltration losses from surface storages can be significant, while subsurface flows may mean limited access to recharged water in upland and medium upland areas. Furthermore, WSD design also needs to consider the landscape of the region being treated in addition to the social factors that can hinder the effectiveness of the WSD.

On the other hand, at the mesoscale (~ 100 km^2), in-stream resources become more important; as a result, there is a shift toward larger structures (e.g., check dams) as a means for retaining surface water resources, as well as for providing enhanced recharge to local aquifers. However, plot-scale techniques are still important. Furthermore, to adequately explore the downstream impacts of deep groundwater use, studies at a larger scale are needed.

In the drier areas where access to cheap electricity has allowed economic access to deep groundwater resources, irrigated agriculture has expanded, placing greater reliance on groundwater resources. However, there must be limits placed on the use of this resource, as overdepletion leads to difficulty in accessing the groundwater, which results in economic, social, and environmental concerns.

REFERENCES

[1] Joy KJ, Kirankumar AK, Lele R, Adagale R. Watershed development review: issues and impacts. Centre for Interdisciplinary Studies in Environment and Development. Bangalore: India; 2004. 146 pp.

[2] Kerr J, Pangare G, Pangare VL. Watershed development projects in India: An evaluation. Washington DC: International Food Policy Research Institute; 2002. Research Report 127: 102 pp.

[3] Kiersch B. Land use impacts on water resources: a literature review. In land-water linkages in rural watersheds, FAO land and water bulletin 9; 2002. ISSN 1024-6703, http://www.fao.org/docrep/004/y3618e/y3618e07.htm.

[4] Cornish PS, Karmakar D, Kumar A, Das S, Croke B. Improving crop production for food security on the East India Plateau 1: evaluating rainfall-related risks in rice and opportunities for improved cropping systems. Agric Systems 2014 (submitted).

[5] Pangare V, Karmakar D. Impact on livelihoods: PRADAN's collaboration study of the 5% technology Purulia, West Bengal, India. Poverty-focused smallholder water management: an IWMI research project supported by DFID. Final Report Document 3 of 9. India: International Water Management Institute; 2003. 83 pp.

[6] Jakeman AJ, Littlewood IG, Whitehead PG. Computation of the instantaneous unit hydrograph and identifiable component flows with application to two small upland catchments. J Hydrol 1990;117:275–300.

[7] Croke BF, Jakeman AJ. A catchment moisture deficit module for the IHACRES rainfall-runoff model. Environ Model Software 2004;19(1):1–5. http://dx.doi.org/10.1016/j.envsoft.2003.09.001.

[8] Croke BF, Jakeman AJ. Corrigendum to "A catchment moisture deficit module for the IHACRES rainfall-runoff model". Environ Model Software 2005;20:977. http://dx.doi.org/10.1016/j.envsoft.2004.11.004.

[9] Verdin KL, Greenlee SK. 1996. Development of continental scale digital elevation models and extraction of hydrographic features. In: Proceedings, Third International Conference/Workshop on Integrating GIS and Environmental Modeling. Santa Fe: New Mexico; January 21-26, 1996. National Center for Geographic Information and Analysis, Santa Barbara, California.

[10] So HB, Kirchoff G. Management of clay soils for rainfed lowland rice-based cropping systems. Soil Tillage Res 2000;56:1–2.

[11] Duchon CE, Essenberg GR. Comparative rainfall observations from pit and aboveground rain gauges with and without wind shields. Water Resources Res 2001;37:3253–63.

[12] Croke BFW, Islam A, Ghosh J, Khan MA. Evaluation of approaches for estimation of rainfall and the unit hydrograph. Hydrol Res 2011;42(5):372–85.

[13] Rajeevan M, Bhate J. A high resolution daily gridded rainfall data set (1971-2005) for meso scale meteorological studies; 2008. NCC res. report no.9.

[14] Srivastava AK, Rajeevan M, Kshirsagar SR. Development of a high resolution daily gridded temperature data set (1969-2005) for the Indian region. Atmospheric Science Letters; 2009. http://dx.doi.org/10.1002/asl.232.

[15] Chapman TG. Estimation of Daily Potential Evaporation for Input to Rainfall-Runoff Models. *MODSIM2001*. Integrating Models Nat Resources Manage across Disciplines, Issues Scales 2001;1:293–8. MSSANZ.

[16] Klemes V. Operational testing of hydrological simulation models. Hydrological Sci J 1986;31:13–24.

[17] Bennet ND, Croke BF, Guariso G, Guillaume JH, Hamilton SH, Jakeman AJ, et al. Characterising performance of environmental models. Environ Model Software 2012;40(1):1–20. http://dx.doi.org/10.1016/j.envsoft.2012.09.011. *25 September.*

[18] Cornish P, Croke BF, Kumar S, Karmakar D. Water harvesting and better cropping systems for smallholders of the East India Plateau. Canberra: ACIAR; 2012.

Chapter 6

Sustainable Watershed Development Design Methodology

K.V. Rao [*], Pratyusha Kranti [*], Hamsa Sandeep [*], P.D. Sreedevi [§] and Shakeel Ahmed [§]

[*] Central Research Institute for Dryland Agriculture, Hyderabad, India, [§] CSIR-National Geophysical Research Institute, Hyderabad, India

Chapter Outline

6.1 INTRODUCTION

Watershed programs are aimed at improving the livelihoods of the rural population through sustainability of rainfed production systems. This is done by creating opportunities for better water use with appropriate natural resource conservation and management interventions. Achieving this goal requires

Integrated Assessment of Scale Impacts of Watershed Intervention
http://dx.doi.org/10.1016/B978-0-12-800067-0.00006-2.
149

careful planning by matching the production system requirements with opportunities available at watershed level without degrading the natural resources.

The existing watershed development program prescribes suitable interventions that are approved by the village watershed committee (WC). These interventions are implemented under the guidance of the Project Implementing Agency (PIA) functionaries and supervised by WC members. The interventions in watersheds typically consist of both *in situ* and *ex situ* interventions. It has been often assumed that watersheds would be successful with proper implementation coupled with the participation of communities.

It has been observed many times that watersheds are planned and implemented without holistic consideration. Often, it is assumed by PIA/WCs that the resources are abundant in the watershed and that by implementing a WSD program, the resources can be used effectively and interventions are made based on the physical feasibility. However, it needs to be recognized that watersheds are located within a physical setting—with varying elevations, streams, soils, vegetation types, and water bodies—in a climatic zone represented by varying rainfall and temperature.

The common watershed guidelines prescribe the development of a mesoscale watershed (3000–5000 ha) as a single unit for implementations larger than the earlier watershed area of 500 ha. Under the Indian context, a 5000 ha watershed would spread over a number of villages (about four to six) with many households dependent on them. Since the village is the smallest level of administrative intervention and has its own elected governing body, there is the possibility of conflict between the WC and the local governance systems. With the promise of improved water availability from a watershed program, there is competition for access among different village communities.

Under these changing dimensions, the PIA needs to plan for appropriate interventions suited to the physical setting of the watershed. As every watershed is different in its characteristics, planning interventions must be based on location-specific requirements. Excessive importance on one type of intervention without considering the overall watershed needs leads to nonrealization of the intended benefits, wastage of money, and creation of new problems.

In this chapter, the study sites are analyzed with respect to different biophysical parameters within these watersheds. It aims to build a mechanism that could be used as a general approach while designing the watershed program.

6.2 METHODOLOGY AND APPROACH

The biophysical aspects of the basic driving factors for watershed programs were analyzed, particularly from a mesoscale perspective of the influence on upstream, midstream, and downstream locations.

The study sites represent low to medium rainfall zones in Andhra Pradesh, and the study focuses on the kharif (June to September) monsoon season in

Anantapur and Kurnool districts—where groundnut is the major cropping system—and the kharif and rabi monsoon seasons in the interior parts of Andhra Pradesh (Prakasam District). The analysis focuses on the variability of rainfall and rainy days over a period of 11 years at annual scale as well as characterization of storm intensities under excess (above normal or wet), normal, and deficit monsoon rainfall years. The analysis is useful in understanding the pattern of rainfall and to suggest suitable interventions.

Availability of natural resources (particularly soil resources) and their distribution within the upstream, midstream, and downstream locations was assessed to understand the possibility of successful crop production within watersheds. Assessment of land use distribution coupled with available soil resources over a time offers valuable insight into the cultivation practices across the landscape. The requisite data for this type of analysis are often available in the public domain. In the present study, similar data sources have been used to develop a new methodology for better watershed planning.

Although the guidelines prescribe mesoscale watershed development, the higher order mesoscales are further divided into similar manageable hydrological units (HUNs) rather than administrative boundaries during implementation, because interventions should be based on hydrological issues. With the availability of digital information on elevation at 30 m resolution (Advanced Spaceborne Thermal Emission and Reflection Radiometer, ASTER) and 90 m resolution (Shuttle Radar Topography Mission, SRTM), and Geographic Information System (GIS) software at an affordable price or for free, the mesoscale watersheds could be further divided into smaller HUNs for prioritization. Since resource conservation is the key issue for watersheds, parameters such as hypsometric integral (HI) and drainage density are used to characterize the HUNs and their variability at upstream, midstream, and downstream locations. The HI depicts the proportion of area under a proportion of elevation and is often used as proxy for erosion susceptibility. Drainage density (the length of drains and/or streams per unit area) is a proxy for runoff potential within the watershed. Combined use of these parameters helps to prioritize the watersheds and the type of interventions (*ex situ* or *in situ*). Identification of stream networks—along with their length, order, and distribution—prevailing in a watershed can provide valuable information. This stream network information was generated through GIS software.

Estimation of water availability is the key requirement for watershed planning. Interventions such as check dams and percolation tanks built on streams are not useful when there is no surface runoff in the streams. This could be due to various reasons such as the physical setting of the watershed, porous soils, good vegetative cover, and the management practices followed at farms (soil, land management, small-scale water-harvesting systems, etc.). Peninsular India, which receives lower rainfall compared with the high rainfall regions of Orissa, Jharkhand, Madhya Pradesh, etc., has had tank systems for a long time.

Although the physical setting of watersheds and drainage density information provide a clue for *ex situ* interventions, they need to be carefully assessed under the prevailing climatic zone considering the rainfall information (low, medium, or high) and the existing storage capacity through available tanks, and so on. Hence, there is no prescribed guideline for estimating the water availability at either farm or watershed level. Practitioners estimate the peak discharge to ensure the structural stability, but they do not take into account interventions planned at the farm level when estimating available water for harvesting.

In this study, water availability information was generated for three scenarios: (1) no watershed interventions at farm level, i.e., for the existing land use under a particular soil type; (2) a hypothetical quantity of 50 m^3 of water harvesting; and (3) a hypothetical quantity of 100 m^3 of water harvesting. Some of the watershed programs managed by organizations like the National Bank for Agriculture and Rural Development (NABARD), etc., prescribe *in situ* conservation measures such as farm bunding, continuous contour trenches, and individual farmer-based farm ponds, which retain a large part of the runoff generated at the plot/farm level; these practices are mainstreamed into integrated watershed management programs. Although the capacity of each intervention to retain various quantities of runoff is different, 50 and 100 m^3/ha capacities were chosen as these are the smallest possible capacities that can be created at the farm level. A one-dimensional, root-zone water balance model was used to estimate the runoff and recharge (deep percolation beyond root-zone) for different land uses under various soil types and depths.

This model works on a daily scale starting from June onward. The runoff is estimated through the Soil Conservation Service (SCS) soil moisture accounting method. Based on the land use information and soil type, curve numbers are assigned to different land uses and soil types to generate runoff and recharge information. The program was further modified to account for various quantities of water harvesting. Thus, if dry conditions prevail, the runoff would be retained following rainfall; similarly, if the runoff is less than 5 mm/day or 10 mm/day there will be infiltration into the soil thus increasing the available soil moisture. On the other hand, when there is continuous rain, the soil is saturated, and the runoff would be available on a large scale, although it could be only 5 mm/day. Aggregation of such information at the watershed level would provide the requirement of additional storage at *ex situ* locations with and without watershed interventions.

6.3 CHARACTERIZATION OF BIOPHYSICAL RESOURCES OF THE STUDY SITES

6.3.1 Rainfall (Temporal and Spatial Analysis)

Rainfall recording stations (six in number) were identified for both HUNs, and they represent upstream, midstream, and downstream locations. These stations have records of daily rainfall for long periods of time, with some dating back to 1963. Long-term average rainfall and average rainfall for the recent period are given in Table 6.1.

The time periods considered for estimating average rainfall for different stations are as follows: for Besthavaripeta (HUN1), Komarolu (HUN1), and Thuggali (HUN2) the average rainfall corresponds to 1989−1999; for Peapily and Gooty located in HUN2, the average rainfall corresponds to 1963−1999; and for Racherla (HUN1), the average rainfall corresponds to 2000−2010. The average annual rainfall from 2000 to 2010 for different stations of each HUN is shown in Figures 6.1 and 6.2.

We see that deviations between long-term average rainfall and average rainfall from 2000 to 2010 is more pronounced in HUN2 (Anantapur) compared with HUN1 (Prakasam); stations representing upstream, midstream, and downstream locations in HUN2 recorded higher average rainfall of 28.4, 17, and 34.5%, respectively (Table 6.2).

Eleven years of data were analyzed to understand the quantum of average rainfall received during excess, normal, and deficit rainfall years. The India Meteorological Department (IMD) classification was followed to categorize the year as excess/normal/mild drought/moderate drought. In HUN1, during the above normal rainfall years, the average rainfall recorded in the station for upstream stations is 10% higher than the average rainfall recorded at both midstream and downstream stations. During normal and mild drought years, the difference in rainfall amounts recorded at various stations is only ∼5−8%, and during moderate drought years, approximately 20 and 10% difference is observed between the downstream and upstream stations and between downstream and midstream stations, respectively. On the other hand, in HUN2, the downstream recorded ∼5% higher rainfall during above normal rainfall years and received <5% rainfall during normal years compared with other stations.

While maximum variability of annual rainfall on the deficit side is observed to be increasing from upstream to downstream locations in HUN1 (−31 to −41), in HUN2 the maximum variability on the deficit side could be observed at midstream locations (−37%). Furthermore, in case of excess rainfall over long-term average, while the upstream station recorded higher rainfall in HUN1, the downstream station recorded higher rainfall in HUN2.

An assessment was made to characterize individual years as deficit or excess, following the IMD classification for different stations representing both HUNs (Table 6.3). Since the 2000−2010 time period coincides with higher rainfall compared with the long-term averages, more years have been categorized as above normal years. For above normal rainfall years, high rainfall was received at all stations except for 2001 in HUN2 and the 2008 in HUN1. However, in deficit rainfall years, variability could be observed between stations located in upstream, midstream, and downstream locations of both HUNs, and the pattern is not uniform.

TABLE 6.1 Station-wise average rainfall information for both HUNs

HUN no.	District	Station name	Location	Average rainfall in mm	Average rainfall in mm (2000–2010)	Percentage deviation (range) of annual rainfall during 2000–2010
1	Prakasam	Komarolu	Upstream	704	757	−31 to 77
		Racherla	Midstream	687	687	−34 to 40
		Besthavaripeta	Downstream	661	674	−41 to 54
2	Anantapur/ Kurnool	Peapily	Upstream	528	678	−12 to 77
		Thuggali	Midstream	528	618	−37 to 74
		Gooty	Downstream	472	635	−10 to121

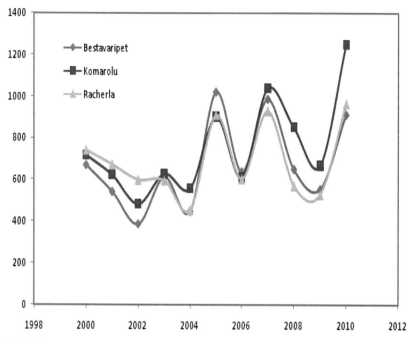

FIGURE 6.1   Temporal distribution of annual rainfall for stations in HUN1.

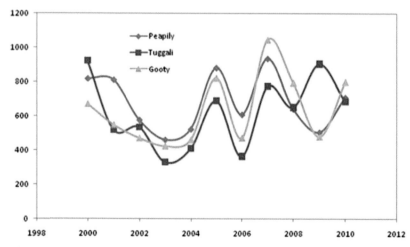

FIGURE 6.2   Temporal distribution of annual rainfall for stations in HUN2.

TABLE 6.2 Average rainfall during normal, above normal, and deficit years

HUN	Station	Rainfall scenario			
		Above normal	Normal	Mild	Moderate
HUN1	Komarolu	1009	650	558	484
	Racherla	932	630	528	453
	Besthavaripeta	971	611	—	418
HUN2	Peapily	799	534	—	—
	Thuggali	773	528	410	348
	Gooty	827	477	—	—

The daily rainfall data were analyzed to understand the daily intensities across different years for all stations representing upstream, midstream, and downstream sections of HUNs (Table 6.4).

We observe that approximately 19—24% of the total rainfall is contributed by events of >50 mm/day across both HUNs, while approximately 45—50% rainfall is contributed by rainy events of >25 mm/day. Little difference could be observed across stations within both HUNs in this context (Table 6.5).

It is observed that a greater number of rainy days is recorded in Komarolu station located at the upstream of HUN1. For the other two stations located at the midstream and downstream, the number of rainy days recorded is observed to be the same. On the other hand, in HUN2, little difference could be observed across stations representing different parts of the watershed.

In terms of distribution of rainy events, approximately 4—6% of the rainy events are recorded with daily intensities of >50 mm/day contributing ~20% of the average rainfall. A total of ~20% of the rainy events has a daily intensity of >25 mm/day. These contribute to ~50% of the rainfall.

Further, an analysis of intense events during different years representing excess, normal, and drought years was performed (Tables 6.6 and 6.7).

It is observed that in HUN1, a high percentage (33%) of rainfall was recorded during excess rainfall years through storms of >50 mm/day at all locations, and during excess rainfall years, ~55% of the annual rainfall is received through intense storms of >25 mm/day at all locations. During normal rainfall years, ~20% of the rainfall is received through storms of >25 mm/day at upstream and downstream locations, while 13% is received at midstream locations. Storm intensities of >25 mm/day contributed to approximately 42—49% of the annual rainfall during normal years at upstream, midstream, and downstream locations. Further, midstream locations received ~27% of the rainfall during moderate drought years through storm

TABLE 6.3 Temporal characterization of rainfall

	HUN1			HUN2		
	Upstream	Midstream	Downstream	Upstream	Midstream	Downstream
Year	Komarolu	Racherla	Besthavaripeta	Peapily	Thuggali	Gooty
2000	Normal	Normal	Normal	Above normal	Above normal	Above normal
2001	Normal	Normal	Normal	Above normal	Normal	Normal
2002	Moderate	Normal	Moderate	Normal	Normal	Normal
2003	Normal	Normal	Normal	Normal	Moderate	Normal
2004	Mild	Moderate	Moderate	Normal	Mild	Normal
2005	Above normal	Above normal	Above normal	Above normal	Above normal	Above normal
2006	Normal	Normal	Normal	Normal	Moderate	Normal
2007	Above normal	Above normal	Above normal	Above normal	Above normal	Above normal
2008	Above normal	Normal	Normal	Above normal	Above normal	Above normal
2009	Normal	Mild	Normal	Normal	Above normal	Normal
2010	Above normal	Above normal	Above normal	Above normal	Above normal	Above normal

TABLE 6.4 Daily rainfall intensity characterization for both HUNs

HUN	Station	Percentage rainfall				Average Rainfall, mm
		<2.5 mm/day	2.5–25 mm/day	26–50 mm/day	>50 mm/day	
HUN1	Komarolu	0.4	52.0	26.7	20.9	757
	Racherla	1.2	52.0	26.7	20.1	687
	Besthavaripeta	1.4	51.8	27.1	19.8	674
HUN2	Peapily	1.9	49.5	27.5	21.0	678
	Thuggali	1.3	56.4	18.7	23.6	618
	Gooty	3.0	45.5	30.8	20.6	635

TABLE 6.5 Characterization of rainy days information based on daily intensity

HUN	Station	Percentage of rainy days				Average number of rainy days
		<2.5 mm/day	2.5–25 mm/day	26–50 mm/day	>50 mm/day	
HUN1	Komarolu		84.2	11.5	4.3	51
	Racherla		82.5	12.8	4.7	42
	Besthavaripeta		83.2	12.7	4.1	42
HUN2	Peapily		81.8	13.3	4.9	40
	Thuggali		86.0	7.7	6.3	40
	Gooty		80.8	14.5	4.7	38

intensities of >50 mm/day. In terms of rainy days, little variability is observed between the upstream, midstream, and downstream locations across excess, normal, or deficit rainfall years. Furthermore, ~80% of the rainy days have daily intensities of <25 mm/day, and with decrease in annual rainfall, less rainy days were observed with storm intensities of >25 mm/day. This phenomenon has been observed across locations in HUN1.

Similarly, it is observed that in HUN2 the midstream locations received ~35% of the annual rainfall during excess rainfall years with storm intensities

TABLE 6.6 Distribution of rainy days and quantum of rainfall in HUN1 (Prakasam District)

Rainfall Met. category	Upstream locations Komarolu Percentage				Midstream locations Racherla Percentage				Downstream locations Besthavaripeta Percentage			
	<2.5 mm/day	2.5–25 mm/day	26–50 mm/day	Total	<2.5 mm/day	2.5–25 mm/day	26–50 mm/day	Total	<2.5 mm/day	2.5–25 mm/day	26–50 mm/day	Total
Above normal	49	20	31	1009	41	23	35	932	42	24	33	971
Normal	50	30	19	650	57	29	13	630	51	27	20	611
Mild	58	30	12	558	49	38	10	528	68	30	0	418
Moderate	69	31	0	485	58	14	27	454				
No. of rainy days												
Above normal	84	10	6	61	78	13	8	48	80	13	7	51
Normal	83	13	4	46	84	13	3	41	83	13	4	41
Mild	87	11	2	45	83	14	3	36	88	12	0	34
Moderate	91	9	0	43	88	6	6	32				

TABLE 6.7 Distribution of rainy days and quantum of rainfall for HUN2 (Anantapur and Kurnool districts)

Rainfall	Upstream Peapily				Midstream Thuggali				Downstream Gooty			
	Percentage			Total	Percentage			Total	Percentage			Total
Met. category	<2.5 mm/day	2.5–25 mm/day	26–50 mm/day		<2.5 mm/day	2.5–25 mm/day	26–50 mm/day		<2.5 mm/day	2.5–25 mm/day	26–50 mm/day	
Above normal	45	25	28	799	50	14	35	773	43	34	22	827
Normal	55	31	12	534	54	26	20	528	48	28	20	477
Mild					61	23	13	410				
Moderate					76	22		348				
No. of rainy days												
Above Normal	80	13	7	44	84	6	9	47	78	17	5	47
Normal	84	13	3	36	84	11	5	34	83	12	4	31
Mild					87	10	3	31				
Moderate					97	3	0	30				

of >50 mm/day, compared with 22 and 28% rainfall received, respectively, at the downstream and upstream locations. Furthermore, approximately 49−56% of the rainfall is recorded with storm intensities of >25 mm/day at the upstream, midstream, and downstream locations. During normal rainfall years, ~20% of the rainfall is received through storm intensities of >50 mm/day at the midstream and downstream locations compared with 12% at the upstream locations. However, this is compensated by a higher quantum of rainfall received through storm intensities of >25 mm/day. During excess rainfall years, ~5% more rainy days were recorded at midstream and downstream locations compared with the upstream locations. During normal rainfall years, less rainy days were observed at downstream locations, compared with the upstream locations. During excess rainfall years, ~20% of the rainy days at the upstream and downstream locations recorded storm intensities of >25 mm/day compared with 15% at the midstream locations. However, during normal rainfall years, ~15% of the rainy days are recorded with storm intensities of >25 mm/day across all locations. During deficit rainfall years, >85% of rainy days recorded storm intensities of <25 mm/day.

6.3.2 Rainfall Projections From 2020 to 2030

Watershed programs are implemented to bring sustainability within the cropping/production system. However, while designing the program and in the process of identification of suitable interventions, often the available data until that period are utilized. However, with increasing climate variability/change, it is necessary to find adequate information on future climate scenarios for near term period; i.e., for approximately 15−20 years ahead from the planning time as the program benefits are expected to last until this period. An attempt was made to get the future climate datasets from 2020 to 2030, representing 2015−2025 and 2025−2035, respectively. MarkSimGCM was used to get the downscaled datasets for the A1b scenario for the ECHAMM5 model for both HUNs. The MarkSimGCM downscales the weather data using GCM datasets in conjunction with the MarkSim weather generator for the location of interest at daily scale, which could be readily used in crop models. An analysis of data indicates that the annual variability of rainfall is from −2.5 to 15% over the long-term average from 1950 to 2000 (WorldClim datasets) during the 2020s and from −1.5 to 20% during the 2030s. The variability expected is within the normal rainfall scenarios only.

6.3.3 Land Use Information

Land use information was collected from the National Remote Sensing Centre from 2004 onward. Land use information was generated for each village within each HUN, for each sub-HUN, i.e., Vajralavanka−Maruvavanka, for 2004−2011, and is presented in the form of tables. The spatial distribution of land use maps indicated degraded forest and scrub lands on hills, and the agricultural area is located in the valley, which could be observed all along the HUN. Based on rainfall in a particular year, current fallows are brought under cultivation.

6.3.3.1 HUN1: Prakasam District

On average, ~16% of the area is under kharif crop cultivation in the downstream location, and ~7% is under kharif crop cultivation in the midstream and upstream locations (Figure 6.3). A similar pattern could be observed for rabi crops. Since this HUN receives rain through northeast monsoon and with the residual moisture, lands are brought under rabi cultivation with crops with a low water requirement. Scrub and other wastelands occupy ~19% of the downstream locations, compared with 61 and 49%, respectively, in the midstream and upstream locations. Current fallow lands are more (25%) at downstream locations compared with midstream (15%) and upstream locations (24%; see Table 6.8).

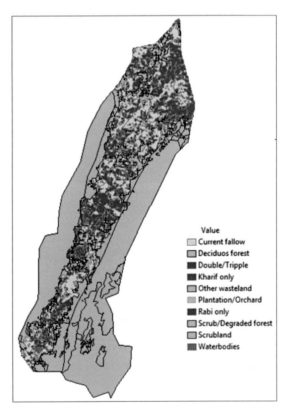

Value
- ☐ Current fallow
- ☐ Deciduos forest
- ■ Double/Tripple
- ■ Kharif only
- ☐ Other wasteland
- ▨ Plantation/Orchard
- ■ Rabi only
- ☐ Scrub/Degraded forest
- ☐ Scrubland
- ▦ Waterbodies

FIGURE 6.3   Land use information at HUN1.

TABLE 6.8 Distribution of land use across different villages in HUN1

Land use type	Besthavaripeta	Mokshagundam	Papaipalli	Pusalapadu	Thaticherla	Vendutla
			Average percent of land use			
Kharif only	16.30	6.74	11.97	6.26	7.14	4.68
Rabi only	10.67	8.31	13.85	26.01	9.00	11.80
Double/triple	19.60	7.94	10.55	10.44	5.81	6.13
Current fallow	25.36	14.59	26.15	26.91	24.03	19.70
Other wasteland	5.14	8.70	10.90	8.46	15.43	17.35
Scrub land	13.76	53.73	15.27	21.90	36.70	40.34

The land use statistics indicate large variability in the kharif cropped area (1.8–13.5% in upstream, 1.5–17% in midstream, and 1.5–45% in downstream villages).

Temporal analysis of land use information from selected villages indicated that a maximum of 15% of the area is cultivated during kharif season in the upstream villages compared with 17% in the midstream and 45% in the downstream villages. Although 2007 is categorized as an excess rainfall year, land use information indicated the lowest amount of area cultivated across kharif and rabi seasons. A high amount of rainfall in June followed by much less rainfall in the following months could be a likely reason for less crops under cultivation. The year 2009 offers a contracting scenario characterized by less rainfall during the kharif season with optimum rains during rabi, which resulted in a higher percentage of crop area during the rabi season (Table 6.9).

6.3.3.2 HUN2: Anantapur Kurnool Districts

Spatial analysis of the land use information in this HUN indicates that kharif cultivated area prevails at parts of upstream villages and downstream villages. Visual interpretation (Figure 6.4) indicates the existence of scrub land and other wastelands at midstream and upstream locations and that current fallows dominate downstream villages.

The village-wise average land use is presented in Table 6.10. Kharif crop cultivation is observed in approximately five villages, with an average of 20% in downstream villages. Current fallows occupy 40–72%, with an average of 62%. Scrub and wastelands occupy ~12% on average in downstream locations. In midstream villages, area under kharif, current fallow, and scrub and wastelands occupies an average of 21, 61, and 16%, respectively. In upstream villages, the area under kharif cultivation, current fallows, and scrub and wastelands occupies approximately 17, 42, and 39%, respectively.

Temporal land use information was analyzed for both sub-HUNs located in this study site: Kharif cultivated area is approximately 21 and 17%, respectively, across the downstream HUN (Maruvavanka) and upstream and midstream HUNs (Vajralavanka). Under severe deficit rainfall during June and July 2006, no kharif area was observed, and current fallows dominated both HUNs. Scrub and other wastelands occupy ~40% of the area in the upstream HUN compared with 7% in the downstream HUN (Table 6.11).

6.3.3.3 Temporal Land use Across Upstream, Midstream, and Downstream Villages

Data on land use was generated (Table 6.12) for each village across the years and variability was observed under kharif cropped area with some years registering almost 0% area. The year 2006 recorded very low rainfall in the June/July months of kharif monsoon resulting in negligible area under cultivation and current fallows increased to 80% in both midstream and

TABLE 6.9 Temporal distribution of land use in upstream, midstream, and downstream villages of HUN1

	2004	2005	2006	2007	2008	2009	2010	2011	Average
Upstream Village: Thaticherla									
Kharif only	13.5	6.1	15.1	3.3	4.4	1.7	11.4	1.8	7.2
Rabi only	2.8	7.1	6.6	6.3	12.1	19.2	4.4	14.0	9.1
Double/triple	0.4	5.6	8.6	6.7	2.8	6.1	14.2	2.4	5.9
Current fallow	29.2	28.7	15.1	29.0	27.0	20.2	15.9	28.4	24.2
Plantation/orchards	0.1	0.2	0.1	0.2	0.1	0.1	0.1	0.1	0.1
Deciduous forest	0.5	0.5	0.6	0.5	0.4	0.5	0.5	0.4	0.5
Scrub/deg. forest	0.6	0.5	0.6	0.6	0.6	0.5	0.6	0.6	0.6
Other wastelands	15.6	15.7	15.7	16.0	15.0	15.2	15.9	15.3	15.5
Scrub lands	37.2	35.6	37.6	37.4	37.6	36.4	37.0	36.9	37.0
Midstream Village: Mokshagundam									
Kharif only	16.9	4.9	11.9	2.6	9.0	1.5	5.6	1.5	6.7
Rabi only	2.0	9.8	4.3	7.0	4.6	16.1	5.0	17.8	8.3
Double/triple	1.9	9.4	13.9	6.5	10.3	7.8	10.3	3.4	7.9
Current fallow	16.2	13.5	6.5	21.5	14.6	13.0	17.0	14.4	14.6
Plantation/orchards	0.0	0.0	0.0	0.0	0.0	0.0	0.0	0.0	0.0

(Continued)

TABLE 6.9 Temporal distribution of land use in upstream, midstream, and downstream villages of HUN1—cont'd

	2004	2005	2006	2007	2008	2009	2010	2011	Average
Deciduous forest	0.0	0.0	0.0	0.0	0.0	0.0	0.0	0.0	0.0
Scrub/deg. forest	0.0	0.0	0.0	0.0	0.0	0.0	0.0	0.0	0.0
Other wastelands	8.6	8.7	8.9	8.3	8.7	9.2	8.9	8.4	8.7
Scrub lands	54.3	53.7	54.6	54.1	52.9	52.4	53.2	54.5	53.7
Downstream Village: Besthavaripeta									
Kharif only	45.9	20.6	31.9	5.0	8.7	4.5	12.4	1.5	16.3
Rabi only	5.4	12.4	5.3	6.8	20.6	19.0	1.2	14.5	10.7
Double/triple	5.6	29.1	27.1	26.2	6.8	17.8	26.5	18.4	19.6
Current fallow	15.4	11.0	8.4	32.5	37.0	29.1	32.3	37.8	25.4
Plantation/orchards	0.0	0.0	0.0	0.0	0.0	0.0	0.0	0.0	0.0
Deciduous forest	0.0	0.0	0.0	0.0	0.0	0.0	0.0	0.0	0.0
Scrub/deg. forest	0.0	0.0	0.0	0.0	0.0	0.0	0.0	0.0	0.0
Other wastelands	5.4	4.5	5.3	6.1	4.3	5.7	5.7	4.2	5.1
Scrub lands	13.3	13.5	13.1	14.3	13.8	15.0	12.9	14.3	13.8

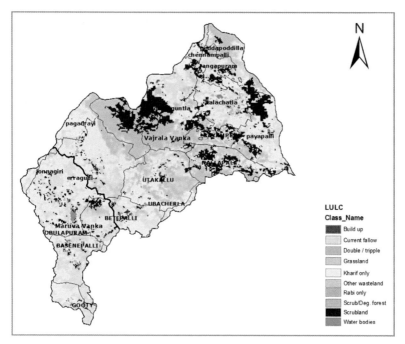

FIGURE 6.4   Land use information at HUN2.

downstream villages. During 2011, downstream villages recorded the lowest kharif cropped area. Midstream and downstream villages showed similar patterns of land use compared with upstream villages where scrub and other wastelands are high.

Land use information for selected villages was collected and is presented in Table 6.13. These villages also were chosen for socioeconomic survey purposes. The area under kharif crops shows clear variation across upstream, midstream, and downstream villages (15, 17, and 22%, respectively). In complete deficit years such as 2006, 2009, and 2011, negligible kharif cropped area was observed across three villages and much of the regular area was left fallow. With negligible kharif area during drought years, current fallows reached a maximum. Approximately 35% of the area was observed to be under scrub and other wastelands in midstream villages, contrary to the aggregate observation.

6.3.4 Soil Resources

Land and soil management options play an important role in the watershed development program. In rainfed areas, reduction in cost of cultivation with enhanced productivity is often seen as a win—win strategy for resource

TABLE 6.10 Average land use scenario in different villages of HUN2

Village	Built-up	Kharif only	Rabi only	Double/ triple	Current fallow	Other wastelands	Scrub land	Water bodies	Total area (ha)	Area location
Downstream villages										
Basinepalle	0	22	1	1	70	3	4	0	800	
Gooty	6	5	2	1	40	33	3	11	774	
Kojjepalli	0	21	1	1	72	3	1	1	429	
Obulapuram	0	32	0	0	68	0	0	0	22	2025
Average	2	20	1	1	62	10	2	3		
Midstream villages										
Basinepalle	0	18	1	2	66	7	6	0	1341	
Erragudi	0	22	1	1	64	8	4	0	3063	
Jonnagiri	0	27	0	1	66	6	0	0	159	
Pagadrayi	0	25	0	0	63	11	1	0	407	
Ubacherla	0	15	0	1	63	17	2	2	1456	
Utakallu	0	17	0	1	47	32	2	0	1401	7826
Average	0	21	1	1	61	13	3	0		

Upstream villages									
Chennampalli	0	19	0	0	66	4	12	0	142
Chetnepalli	8	13	0	1	69	7	2	0	1218
Kadamaguntla	0	12	0	0	31	30	26	0	1888
Kalachatla	0	28	2	5	49	5	12	0	972
Nallapalli	0	15	1	1	19	42	22	0	873
Payapalli	0	19	0	1	32	23	26	0	1844
Peddapodilla	0	13	0	0	35	41	11	0	646
Potedoddi	0	19	0	0	43	34	3	0	1482
S. Rangapuram	0	15	0	0	58	10	17	0	343
Rajampeta	1	17	0	1	20	41	24	0	406
Average					42	24	15	0	10747

TABLE 6.11 Temporal distribution of land use across two sub-HUNs in HUN2

Year	2004	2005	2006	2007	2008	2009	2010	2011	Average
Downstream: Maruvavanka									
Kharif only	35	52	0	34	11	5	32	2	21
Rabi only	0	1	2	0	1	2	1	3	1
Double/triple	1	1	0	1	1	2	1	1	1
Current fallow	57	39	91	58	80	84	59	87	69
Other wastelands	2	2	2	2	2	2	3	3	2
Scrub lands	5	5	5	5	5	5	5	5	5
Upstream and Midstream: Vajralavanka									
Kharif only	25	34	0	28	9	3	27	10	17
Rabi only	0	1	1	0	0	1	0	1	1
Double/triple	0	1	1	1	1	1	1	1	1
Current fallow	35	24	58	31	50	55	32	49	42
Other wastelands	26	26	26	26	26	26	26	26	26
Scrub lands	14	14	14	14	14	14	14	14	14

management and sustainable farm production. Availability of good soil resources within the watershed can either ensure sustainability or enhance the cropping intensity with appropriate interventions in different rainfall regimes. Soil type, depth, and drainage characteristics, as well as their distribution, influence the outcome for a watershed program. Thus, understanding the available soil resources is a key requirement.

The soil resource map from the National Bureau of Soil Survey and Land Use Planning was utilized to generate the maps for both HUNs and the details are presented in the following sections.

6.3.4.1 HUN1: Prakasam District

Deep soils (>75 cm depth) are available in downstream locations and shallow soils are located all along the length of the watershed (Figure 6.5). This is a

TABLE 6.12 Temporal land use information across upstream, midstream and downstream villages

Year	2004	2005	2006	2007	2008	2009	2010	2011	Average
Upstream villages									
Kharif only	25.03	30.07	0.28	24.22	13.25	2.81	26.91	10.58	16.6
Rabi only	0.14	0.50	0.54	0.13	0.19	0.90	0.27	0.52	0.4
Double/triple	0.04	0.87	0.70	1.32	1.13	1.24	0.48	0.43	0.8
Current fallow	34.55	28.54	58.23	35.02	45.42	55.13	32.33	48.82	42.3
Other wastelands	24.15	23.80	23.69	23.58	23.78	23.47	23.73	23.50	23.7
Scrub lands	15.16	15.30	15.61	14.69	15.35	15.57	15.40	15.29	15.3
Midstream villages									
Kharif only	32.03	46.77	0.04	39.87	8.79	2.81	29.72	5.41	20.7
Rabi only	0.30	0.61	0.56	0.27	0.41	1.20	0.36	1.06	0.6
Double/triple	0.18	0.91	0.37	2.37	0.75	0.60	0.70	0.44	0.8
Current fallow	50.35	35.19	82.56	40.47	73.90	79.48	52.88	76.90	61.5
Other wastelands	14.39	13.39	13.26	13.50	12.97	12.91	13.11	13.04	13.3
Scrub lands	2.56	2.62	2.70	2.50	2.66	2.74	2.79	2.74	2.7

(Continued)

TABLE 6.12 Temporal land use information across upstream, midstream and downstream villages—cont'd

Year	2004	2005	2006	2007	2008	2009	2010	2011	Average
Downstream villages									
Kharif only	21.88	46.99	0.00	38.08	17.80	4.03	31.29	0.61	20.09
Rabi only	1.23	1.33	1.26	0.17	1.31	1.33	0.11	0.60	0.92
Double/triple	0.42	1.15	0.25	1.59	0.89	1.27	0.22	0.41	0.77
Current fallow	63.41	37.48	85.09	43.27	62.46	76.01	51.20	80.67	62.45
Other wastelands	9.34	9.00	9.13	7.87	7.81	11.84	11.13	11.40	9.69
Scrub lands	2.03	2.02	1.95	1.87	2.12	1.99	1.79	1.88	1.96

TABLE 6.13 Land use information for selected villages in HUN2

Year	2004	2005	2006	2007	2008	2009	2010	2011	Average
Upstream: S. Rangapuram									
Kharif only	25	26	0	19	15	2	28	2	15
Rabi only	0	0	0	0	0	0	0	0	0
Double/triple	0	0	0	0	0	0	0	0	0
Current fallow	47	47	72	57	56	72	45	73	59
Other wastelands	12	9	10	10	10	8	9	9	10
Scrub lands	15	18	18	14	18	18	18	16	17
Midstream: Utakallu									
Kharif only	28	33	0	31	10	2	28	6	17
Rabi only	0	0	0	0	0	0	0	1	0
Double/Triple	0	1	1	1	0	1	1	1	1
Current fallow	35	31	64	32	55	64	37	59	47
Other wastelands	34	33	32	32	31	30	30	30	32
Scrub lands	2	2	2	2	2	3	3	3	3
Downstream: Basinepalle									
Kharif only	40	53	0	26	15	6	31	2	22
Rabi only	1	1	2	0	0	1	0	2	1
Double/Triple	1	1	0	1	1	2	1	1	1
Current fallow	52	39	91	66	77	84	62	88	70
Other wastelands	3	2	3	2	2	3	3	3	3
Scrub lands	4	4	4	4	4	4	3	4	4

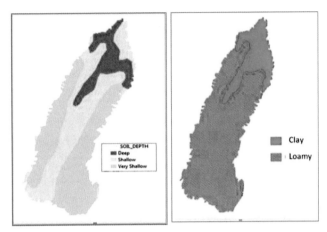

FIGURE 6.5   Spatial distribution of soil properties (soil depth and type) for HUN1.

typical feature of this HUN as watersheds are surrounded by hills on both sides, and the width of the plain area increases horizontally at the downstream locations. Very shallow soils are located on the slopes of the hills located along the length of the HUN. Clayey and loamy soils are dominant in the HUN and the former occupy a higher percentage of area in downstream locations (Figure 6.6).

Deeper soils coupled with more clay content in the downstream locations offer better prospects for crop production. However, since shallow soils are located all along the watershed with hills on either side, the possibility exists for enhanced crop production through suitable watershed interventions.

6.3.4.2 HUN2: Anantapur and Kurnool Districts

Gravelly loamy soils occupy more area in the upstream locations while clayey and calcareous clayey soils are present in the midstream and downstream locations. Gravelly loamy soils in upstream locations have moderate depth, indicating the constraint on soil resource available for plant growth. Parts of clayey soils located in upstream locations also have moderate depth compared with the availability of deeper soils in the midstream and downstream locations. Since the available resources in the midstream and downstream locations support crop production, it is the quantum and distribution of rainfall that is vital for ensuring optimum farm production.

6.4 DESCRIPTION OF HUNs

6.4.1 Delineation of HUN into Watersheds

6.4.1.1 HUN1: Prakasam District

HUN1 in the Prakasam District was delineated to sub-watersheds in a GIS environment with different threshold areas ranging from 50 to 250 ha.

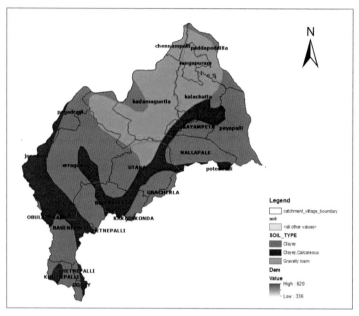

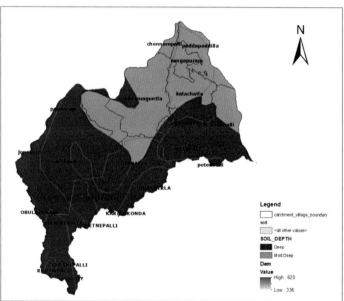

FIGURE 6.6 Spatial distribution of soil properties (soil depth and texture) for HUN2.

Parameters such as HI and drainage density were estimated and presented for each threshold area. With fine resolution thresholds, minor variations could be observed, facilitating informed decisions on interventions.

An HI value of <0.3 indicates a stabilized topography with less chance for further erosion, while an HI value of 0.3–0.6 indicates semi-equilibrium and an HI value of >0.6 indicates an unstable topography representing disequilibrium and chances of erosion. A few sub-watersheds located in the upstream villages are in an unstable condition, and erosion potential was also observed. However, many sub-watersheds in the midstream and downstream locations are under equilibrium conditions, except for those located on the hills on either side, and have more length of drains per unit area, indicating potential water-harvesting systems (Figure 6.7).

Stream network for the HUN was generated in a GIS framework and the stream order was generated following the Strahler procedure (Figure 6.8). A visual interpretation of the drainage network indicates a greater number of first- and second order streams with parallel drainage network, while the 3D perspective of the stream network indicates an opportunity for water harvesting as hills are located along the length of the watershed with first-order or second-order streams directly draining into a fourth-order stream originating at upstream locations.

6.4.1.2 HUN2: Anantapur and Kurnool Districts

A similar procedure to that followed for HUN1 was undertaken to delineate the sub-watersheds and gather information on HI, drainage density, and stream network. Upstream and midstream watersheds are observed to be under a semi-equilibrium state while a majority of the downstream watersheds are under an equilibrium state (Figure 6.9).

The stream network for the HUN was generated in a GIS framework and the stream order was generated following the Strahler procedure (Figure 6.10). A visual interpretation of the drainage network indicates a continuous interaction between the upstream, midstream, and downstream locations, which would influence the interventions. This type of drainage system requires careful planning of interventions, as any excess amount of intervention at a watershed will affect the sub-watersheds located below. The upstream and midstream locations have higher order streams (fifth-order) compared with the downstream locations (fourth-order).

6.5 LAND AND WATER MANAGEMENT INTERVENTIONS THROUGH WATERSHED PROGRAMS

An inventory was made with a field survey of watershed interventions in both HUNs. Data on interventions such as check dams, farm ponds, and rock-fill dams were collected from the field survey and geo-referenced on the watershed map.

(a) Threshold area: 50 ha

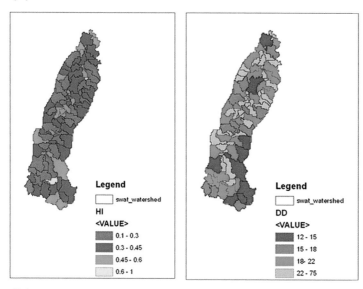

(b) Threshold area: 100 ha

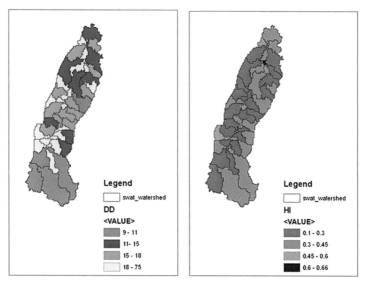

FIGURE 6.7   (a) Threshold area: 50 ha, (b) threshold area: 100 ha, (c) threshold area: 150 ha, and (d) threshold area: 250 ha.

(c) Threshold area: 150 ha

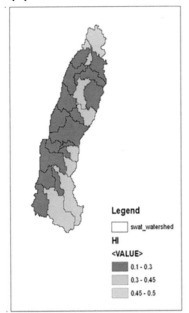
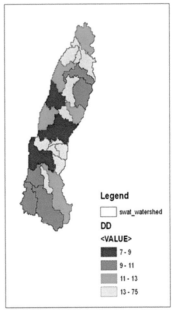

(d) Threshold area: 250 ha

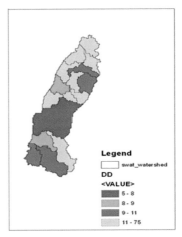

FIGURE 6.7   Continued.

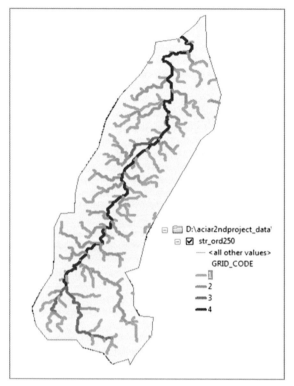

FIGURE 6.8 Stream network and order for HUN1.

6.5.1 HUN1: Prakasam District

Check dam (Figure 6.11), spatial distribution of rock-fill dams (Figure 6.12), and spatial distribution of farm ponds (Figure 6.13) in HUN1.

6.5.2 HUN2: Anantapur and Kurnool Districts

Spatial distribution of check dams and rock-fill dams (Figure 6.14), and spatial distribution of tanks in HUN2 (Figure 6.15).

6.6 ASSESSMENT OF WATERSHED INTERVENTIONS ON HYDROLOGY OF WATERSHEDS

6.6.1 Influence of Watershed Interventions on Resource Conservation at Plot Scale and Watershed Scale

The influence of watershed interventions on resource conservation was assessed through a one-dimensional, root-zone water balance model. This

(a) Threshold area: 50 ha

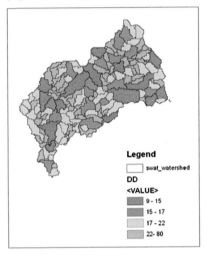
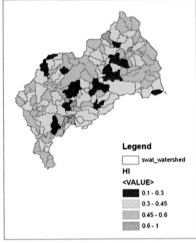

(b) Threshold area: 100 ha

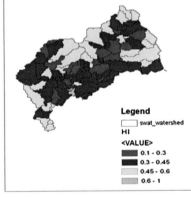
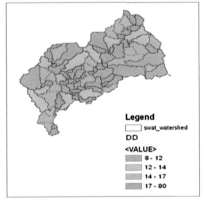

FIGURE 6.9   HI and drainage density for different threshold limits in HUN2. (a) Threshold area: 50 ha, (b) threshold area: 100 ha, (c) threshold area: 150 ha, and (d) threshold area: 250 ha.

model requires daily data on rainfall and soil parameters such as field capacity, wilting point, and crop phonological characters—namely time to reach maturity and crop coefficients. The maximum root growth is restricted to the available soil depth (Figure 6.16).

While reference evapotranspiration is estimated by the temperature-dependent Hargreaves' Method, daily temperature data were extracted from IMD grid datasets, and runoff is estimated through the SCS curve number moisture accounting procedure. For each land use, soil type, and depth, a curve number model was simulated, starting from June onward. The model was run

(c) Threshold area: 150 ha

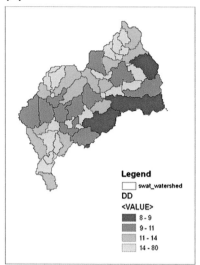 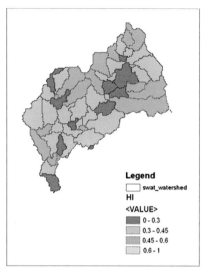

(d) Threshold area: 250 ha

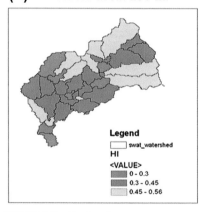 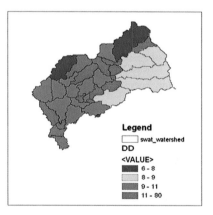

FIGURE 6.9 Continued.

for all combinations of land use, soil type, depth, and rainfall recording stations (three stations in each HUN). Deep percolation quantity, which goes beyond the root zone, is considered as recharge.

The model simulations were made under three scenarios: without watershed interventions, interventions at 50 m³/ha, and interventions at and 100 m³/ha. If watershed interventions are implemented in each parcel of various land uses, as is practiced in the NABARD watershed programs, potential reductions across the landscape could be seen with outflows

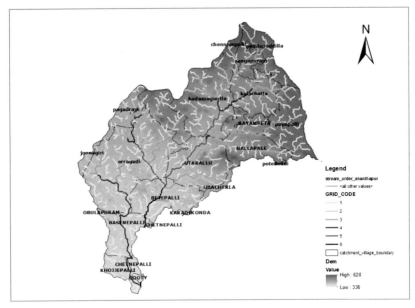

FIGURE 6.10 Stream network and stream order for HUN2.

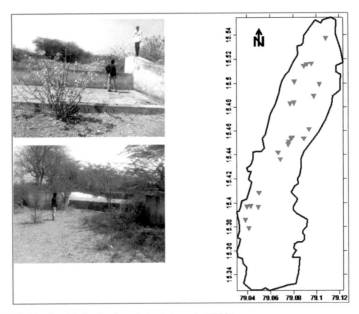

FIGURE 6.11 Spatial distribution of check dams in HUN1.

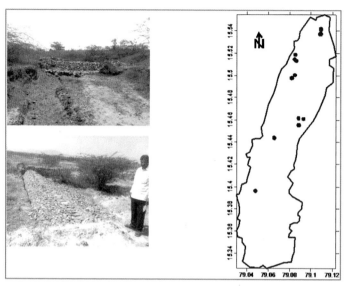

FIGURE 6.12   Spatial distribution of rock-fill dams in HUN1.

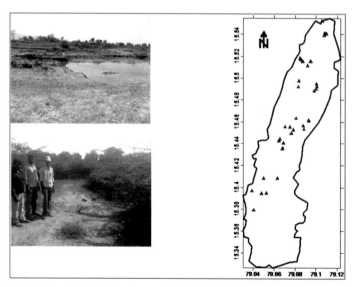

FIGURE 6.13   Spatial distribution farm ponds in HUN1.

FIGURE 6.14   Spatial distribution of Check dams and rock-fill dams in HUN2.

FIGURE 6.15   Spatial distribution of tanks in HUN2.

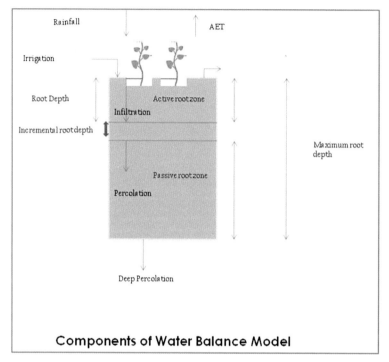

FIGURE 6.16   Schematic diagram of water balance model.

reduced in existing tanks, thus depriving existing livelihoods in the low to medium rainfall zones. The impact would be much higher during deficit rainfall years.

Model output on average runoff percolation, deep percolation (recharge), and actual evapotranspiration for clayey and loamy soils of different depths ranging from very shallow to deep soils is presented in Table 6.14 for without watershed interventions, with different quantities of water harvesting at plot level with rainfall from Komarolu station (upstream) in the Prakasam District.

We see that with watershed interventions at the plot level, average runoff could be reduced by ~30% with increase in actual evapotranspiration (AET) and recharge values in both types of soils. Furthermore, large variability is observed between the AET values for very shallow and deep soils, indicating that the moisture available in the root zone actively contributes to better crop growth as compared with more recharge or deep percolation in shallow soils.

TABLE 6.14 Model output in Komarolu station, Prakasam District

Soil type	Soil depth		Without watershed intervention				50 m³ storage				100 m³ storage			
			Ro	PER	DEEP	AET	Ro	PER	DEEP	AET	Ro	PER	DEEP	AET
Clayey	Very deep	Kharif1	158	52	34	607	112	77	59	626	108	83	66	624
	Deep	Kharif2	156	53	38	587	111	79	64	604	107	85	69	604
	Moderately deep	Kharif3	153	58	47	548	109	86	75	563	105	90	79	564
	Shallow	Kharif4	153	80	75	481	108	115	110	489	102	124	119	487
	Very Shallow	Kharif5	148	140	139	396	103	184	183	397	100	187	186	397
Loamy	Very deep	Kharif25	117	76	58	626	83	98	80	637	79	103	85	635
	Deep	Kharif26	115	78	63	604	82	101	85	614	78	105	90	614
	Moderately deep	kharif27	113	85	74	563	81	108	97	571	75	115	103	570
	Shallow	Kharif28	111	114	109	488	79	144	138	490	74	149	144	490
	Very Shallow	Kharif29	106	182	181	396	75	212	211	396	68	220	219	396

RO, runoff; PER, percolation; DEEP, deep percolation; AET, actual evapotranspiration.

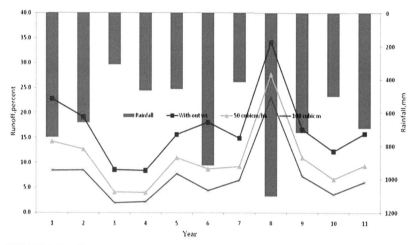

FIGURE 6.17   Change in runoff under different water storage due to conservation.

Conservation of water from runoff reduction in different rainfall years is shown in Figure 6.17 for kharif crop cultivation in deep clayey soils of the Gooty station. The reduction in runoff is observed to vary from 19 to 50% for 50 m^3/ha water conservation and from 33 to 74% for 100 m^3/ha water conservation.

Similar information for loamy soils is presented in Table 6.15. We observe maximum reduction in runoff during low rainfall years, while more reduction is observed with enhanced water harvesting, which needs to be promoted in crop lands at the farm level, thus effectively using the available rainfall. Interventions of a similar nature for other land uses need careful consideration, especially in low to medium rainfall zones. Use of long-term rainfall data to generate model outputs with these interventions would help reduce the uncertainty in runoff reduction during excess and deficit rainfall years.

Based on the assessment of the watershed interventions for resource conservation through surface water balance modeling for different land uses within HUNs of Anantapur and Prakasam districts, the long-term estimations indicate that the resource conservation (Figure 6.18) is higher during drought years and would impact the water availability in midstream and downstream locations.

The temporal distribution of deep percolation from kharif cropped areas is shown in Table 6.16. We observe that recharge is almost doubled with a higher quantity of water conservation at the farm level. During deficit years, the recharge is much higher indicating the effectiveness of watershed interventions.

TABLE 6.15 Temporal variability in runoff with watershed management interventions

Year	Rainfall	Runoff: very deep loamy soil			Runoff (%)			Reduction (%)	
		Without watershed	50 m³/ha	100 m³/ha	Without watershed	50 m³/ha	100 m³/ha	50 m³/ha	100 m³/ha
1	748	110.9	59.14	23.7	14.8	7.9	3.2	47	79
2	660	79.15	46.05	24.24	12.0	7.0	3.7	42	69
3	309	14.73	5.92	0.92	4.8	1.9	0.3	60	94
4	468	19.66	7.74	2.74	4.2	1.7	0.6	61	86
5	458	45.73	29.79	15.36	10.0	6.5	3.4	35	66
6	918	101.5	38.36	23.93	11.1	4.2	2.6	62	76
7	416	37.04	21.67	16.67	8.9	5.2	4.0	41	55
8	1101	295.79	233.98	190.78	26.9	21.2	17.3	21	36
9	720	79.13	47.12	29.56	11.0	6.5	4.1	40	63
10	501	33.98	13.56	3.56	6.8	2.7	0.7	60	90
11	692	68.12	38.21	20.04	9.8	5.5	2.9	44	71
Average	636	81	49	32	11	6	4	47	71

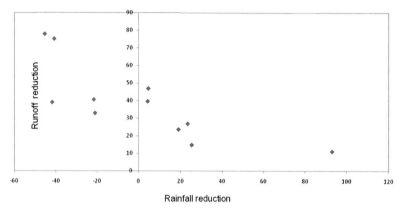

FIGURE 6.18   Reduction in runoff due to watershed interventions at plot scale.

TABLE 6.16 Temporal distribution of deep percolation from kharif cropped areas

Year	Rain-fall	Deep Percolation: Very deep loamy soil			Percolation (%)			Increase (%)	
		Without watershed	50 m³/ ha	100 m³/ ha	Without watershed	50 m³/ ha	100 m³/ ha	50 m³/ ha	100 m³/ ha
1	749	79	106	130	11	14	17	35	66
2	661	0	0	0	0	0	0		
3	310	3	11	16	1	4	5	317	505
4	469	0	0	0	0	0	0		
5	458	0	0	0	0	0	0		
6	918	26	64	75	3	7	8	145	188
7	416	0	0	0	0	0	0		
8	1101	70	100	125	6	9	11	44	79
9	720	0	11	24	0	2	3		
10	501	16	21	21	3	4	4	34	34
11	693	28	49	67	4	7	10	72	136

6.7 CONCLUSIONS

Uniform technological interventions under watershed development programs are implemented across climatic zones without considering the watershed characteristics such as soil resources and the prevailing land use conditions. In the absence of careful planning, especially in low to medium rainfall zones, the purported good the watershed interventions are assumed to provide may in fact create new problems that are visibly seen by PIAs under the mesoscale watershed development programs. Further, interventions on every land parcel, namely "net planning," for water conservation intervention mainly through farm bunding and water absorption trenches for land uses such as scrub lands not only make the investments unproductive in the immediate term but also raise new hydrological issues such as reduced flows into the existing water bodies. This can create conflicts within communities.

To overcome these problems, it is necessary to estimate the water availability under different scenarios, such as with and without watershed interventions. Water conservation efforts, through a certain quantum of water harvesting under a modeling framework, would provide valuable insights into water availability. Based on the available water after water conservation efforts at farm level, additional storage could be planned on streams as *ex situ* conservation interventions, after accounting for the existing storage capacities through tanks. Modern tools, such GIS coupled with the high computing power and publicly available datasets, enhance the capabilities of the PIAs in visualizing the watershed features and key parameters representing erosion status and runoff potential. This helps when making informed decisions for prioritizing the sub-watersheds within the mesoscale HUNs.

Part III

Socio-economic and Livelihood Impacts of Watersheds

Chapter 7

Assessing Livelihood Impacts of Watersheds at Scale: An Integrated Approach

V. Ratna Reddy*, T. Chiranjeevi*, Sanjit Kumar Rout* and M. Sreenivasa Reddy[§]

Livelihoods and Natural Resource Management Institute, Hyderabad, India, [§]Centre for Economic and Social Studies, Hyderabad, India

Chapter Outline

7.1 INTRODUCTION

Impact measurements of developmental initiatives are more often used to correct the type and nature of interventions and implementation modalities. The objective is to improve allocative efficiency of resources and their value. It is important for specific programs like watershed development (WSD) to receive huge budgetary allocations (Rs.250,000 million per year, or more than $4 hundred million). Measuring the watershed impacts in India is becoming more complex in as WSD programs have transformed from a soil and water

Integrated Assessment of Scale Impacts of Watershed Intervention
http://dx.doi.org/10.1016/B978-0-12-800067-0.00007-4.

conservation initiative to a comprehensive rural development and livelihoods program, although the former remains the core. The recent change in the scale of watersheds from micro (500 ha) to meso (5000–10,000 ha) under the Integrated Watershed Management Program (IWMP) is expected to make the impact assessments comprehensive as well as demanding in terms of data and methods of assessment. A larger watershed scale would facilitate the capture of externalities relating to groundwater and surface water flows. Because of this, the impacts of positive and negative externalities across the streams should be considered while assessing the watershed impacts.

Until now, watershed impact assessment studies have focused on the impacts from socioeconomics and natural resources [2,6]. Such assessments are also used to estimate the benefit-cost ratios of the program [2]. With the introduction of the livelihoods component along with a participatory approach to implementation during the late 1990s, impact studies have started to use the sustainable livelihoods (SL) framework to assess the impacts [4,5]. The SL framework is a more comprehensive approach that looks beyond the income and employment aspects of poverty using the five capitals to assess the impacts. The framework incorporates the social, human, and physical (capitals) dimensions of poverty, which are more long term in nature. Despite the fact that the prime objective of WSD is soil and water conservation and thus improving the productivity and resilience of the system, not much has been done to assess the resilience aspects of WSD.

Of late, resilience is gaining prominence as an important attribute of the farming communities, especially in the context of climate change impacts. The SL framework has resilience built in. In most cases, impact studies do not have the backing of proper baseline information. This limits the validity of the impact assessment, as the data generated from the households suffer from memory lapse when "before and after" methods are used, and getting a perfectly matching sample becomes a limitation when "with and without" methods are used. However, adopting a "double difference" method, in which both before and after and with and without approaches are combined, is expected to provide the best proxy in the absence of a baseline [4]. Impact assessments are also influenced by the timing of the study. While impacts are clearly captured in the immediate post-implementation phase, attribution of impacts gets blurred as the gap between implementation and assessment increases. In this context, using resilience as an impact indicator would help in addressing the limitations to a large extent. Resilience, in a way, is directly linked to watershed interventions and could be attributed directly to the current watershed condition. Furthermore, resilience is more long term in nature so it addresses the sustainability aspects of WSD. When resilience is linked to the five capitals, it becomes robust and comprehensive in understanding the impacts in the absence of baseline information.

Given the nature and scope of WSD in the context of rural development in India, two important aspects need to be considered while assessing its impacts:

(1) the impact of WSD on the resilience of the households to cope with drought and climate change at different locations of the watershed (scale), i.e., upstream, midstream and downstream locations and (2) the nature and intensity of the impacts in the context of differing hydrogeological and biophysical conditions that can be internalized in the context of scale. Assessing the resilience at different locations of the watershed would help in assessing the differential impacts of watershed at scale. Comparing this with the control situations (without watershed) would help in assessing the watershed impacts per se, although this has the limitation of getting a matching situation.

This chapter is an attempt to demonstrate that resilience is a better indicator of watershed impacts when compared with the conventional indicators of irrigation such as yield and income. This is more evident in the context of the long gap between implementation and impact assessment. Assessing resilience is relevant in the IWMP context, which incorporates the scale aspects of impact assessment.

The specific objectives of this chapter include: (1) assessing resilience of the households in the context of WSD using the SL framework (five capitals) and comparing the same with the conventional approach of crops, yields, income, etc.; (2) assessing the impact of WSD at upstream, midstream, and downstream locations; and (3) identifying the important indicators of the different capitals that influence resilience.

It is hypothesized that (1) WSD has the inherent potential to enhance the resilience of the system and (2) resilience is a better indicator of WSD when compared with conventional indicators, especially in the absence of baseline data. Related hypotheses include resilience is expected to be (3) linked to average rainfall directly, (4) greater in the downstream locations, and (5) greater among more economically and socially advantaged sections.

This chapter is organized into seven sections. The following section provides the approach and method for impact assessment in the context of WSD. Section 7.3 presents the profile of the study sites. Analysis of the impact of WSD using sustainable rural livelihoods approach (SRL) is discussed in Section 7.4. The effect of WSD on resilience is assessed in Section 7.5. These results are validated further in the context of hydrogeological and biophysical aspects in Section 7.6. Finally, some concluding remarks are made in the last section.

7.2 APPROACH AND METHODS

Defining the resilience of a household is critical for assessing the impact of WSD on resilience. Although resilience has been defined in a number of ways, the following three definitions are the most relevant for our purpose:

1. "The amount of change a system can undergo without changing state" [9].
2. "The ability of a system to recover from the effect of an extreme load that may have caused harm"[8].

3. "Resilience refers to three conditions that enable social or ecological systems to bounce back after a shock. The conditions are: ability to self-organize, ability to buffer disturbance, and capacity for learning and adapting" [7].

These definitions deal with the ability or capacity of a system or individual to deal with the magnitude and intensity of change. The reasons behind the change vary across locations. For instance, drought is a common phenomenon in rainfed regions, and its spread and intensity is expected to rise in the future due to climate change. Therefore, we define the household resilience in the context of drought as: "the number of droughts a household can survive." This captures the magnitude and intensity of droughts, and the households' capacity is assessed in terms of its access to the five capitals. Households are questioned specifically about their perceived capacity to survive drought within the context of each capital (physical, natural, financial, human, and social). Households with a perceived capacity to survive more droughts are considered more resilient than those with a perceived capacity to survive fewer droughts.

The rationale behind using the five capitals framework (SRL) to assess households is that SRL provides a comprehensive understanding of farmers' assets and capabilities to deal with change; it looks beyond income for understanding a household's livelihood strategies. SRL as a framework is comprehensive and has the potential to provide a dynamic assessment (as it ensures sustaining the livelihoods in future) of the livelihood capitals at different scales (as the capitals could be assessed at household, community, and village levels, as well as wider levels).

Different methods and tools are used to assess the capitals and their potential to deal with the changes (such as drought or climate variability) at the household level. These methods and tools range from pure qualitative to pure quantitative methods. Adoption of these are often dictated by the constraints of time and budget apart from the researchers' convictions, and they may have their own limitations and biases. While the SRL framework has its own limitations, such as quantifying the qualitative data and integrating data from different scales (households, community, village, etc.), the methods and tools used in data generation could further complicate or limit the applicability of the framework. Therefore, one has to differentiate between the framework on one hand and the methods and tools used for data generation on the other.

A multilayered approach was adopted for the present study. Focus group discussions were used to assess the potential of the five capitals in dealing with droughts. This information was complemented with the information from the quantitative data generated using two rounds of questionnaires. In addition, case studies were used to understand specific narratives representing different groups. Thus, on the whole, three types of instruments were used to generate data.

Analysis was performed from different angles to make it analytically robust. Given the complex nature of the data generated (qualitative as well as quantitative), one has to be cautious when choosing analytical tools and instruments. Further, it is necessary to understand the limitations of each tool and method—especially for understanding the investigator/respondent sensitivity for the tools. No single tool/method on its own is enough to understand the complexities of the issues at hand, as each tool/method has its own set of advantages and disadvantages.

Impact assessment was performed using the before and after and with and without approaches. Quantitative data on various indicators of five capitals were collected using a detailed household questionnaire, which was canvassed among the sample households in six sample villages that have undergone watershed treatment. All these villages are located within a hydrological (one or interconnected) boundary. Data on various indicators for the five capitals were collected for 2 years, i.e., for 2010−2011 and 2011−2012. The year 2010−2011 was a normal year, while 2011−2012 was a drought year. This gives us an opportunity to assess the watershed impacts during a drought year vis-à-vis a normal one. In other words, the effectiveness of watershed interventions in a drought situation could be assessed.

The resilience information from the sample households was collected during the drought year (2011−2012), as it helps households to contextualize WSD in the event of drought. One control village from each hydrological unit (HUN) was also selected for a detailed comparative assessment. The field work was conducted over a period of four months (December to March) during both years. Qualitative research was conducted in different periods over 4 years, i.e., between 2009 and 2013. The results from the analysis were validated with the village communities in 2013.

7.3 SAMPLE SELECTION AND PROFILE OF THE STUDY SITES

As mentioned earlier, six watersheds spread over three HUNs located in Kurnool/Anantapur and Prakasam districts of Andhra Pradesh were selected after intensive field visits from the technical teams, including experts from hydrogeological, biophysical, and socioeconomic fields. The sample watersheds are located in the upstream, midstream, and downstream locations of the HUNs. These HUNs were formed under the Andhra Pradesh Farm Managed Groundwater Systems (APFMGS) project in partnership with local nongovernmental organizations (NGOs), and were implemented in 650 villages spread over 63 HUNs across seven drought-prone districts of Andhra Pradesh using the hydrological boundaries as an operational unit (Figure 7.1).

Two broad criteria were adopted for selection of the field sites: (1) a technically demarcated HUN and (2) substantial coverage of area under the WSD program implemented by the Department of Rural Development (DRD).

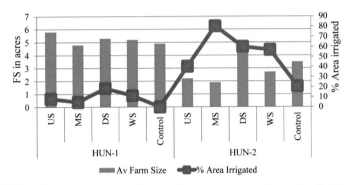

FIGURE 7.1 Natural capital: status of land and water resources in the sample villages.

TABLE 7.1 Selected HUNs and coverage of WSD program

HUN	Vajralavanka	Maruvavanka	Peethuruvagu
District	Anantapur/ Kurnool	Anantapur/ Kurnool	Prakasam
Area (ha)	10594	5025	9425
Villages covered	14	13	14
Watershed covered villages Approximate area (500 ha per village)	7 (3500 ha)	4 (2000 ha)	7 (3500 ha)
Approximate percentage of coverage of DRD watersheds to HUN area	33	40	37
Average rainfall (in mm)	631	654	702

Three HUNs were selected after an elaborate process of assessing their technical aspects under the APFMGS projects and coverage of area under the WSD through the DRD (Table 7.1). The area covered under each HUN ranges between 5000 and 10,000 ha, although the number of villages covered is about 13 or 14, and the coverage of area under each watershed is between 33 and 40%. The average rainfall (over the last 10 years) is higher in the Prakasam District compared with the Anantapur/Kurnool districts.

Initially, two districts, Anantapur and Prakasam, were identified after considering the variations in rainfall and hydrogeological formations, and a few HUNs and villages were identified after assessing the cadastral maps of each HUN. After a rapid appraisal of these HUNs, three were identified for the study (Table 7.2).

TABLE 7.2 Villages visited and the selected sample HUNs

HUN	District	Village	Location in HUN	Program status
Vajralavanka	Kurnool/Anantapur	S. Rangapuram[a]	Upstream	WSD and APFMGS
		Utakallu[a]	Midstream	WSD and APFMGS
Maruvavanka	Anantapur	Basinepalle[a]	Downstream	WSD and APFMGS
		Lachanapalli	Downstream	WSD and APFMGS
Upparavanka	Anantapur	Vennedoddi	Upstream	WSD and APFMGS
Bellamvanka	Anantapur	Mamilla Cheruvu	Downstream	WSD and APFMGS
		Kothur	Downstream/midstream	WSD and APFMGS
		Kottapet	Midstream	WSD and APFMGS
Peddavanka	Anantapur	Dimmaguda	Downstream	WSD and APFMGS
		Kottapalle	Downstream	APFMGS
Peethuruvagu	Prakasam	Thaticherla[a]	Upstream	WSD and APFMGS
		Penchikalapadu[a]	Midstream	WSD and APFMGS
		Vendutla[a]	Downstream	WSD and APFMGS
Upparavanka	Anantapur	a) Karidikonda[a]	Midstream	Control
Uppuvagu	Prakasam	b) Alasandalapalle[a]	Upstream	Control

[a]Selected villages.

We have opted for two instead of one HUN as the HUNs individually are not big enough to fulfill the criteria in the Anantapur/Kurnool districts. These two HUNs are interconnected hydrologically as well as by surface-flow pattern, and provide the upstream/downstream linkages between HUNs. In the Prakasam District, we could find a classic upstream/downstream case in a single HUN with coverage under the WSD program.

From each HUN, three villages were identified—one each at upstream, midstream, and downstream locations. The criteria for the village selection include (1) location and (2) being covered under the watershed program and the APFMGS project. In both sites, upstream villages are located on the mountain slopes and the downstream villages are located in the valley and drain into the major surface water bodies or streams. One of the main differences between the sample villages in the hydrological sites is that the sample villages in the Anantapur/Kurnool districts do not have any surface water bodies (tanks), while all three villages in the Prakasam District have surface water bodies. The Prakasam HUN drains into one of the biggest tanks (Kambam Cheruvu) in the state.

All of the sample villages are covered under the watershed program under different batches and programs. While watersheds in the Anantapur/Kurnool districts are covered under the desert development program (DDP), the other watersheds are covered under the integrated WSD program (IWDP), drought-prone area development program (DPAP), and the Andhra Pradesh Rural Livelihoods Program (APRLP; Table 7.3). These watersheds were implemented between 1995–1996 and 2003–2004. All of the watersheds, except S. Rangapuram, were implemented by government agencies and have an average coverage of 500 ha; S. Rangapuram has more than 800 ha as it extends to the forest and hillocks, which are outside the village area. The size of the villages in terms of the number of households varies between 87 in S. Rangapuram to 425 in Basinepalle in the Anantapur/Kurnool districts. One control village each was selected from the respective locations to assess the impact of the watershed in a with and without context; a comparison between a village that is closely located to the watershed (upstream/midstream) and the control village would be more meaningful.

7.4 IMPACT OF WSD—THE SRL APPROACH

Watershed impacts are assessed using the five capitals: natural, physical, financial, human, and social. Within each capital, different indicators are used to assess the impacts, which are measured in terms of the percentage change over the period of watershed implementation, i.e., before and after comparison. The changes are tested for statistical significance with the help of paired t tests. With and without comparison is made between watershed and non-watershed (control) villages, and the impacts are measured separately for upstream, midstream, and downstream locations as well as between normal and drought years.

TABLE 7.3 Basic features and household sample selection in the sample villages

Name of the watershed	Type of PIA	Scheme of funding	Year of formation (batch)	Area of village (ha)	Watershed area (ha)	Total population[a]	% of SC & ST population	Number of households			
								LL[b]	SMF[b]	LMF[b]	Total[b]
S. Rangapuram	NGO	IWDP	95–96 (I)	339	816	466 (47)	34	10 (5)	11 (7)	66 (42)	87 (54)
Utakallu	GO	DDP	99–00 (V)	1373	500	1523 (47)	14	37 (5)	140 (43)	143 (43)	320 (91)
Basinepalle	GO	DDP	98–99 (IV)	883	500	1955 (49)	29	175 (10)	139 (49)	111 (41)	425 (100)
Thaticherla	GO	DPAP	98–00 (V)	1903	500	1139 (48)	15	45 (10)	206 (85)	14 (06)	265 (101)
Penchikallupadu	GO	APRLP	02–03	974	500	491 (49)	10	22 (05)	87 (52)	05 (03)	114 (60)
Vendutla	GO	DPAP	98–99 (V)	2512	500	552 (48)	24	47 (05)	55 (41)	19 (14)	121 (60)
Kardikonda	Control (Anantapur/Kurnool)		Midstream location	1351	NA	1097 (49)	13	34 (5)	70 (18)	104 (27)	208 50
Alasandalapalle	Control (Prakasam)		Upstream location	1997	NA	581 (47)	06	5 (5)	92 (32)	39 (13)	136 (50)

PIA, Project Implementing Agency; GO, government organization; LL, land less.
[a]Figures in parentheses indicate the proportion of female population as per 2001 Census.
[b]Figures in parentheses are the sample size.

7.4.1 Natural Capital

Under natural capital, land, water (irrigation), and fodder are considered for impact assessment. The average landholding size and proportion of irrigated area in the sample villages indicates that HUN1 (Kurnool/Anantapur) is endowed with better land resources and poor water resources, while the reverse is true for HUN2 (Prakasam; Figure 7.1). This is natural because the average landholding size tends to be lower in the irrigated regions due to population density. The land use pattern is observed to have changed during watershed interventions.

7.4.1.1 Land

Changes in land use pattern, land quality, and cropping pattern are considered as impact indicators of land; land quality indicators are measured in terms of monetary benefits perceived by the farmers. Significant changes are observed in the net sown area and current fallows (Table 7.4). The watershed villages have experienced a significant decline in the net sown area, although the

TABLE 7.4 Changes in land use pattern in the sample villages (average per HH)

Location	HUN1 (Anantapur/Kurnool)			HUN2 (Prakasam)		
	Net sown area	Wasteland	Current fallow	Net sown area	Wasteland	Current fallow
Normal Year (2010–2011)						
Upstream	−13	NA	NA	−9[a]	+ve	208[b]
Midstream	−18[b]	NA	NA	−1	NA	NA
Downstream	−1	NA	NA	−4	NA	108
Watershed	−10[b]	NA	NA	−5[b]	+ve	143[c]
Control	0.0	NA	NA	4	NA	NA
Drought Year (2011–2012)						
Upstream	−24[a]	NA	+ve	−4[b]	NA	470[b]
Midstream	−5[c]	−75	+ve	−12[b]	NA	NA
Downstream	−10[c]	−100	−25	−10	NA	NA
Watershed	−12[a]	−83	70	−8[b]	NA	1195[c]
Control	0.1[c]	0	NA	−6	NA	NA

[a]*Level of significance at 1% level.*
[b]*Level of significance at 5% level.*
[c]*Level of significance at 10% level.*

decline is greater in HUN1. The decline in the net sown area is significant in the midstream locations of HUN1 and upstream locations in HUN2. In HUN2, the area under current fallows has gone up significantly in the upstream locations. As expected, the decline in net sown area is greater during the drought year when compared with the normal year. No land use changes are observed in the control villages. However, this may not necessarily be attributed to watershed interventions. At the same time, the impact of watershed interventions on improving *in situ* moisture and irrigation is expected to increase the net sown area, but this is not actually the case.

The impacts related to land quality are measured in terms of soil moisture conservation (SMC), rainwater harvesting (RWH), and dryland horticulture (DH). These impacts appear to be substantial in both HUNs, according to the farmers (Figure 7.2). These benefits range between Rs.1000 ($15) to Rs.3000 ($45) per acre. The impacts are marginally higher in HUN1, and the farmers have benefited the most from DH, followed by SMC. This is mainly because horticulture plants were distributed free of cost with some support for their maintenance during the initial years. Similarly, in SMC, the farmers benefited directly from land clearance and improvement activities. Across the streams, the impacts are more positive in the upstream and midstream villages of HUN1 and in the downstream villages of HUN2.

Along with the land use changes, the cropping pattern also changed. The decline in net sown area is associated more with the decline area under cereal crops during both normal and drought years, although the decline is sharper during the drought year (Figures 7.3 and 7.4). Further, the decline in cereal crops is more in upstream villages. During the normal year, the area under cash

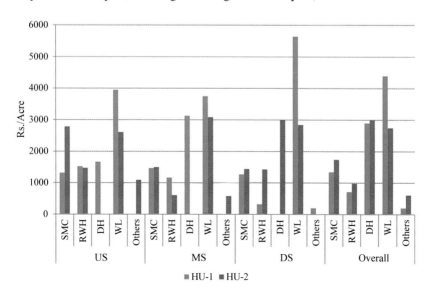

FIGURE 7.2   Land-related benefits from WSD. (WL denotes Wage Labour).

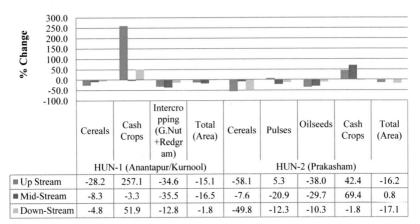

	Cereals	Cash Crops	Intercropping (G.Nut +Redgram)	Total (Area)	Cereals	Pulses	Oilseeds	Cash Crops	Total (Area)
			HUN-1 (Anantapur/Kurnool)				HUN-2 (Prakasham)		
■ Up Stream	-28.2	257.1	-34.6	-15.1	-58.1	5.3	-38.0	42.4	-16.2
■ Mid-Stream	-8.3	-3.3	-35.5	-16.5	-7.6	-20.9	-29.7	69.4	0.8
■ Down-Stream	-4.8	51.9	-12.8	-1.8	-49.8	-12.3	-10.3	-1.8	-17.1

FIGURE 7.3   Stream-wise changes in cropping pattern (kharif): normal year.

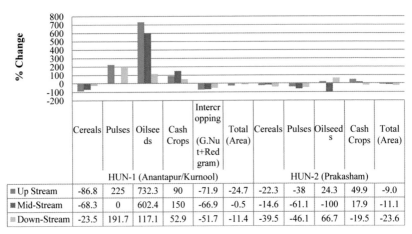

	Cereals	Pulses	Oilseeds	Cash Crops	Intercropping (G.Nut+Redgram)	Total (Area)	Cereals	Pulses	Oilseeds	Cash Crops	Total (Area)
				HUN-1 (Anantapur/Kurnool)					HUN-2 (Prakasham)		
■ Up Stream	-86.8	225	732.3	90	-71.9	-24.7	-22.3	-38	24.3	49.9	-9.0
■ Mid-Stream	-68.3	0	602.4	150	-66.9	-0.5	-14.6	-61.1	-100	17.9	-11.1
■ Down-Stream	-23.5	191.7	117.1	52.9	-51.7	-11.4	-39.5	-46.1	66.7	-19.5	-23.6

FIGURE 7.4   Stream-wise changes in cropping pattern (kharif): drought year.

crops (cotton) increased in both HUNs. On the other hand, during the drought year, while the area under pulses and oilseeds along with cash crops increased in HUN1, the area under cash crops and oilseeds increased in HUN2. However, such changes in cropping pattern have also been observed in the control village, which reported greater and significant decline in area under cereals during the normal year in HUN1 and the reverse during the drought year. For cash crops, the control village reported higher growth in area when compared with watershed villages, while with oilseeds, the control village reported positive growth during normal years while watershed villages reported negative growth. However, in most of the other cases, i.e., between streams and farm sizes, the differences are not statistically significant.

7.4.1.2 Irrigation

The proportion of area under irrigation is found to be greater in HUN2 (53%) when compared with HUN1[1] (14%) before implementation of the watershed programs in the watershed villages during the normal year. On the other hand, during the drought year, while the proportion of area under irrigation has decreased to 9% in HUN1, the proportion of irrigated area in HUN2 remained unchanged. The extent of irrigation is the lowest in the upstream villages (10%), followed by midstream (16%) and downstream (22%) villages in HUN1, while in HUN2, the extent of irrigation is the highest in the midstream village (73%), followed by downstream (68%) and upstream (43%) villages. During the drought years the pattern is similar also across the streams, although the extent of irrigation decreased in all of the villages.

The area under irrigation remained constant during watershed interventions. Very few households have reported an increase in access to irrigation, although the performance is better in HUN2 (Table 7.5). Similarly, in terms of actual irrigation no significant changes have taken place in the HUNs, across the streams as well as between the drought and normal years (Table 7.6); source-wise irrigation also does not show any significant positive impact. However, the negative impacts appear to be greater in HUN1 when compared with HUN2 (Table 7.7).

During the normal year, between the size classes, larger farmers seem to benefit more when compared with the small and marginal farmer (SMF) households (HHs) in terms of percentage changes in their irrigated area in HUN1, except in the upstream village. On the other hand, a reverse trend was observed in the SMF HHs in HUN2—these households received more benefit compared with their large and medium farmer (LMF) counterparts, except in the control village during this time (Figure 7.5).

During the drought years, while the SMF HHs in upstream and midstream villages of HUN1 appear to be experiencing no change in their irrigated area, the percentage changes in the irrigated area for the LMF HHs shows a declining trend. In the downstream village of HUN1, the LMF HHs gained (increase in irrigated area) when compared with their SMF counterparts (decline in irrigated area; Figure 7.6). However, the changes are not significant between the size classes (Table 7.8), meaning that the impact of watershed interventions is not biased as far as irrigation is concerned.

7.4.1.3 Fodder

Four indicators of fodder availability are considered for assessing the impact: quantity, share of own field, share of common property resources (CPRs), and share of purchased fodder. Quantity of fodder is an indicator of improved natural resource base and the share of CPRs indicates improved common pool resources.

---

1. Control village of HUN1 has no irrigation facility.

TABLE 7.5 Natural capital: access to irrigation facility (% of HHs)

Type of watershed	Year	HUN1			HUN2		
		Declined	No change	Increased	Declined	No change	Increased
Upstream	Normal	4	88	8	1	94	5
	Drought	7	91	2	2	94	4
Midstream	Normal	5	93	2	2	85	13
	Drought	1	98	2	8	88	4
Downstream	Normal	4	92	3	11	80	9
	Drought	3	92	5	19	77	4
Control	Normal	0	100	0	4	91	5
	Drought	0	100	0	6	90	4

TABLE 7.6 Natural capital: changes in area under irrigation across streams (% area irrigated)

	HUN1 (Anantapur/Kurnool)					HUN2 (Prakasam)				
Year	US	MS	DS	WS	Control	US	MS	DS	WS	Control
Normal	20	−58	−4	−18	NA	8	8	−10	−1	2
Drought	−38	−8	7	−8	NA	2	−9	−23	−14	−5

US, upstream; MS, midstream; DS, downstream; WS, watershed.

Fodder consumption in absolute quantity has increased in both HUNs, although the increase is significant only in HUN2 (Table 7.9). It appears that the increase is because of the increase in demand for fodder and not due to the increased availability in HUN2, since the share of fodder from each field has not shown any significant increase in HUN2. On the other hand, the share of self-grown fodder has increased significantly in HUN1.

Between watershed and control villages, no clear difference is observed, although the share of CPRs as well as purchase of fodder has increased significantly in HUN1.

The impacts are clearer when the normal year is compared with the drought year: During the drought year, there is a significant increase in the quantity of fodder consumed in the watershed villages, which is mainly due to the increase in livestock as a drought-coping mechanism. This is greater in the upstream and midstream villages when compared with the downstream villages.

The decline in the share of self-grown fodder and fodder from CPRs is greater in the control villages when compared with the watershed villages.

7.4.2 Physical Capital

Livestock and irrigation assets are seen as indicators of physical capital. Livestock is grouped under big ruminants and small ruminants—the small ruminants are considered a drought-coping strategy, i.e., households tend to sell small ruminants during drought years; these two groups are standardized as total livestock units (TLUs).

7.4.2.1 Livestock

Except for the midstream villages of HUN2, the population of big ruminants has declined in watershed as well as control villages of both HUNs during normal as well as drought years (Figures 7.7 and 7.8). While the decline is greater during the drought year than during the normal year in HUN1 (statistically significant differences were found in upstream and midstream villages and also at the aggregate level in HUN1), in HUN2 the decline is

TABLE 7.7 Natural capital: changes in source-wise irrigation

Source of irrigation	Year	HUN1 (Anantapur/Kurnool)		HUN2 (Prakasam)				
		US–MS	US–DS	US–MS	US–DS	MS–DS	WS–control	HUN1–HUN2
Open wells	Normal	(19.5)–(−75.3)[b]	(19.5)–(−29.4)	(0)–(NA)	(0)–(−15.8)	(NA)–(−15.8)	(24.0)–(−100)[c]	(−37.6)–(−4.8)[c]
	Drought	(−47.6)–(−63.6)	(−47.6)–(−52.4)	(NA)–(−50.0)	(NA)–(NA)	(−50.0) (NA)	(−50.0)–(NA)	(−51.2)–(−50.0)
Bore wells	Normal	(NA)–(166.7)	(NA)–(82.9)	(11)–(−0.1)	(11)–(−8.9)	(−0.1)–(−8.9)	(−1.1)–(12.4)	(94.9)–(0.9)
	Drought	(+ve)–(36.4)	(+ve)–(14.4)	(3.1)–(−6.7)[c]	(3.1)–(−22.9)[c]	(−6.7)–(−22.9)	(13.5)–(−5.4)	(21.9)–(−12.7)
Tanks	Normal	(NA)–(−100)	(NA)–(−100)	(−15.8)–(0)	(−15.8)–(−100)[b]	(0)–(−100)	(−18.9)–(0)[b]	(−100)–(−15.2)[a]
	Drought	(NA)–(NA)	(NA)–(NA)	(−4.5)–(−24.8)	(−4.5)–(+ve)	(−24.8) (+ve)	(−5.4)–(0)	(NA)–(−5.1)
Total	Normal	(19.5)–(−58.2)	(19.5)–(−3.5)	(7.5)–(7.5)	(7.5)–(−10.2)[b]	(7.5)–(−10.2)[b]	(−0.6)–(2.1)	(−18.2)–(−0.1)[b]
	Drought	(−38.1)–(−7.6)[c]	(−38.1)–(6.6)[b]	(1.7)–(−9.3)[c]	(1.7)–(−22.7)[c]	(−9.3)–(−22.7)	(−13.6)–(−5.3)	(−7.6)–(−12.8)

[a] Level of significance at 1% level.
[b] Level of significance at 5% level.
[c] Level of significance at 10% level.

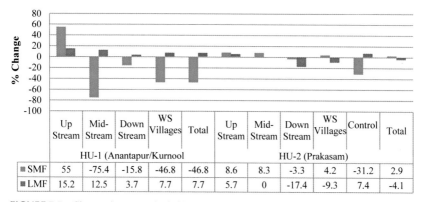

	Up Stream	Mid-Stream	Down Stream	WS Villages	Total	Up Stream	Mid-Stream	Down Stream	WS Villages	Control	Total
	HU-1 (Anantapur/Kurnool)					HU-2 (Prakasam)					
■ SMF	55	-75.4	-15.8	-46.8	-46.8	8.6	8.3	-3.3	4.2	-31.2	2.9
■ LMF	15.2	12.5	3.7	7.7	7.7	5.7	0	-17.4	-9.3	7.4	-4.1

FIGURE 7.5 Changes in area under irrigation across streams and size classes (% area irrigated): normal year.

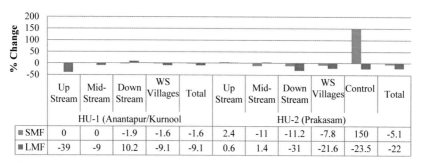

	Up Stream	Mid-Stream	Down Stream	WS Villages	Total	Up Stream	Mid-Stream	Down Stream	WS Villages	Control	Total
	HU-1 (Anantapur/Kurnool)					HU-2 (Prakasam)					
■ SMF	0	0	-1.9	-1.6	-1.6	2.4	-11	-11.2	-7.8	150	-5.1
■ LMF	-39	-9	10.2	-9.1	-9.1	0.6	1.4	-31	-21.6	-23.5	-22

FIGURE 7.6 Changes in area under irrigation across streams and size classes (% area irrigated): drought year.

greater in the downstream and control villages during the normal year when compared with the drought year (however, this difference is significant only in case of the control village; Table 7.10). Moreover, the declines are more prominent in HUN2 (Prakasam) than in HUN1 (Anantapur/Kurnool). While the changes are statistically significant in HUN2 in both normal and drought years, they are significant in HUN1 only during the drought year.

When the changes in big ruminant population are compared between different streams, the changes between the midstream and downstream villages turned out to be statistically significant in both HUNs during the normal year, whereas during the drought year, statistically significant differences were observed between upstream and midstream villages in HUN1 and midstream and downstream villages in HUN2.

Similarly, the population of small ruminants also declined in the downstream villages in both HUNs during the normal year, whereas during the drought year their population increased. However, while big ruminant population increased in the midstream villages of HUN2 during normal and drought years, the small ruminant population declined during the normal year and

TABLE 7.8 Impact of WSD on access to irrigation facility (% of HHs)

Type of watershed (WS)	Category of HHs	HUN1 (Anantapur/Kurnool)			HUN2 (Prakasam)		
		Declined	No change	Increased	Declined	No change	Increased
Upstream	Total	4	89	7	1	95	4
	SMF–LMF	0–6	93–86	7–9	1–0	95–88	4–13
Midstream	Total	4	93	2	2	87	12
	SMF–LMF	6–3	92–94	2–3	2–0	85–100	13–0
Downstream	Total	4	93	3	10	82	8
	SMF–LMF	3–7	95–86	2–7	9–18	82–73	9–9
WS villages	Total	4	92	4	4	89	7
	SMF–LMF	4–5	94–89	2–6	3–10	89–81	8–10
Control village	Total	0	100	0	4	92	4
	SMF–LMF	0–0	100–100	0–0	7–0	93–87	0–13
	SMF–LMF	3–4	95–91	2–5	4–6	89–83	7–11

TABLE 7.9 Changes in availability of fodder in the sample villages

% Change	HUN1 (Anantapur/Kurnool)					HUN2 (Prakasam)				
	Upstream	Midstream	Downstream	Watershed villages	Control villages	Upstream	Midstream	Downstream	Watershed villages	Control villages
Normal year										
Quantity consumed	13	20	65	26	79	-13	52[b]	43[a]	22[a]	7[a]
% Share of own field	23 (86)	15 (88)	13 (99)	17[b] (89)	53 (51)	-15 (8)	103[b] (31)	-28 (13)	30 (18)	-4[c] (32)
% Share of CPRs	-47[c] (4)	15[c] (5)	-55[c] (1)	-22 (4)	35[c] (36)	33 (29)	62[a] (26)	9 (20)	36[a] (25)	-10 (14)
% Share of purchased fodder	-16 (10)	-7 (7)	-26 (0)	-12 (7)	99[c] (13)	-20 (64)	48[b] (43)	4 (67)	1 (57)	-23 (54)
Drought year										
Quantity consumed	68[b]	48	43	55[a]	59	2	22[b]	38	19[b]	-3
% Share of own field	-53[a] (25)	-37[a] (37)	-26[c] (41)	-38[a] (35)	-75[a] (24)	-22[b] (25)	-3 (30)	-7 (47)	-11 (36)	-52[b] (37)
% Share of CPRs	-7 (38)	-38[a] (32)	-49[c] (9)	-29[a] (25)	-42[c] (30)	-8 (39)	35[b] (46)	-19[c] (22)	-3 (33)	-32[b] (20)
% Share of purchased fodder	63[b] (37)	37[c] (32)	35 (50)	42[b] (40)	17 (47)	-8 (35)	90[b] (24)	97 (31)	36 (31)	156[c] (43)

Figures in parentheses indicate the actual shares.
[a]Level of significance at 1% level.
[b]Level of significance at 5% level.
[c]Level of significance at 10% level.

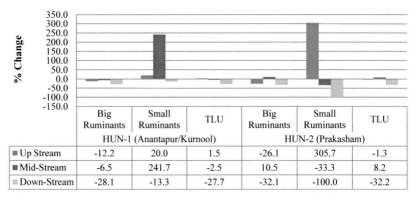

	Big Ruminants	Small Ruminants	TLU	Big Ruminants	Small Ruminants	TLU
	HUN-1 (Anantapur/Kurnool)			HUN-2 (Prakasham)		
■ Up Stream	-12.2	20.0	1.5	-26.1	305.7	-1.3
■ Mid-Stream	-6.5	241.7	-2.5	10.5	-33.3	8.2
▫ Down-Stream	-28.1	-13.3	-27.7	-32.1	-100.0	-32.2

FIGURE 7.7   Changes in size and composition of livestock in the sample households: normal year.

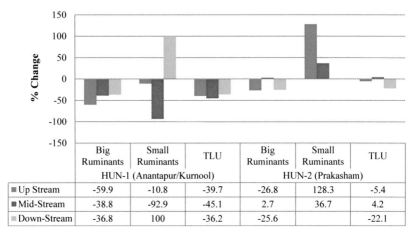

	Big Ruminants	Small Ruminants	TLU	Big Ruminants	Small Ruminants	TLU
	HUN-1 (Anantapur/Kurnool)			HUN-2 (Prakasham)		
■ Up Stream	-59.9	-10.8	-39.7	-26.8	128.3	-5.4
■ Mid-Stream	-38.8	-92.9	-45.1	2.7	36.7	4.2
▫ Down-Stream	-36.8	100	-36.2	-25.6		-22.1

FIGURE 7.8   Changes in size and composition of livestock in the sample households: drought year.

increased during the drought year; furthermore, there was an increase in the small ruminant population during both normal and drought years in the upstream village of this HUN. On the other hand, in HUN1, the small ruminant population increased in both upstream and midstream villages during the normal years, and declined in these two villages during the drought year. When the changes in small ruminant population during normal and drought years are compared, the changes in small ruminant population are observed to be more prominent (difference is significant) as seen in the midstream village in HUN1 (increased during normal year and declined during drought year).

The standardized TLUs are found to be increasing in the upstream village of HUN1 and the midstream village of HUN2 during the normal year, whereas, during the drought year, all villages except the midstream village of HUN2 are

TABLE 7.10 Changes in livestock holdings

% Change	HUN1 (Anantapur/Kurnool)					HUN2 (Prakasam)				
	US	MS	DS	WS villages	Control village	US	MS	DS	WS villages	Control village
Normal year										
BR	-12	-7	-28	-15	-36[c]	-26[b]	11	-32[b]	-18[b]	-43[a]
SR	20	242[b]	-13	25	-11	306	-33	-100	180	5
TLU	2	-3	-28	-7	-32[c]	-1	8	-32[b]	-10	-30[b]
Drought year										
BR	-60[a]	-39[a]	-37[b]	-45[a]	-68[b]	-27[a]	3	-26[c]	-16[b]	-18[b]
SR	-11	-93	100	-23	-40	128[c]	37	+ve	124[b]	-29
TLU	-40[b]	-45[a]	-36[b]	-40[a]	-66[b]	-5	4	-22	-7	-23[b]

US, upstream; MS, midstream; DS, downstream; WS, watershed; BR, big ruminants; SR, small ruminants.
[a]*Level of significance at 1% level.*
[b]*Level of significance at 5% level.*
[c]*Level of significance at 10% level.*

observed to decline. Further, while the differences between the upstream/downstream villages and midstream/downstream villages in HUN1 are statistically significant during the normal year, in HUN2 the difference between the midstream/downstream villages is found to be statistically significant during the drought year.

Among the watershed villages, the decline in the TLUs is observed to be greater in the downstream villages of both HUNs during the normal year, while this trend is observed only in HUN2 during the drought year.

When changes in TLUs during normal and drought years are compared, statistically significant differences were found in upstream and midstream villages and also at the aggregate level in HUN1. In HUN2, on the other hand, the difference was found only in the control village.

On the whole, the big ruminant population declined in the watershed as well as control villages in both HUNs during the normal as well as drought years; the decline is greater in the control villages than in the watershed villages in their respective HUNs. However, the differences (between watershed and control villages) are statistically significant only during the normal year. It is further observed that while small ruminant population increased both in watershed and control villages of HUN2 during normal as well as drought years, increase in the population of small ruminants is observed only during the normal year in the watershed villages of HUN1. Further, although the extent of decline is greater in control than in watershed villages, the differences are not statistically significant. The TLUs are declining in watershed and control villages in both HUNs in normal as well as in drought years. However, the declines in TLUs are more in the control village than in the watershed villages in their respective HUNs, especially in HUN2, where the differences turned out to be statistically significant.

A comparison between SMF and LMF HHs within their respective watersheds in HUN1 during the normal year reveals an increase in big ruminant population among the LMF HHs and a decline in the same among SMF HHs (Figure 7.9). On the other hand, in HUN2, both SMF and LMF HHs experienced a decline in big ruminant population, except in the midstream village where their population increased in the SMF HHs during the normal year (Figure 7.10). Further, a decline in big ruminant population is observed for both SMF and LMF categories of HHs during the drought years in both HUNs, except in the downstream village in HUN1 and midstream village in HUN2 (Figures 7.11 and 7.12). It is interesting to note here that while big ruminant population among the LMF HHs declined in the midstream village of HUN2 during the normal year, it was observed to increase during the drought year.

Small ruminants are generally found to be associated more with the SMF HHs than with the LMF HHs in both HUNs during the normal year. While the small ruminant population of SMF HHs increased in HUN1, except in the downstream village, they declined in HUN2, except in the upstream village.

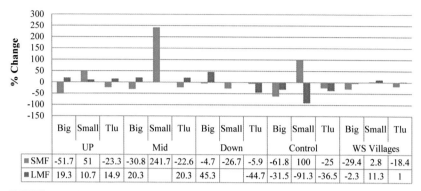

FIGURE 7.9 Changes in size and composition of livestock by farm size: normal year (HUN1).

	Big	Small	Tlu	Big	Small	Tlu	Big	Small	Tlu	Big	Small	Tlu	Big	Small	Tlu
		UP			Mid			Down			Control			WS Villages	
■ SMF	-51.7	51	-23.3	-30.8	241.7	-22.6	-4.7	-26.7	-5.9	-61.8	100	-25	-29.4	2.8	-18.4
■ LMF	19.3	10.7	14.9	20.3		20.3	45.3		-44.7	-31.5	-91.3	-36.5	-2.3	11.3	1

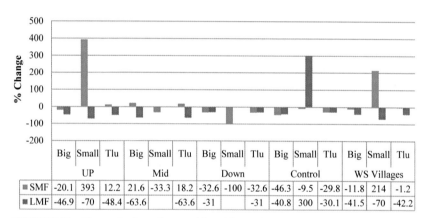

FIGURE 7.10 Changes in size and composition of livestock by farm size: normal year (HUN2).

	Big	Small	Tlu	Big	Small	Tlu	Big	Small	Tlu	Big	Small	Tlu	Big	Small	Tlu
		UP			Mid			Down			Control			WS Villages	
■ SMF	-20.1	393	12.2	21.6	-33.3	18.2	-32.6	-100	-32.6	-46.3	-9.5	-29.8	-11.8	214	-1.2
■ LMF	-46.9	-70	-48.4	-63.6		-63.6	-31		-31	-40.8	300	-30.1	-41.5	-70	-42.2

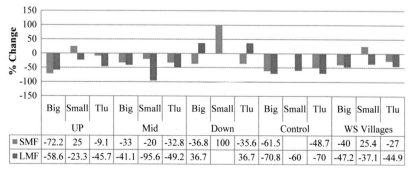

FIGURE 7.11 Changes in size and composition of livestock in the sample households: drought year (HUN1).

	Big	Small	Tlu	Big	Small	Tlu	Big	Small	Tlu	Big	Small	Tlu	Big	Small	Tlu
		UP			Mid			Down			Control			WS Villages	
■ SMF	-72.2	25	-9.1	-33	-20	-32.8	-36.8	100	-35.6	-61.5		-48.7	-40	25.4	-27
■ LMF	-58.6	-23.3	-45.7	-41.1	-95.6	-49.2	36.7		36.7	-70.8	-60	-70	-47.2	-37.1	-44.9

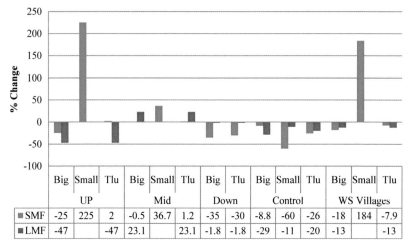

	Big	Small	Tlu	Big	Small	Tlu	Big	Tlu	Big	Small	Tlu	Big	Small	Tlu
	UP			Mid			Down		Control			WS Villages		
SMF	-25	225	2	-0.5	36.7	1.2	-35	-30	-8.8	-60	-26	-18	184	-7.9
LMF	-47		-47	23.1		23.1	-1.8	-1.8	-29	-11	-20	-13		-13

FIGURE 7.12 Changes in size and composition of livestock in the sample households: drought year (HUN2).

On the other hand, among the LMF HHs the small ruminant population is increased in the upstream village of HUN1 and declined in the upstream village of HUN2, where they were found to be rearing small ruminants. On the other hand, during the drought year, the population of small ruminants in the SMF HHs increased in both HUNs, except in the midstream village of HUN1; among the LMF HHs, their population declined in upstream and midstream villages of HUN1, where they were found to be rearing small ruminants.

The TLUs of SMF HHs declined both in normal as well as in drought years in HUN1, while in HUN2, the TLUs of the SMF HHs increased, except in the downstream village during the same period. Similarly, the TLUs of the LMF HHs, except in HUN1, increased during the normal year and declined during the drought year; whereas in HUN2, the TLUs of LMF HHs in all of the villages, except in the midstream village, declined during the normal as well as drought years.

7.4.2.2 Irrigation Assets

Open wells, bore wells, and motors are considered part of irrigation assets. There is widespread decline in the open wells across the villages during normal as well as drought years (Table 7.11). This is a common phenomenon across the regions where open wells are replaced by bore wells to meet the increasing demand for water.

There is a significant increase in the number of bore wells in the watershed villages of HUN1 during the drought year; this is mainly to cope with the dwindling water table. On the contrary, in HUN2, the control village has reported a significant increase in the number of bore wells while the increase is not significant in the watershed villages.

TABLE 7.11 Physical capital: irrigation assets

	Anantapur/Kurnool (HUN1)					Prakasam (HUN2)				
	US	MS	DS	WS villages	Control village	US	MS	DS	WS villages	Control village
Normal year										
OW	6	−44[b]	−100[a]	−51[a]	−100.0	−75[a]	−25	−48[a]	−50[a]	−61[a]
BW	33	−39	67[b]	30	NA	33[a]	35[a]	−15	9	17
Motors	−14	−50	−19	−20	NA	42[a]	54[a]	28[b]	39[a]	23
Drought year										
OW	−36	−60	−81[a]	−63[a]	NA	−100[b]	−86[b]	−100[a]	−97[a]	−100[b]
BW	NA	400	18	35[c]	NA	14	16[b]	−3	7	38[c]
Motor	−38[c]	200	64[a]	47[a]	NA	10	16[a]	2	9[b]	38[c]

US, upstream; MS, midstream; DS, downstream; WS, watershed; OW, open well; BW, bore well.
[a]Level of significance at 1% level.
[b]Level of significance at 5% level.
[c]Level of significance at 10% level.

7.4.3 Financial Capital

Three indicators, yield rates, household income, and debt-saving ratio, are used to assess the impact of WSD on financial capital.

7.4.3.1 Yield Impacts

Watershed interventions are expected to have a direct impact on crop yields through improved soil quality, *in situ* moisture, and irrigation. Changes in crop yields are assessed for kharif as well as rabi crops. Only paddy, groundnut, and groundnut + red gram (mixed) are grown in the kharif season in HUN1, while a number of crops are grown in HUN2 (Table 7.12).

Kharif crop yields have increased significantly in both HUNs for most crops: while groundnut + red gram recorded 20% growth in per acre yield in HUN1, the increase in yield rates, which ranged between 47 (red gram) and 155% (jowar) for all kharif crops, except castor, was significant in HUN2. Similarly, while midstream villages in HUN2 recorded higher growth in yield, the differences between streams are not much in HUN1. It is further observed that during the normal year, watershed villages have done better when compared with control villages. On the other hand, during the drought year, yield rates of all crops have experienced a significant decline. Similar impacts are observed in rabi crops in both HUNs (Table 7.13).

7.4.3.2 Income

In absolute terms, the total household income has increased substantially in both HUNs (except in the upstream village of HUN1, which has reported a decline in the total income) during both normal and drought years (Figures 7.13 and 7.14). The percentage change in the overall income during the normal year in HUN2 (148%) is found to be significantly more than that of HUN1 (36%; Table 7.14); that is, households from HUN2 had higher income than those from HUN1 during the normal as well as drought years, although the difference is not much during the drought year. The increased income in HUN2 is significant in all villages including control villages, while in HUN1, it is significant only in downstream and control villages during the normal year.

Further, the downstream village in HUN1 reported greater impact both during normal and drought years; on the other hand, in HUN2, the midstream village has shown greater impact, followed by the upstream and downstream villages during the normal year, while the upstream village performed better, followed by downstream and midstream villages during the drought year. However, the stream-wise differences are significant only in HUN1 during the drought year (Table 7.15).

Among farm sizes, the SMF category revealed greater impact on income when compared with their LMF counterparts in HUN2 during the normal year; while in HUN1, barring the midstream village, the same trend is observed.

TABLE 7.12 Changes in yield rates of kharif crops

	HUN1		HUN2							
	Paddy	Groundnut + Red gram	Groundnut	Paddy	Red gram	Jowar	Bajra	Cotton	Castor	Sunflower
Normal year										
US	26^c	20^b	NA	38^c	10	80^c	19	150^a	NA	73^c
MS	26^c	18^a	NA	66^b	51^a	196^c	143^a	0	NA	NA
DS	3	23^a	27	50	120^a	174^a	131^c	85^a	100	113^b
WS	9	20^a	27	53^a	47^a	155^a	103^a	79^a	100	93^a
CON	NA	NA	55^a	21	1	26	27	NA	23^c	NA
Drought year										
US	−24	$−71^a$	−75	$−20^a$	$−65^a$	$−29^b$	—	—	—	—
MS	NA	$−69^a$	$−67^b$	−24	−22	$−23^b$	—	—	—	—
DS	$−20^c$	$−64^a$	$−75^c$	$−20^a$	$−55^c$	$−44^b$	—	—	—	—
WS	21^c	$−67^a$	$−71^c$	$−21^a$	$−56^a$	$−30^a$	—	—	—	—
CON	NS	$−67^a$	NA	NA	$−49^a$	$−20^a$	—	—	—	—

US, upstream; MS, midstream; DS, downstream; WS, watershed; CON, control.
[a]Level of significance at 1% level.
[b]Level of significance at 5% level.
[c]Level of significance at 10% level.

TABLE 7.13 Changes in yield rates of rabi crops

| | Normal year | | | Drought year | |
| | HUN1 | HUN2 | | HUN2 | |
Location	Groundnut	Sunflower	Bengal gram	Sunflower	Vegetables
Upstream	NA	33	NA	-29^b	NA
Midstream	NA	49^a	NA	NA	NA
Downstream	85^a	157	NA	NA	NA
Watershed	85^a	54^a	NA	-29^b	-15^b
Control	NA	100	-34^b	NA	NA

[a]Level of significance at 1% level.
[b]Level of significance at 10% level.

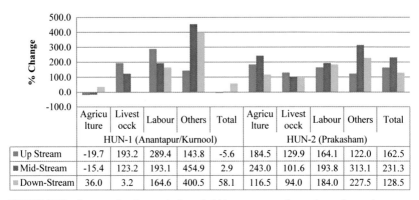

	Agriculture	Livestock	Labour	Others	Total	Agriculture	Livestock	Labour	Others	Total
	HUN-1 (Anantapur/Kurnool)					HUN-2 (Prakasham)				
Up Stream	-19.7	193.2	289.4	143.8	-5.6	184.5	129.9	164.1	122.0	162.5
Mid-Stream	-15.4	123.2	193.1	454.9	2.9	243.0	101.6	193.8	313.1	231.3
Down-Stream	36.0	3.2	164.6	400.5	58.1	116.5	94.0	184.0	227.5	128.5

FIGURE 7.13   Source-wise changes in household income: normal year (gross income).

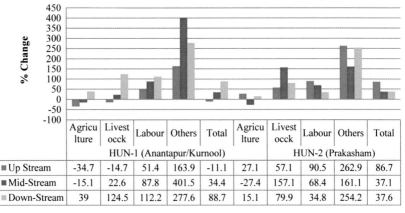

	Agriculture	Livestock	Labour	Others	Total	Agriculture	Livestock	Labour	Others	Total
	HUN-1 (Anantapur/Kurnool)					HUN-2 (Prakasham)				
Up Stream	-34.7	-14.7	51.4	163.9	-11.1	27.1	57.1	90.5	262.9	86.7
Mid-Stream	-15.1	22.6	87.8	401.5	34.4	-27.4	157.1	68.4	161.1	37.1
Down-Stream	39	124.5	112.2	277.6	88.7	15.1	79.9	34.8	254.2	37.6

FIGURE 7.14   Source-wise changes in household income: drought year (gross income).

TABLE 7.14 Changes in source-wise impact on income across streams (% change)

	HUN1 (Anantapur/Kurnool)					HUN2 (Prakasam)				
	Upstream	Midstream	Down-stream	WS villages	Control village	Upstream	Midstream	Down-stream	WS villages	Control village
Normal year										
Agricultural income	−20	−15	36[a]	3.0	283[a]	185[a]	243[a]	117[a]	145[a]	105[a]
Livestock income	193	123[a]	3.	131	40[b]	130[a]	102[a]	94[a]	110[a]	118
Labor income	289[a]	193[a]	165[a]	195[a]	319[a]	164[a]	194[a]	184[a]	177[a]	172[a]
Other income	144[c]	455	401[a]	372[a]	1290[a]	122[a]	313[c]	228[a]	190[a]	173[c]
Total income	−5.6	3	58[a]	22[b]	332[a]	163[a]	231[a]	129[a]	154[a]	121[a]
Drought year										
Agricultural income	−35[a]	−15[c]	39[c]	1	−21[c]	27[b]	−27	15	11.2	40[c]
Livestock income	−15	23	125[b]	39[b]	−29	57[a]	157[a]	80[b]	82[a]	57[b]
Labor income	51[a]	88[a]	112[a]	90[a]	103[a]	91[a]	68[a]	35[c]	73[a]	69[a]
Other income	164[a]	402[a]	278[a]	287[a]	189[a]	263[a]	161[a]	254[a]	235[a]	374[a]
Total income	−11	34[a]	89[a]	45[a]	37[a]	87[a]	37[b]	38[b]	50[a]	78[a]

[a] Level of significance at 1% level.
[b] Level of significance at 5% level.
[c] Level of significance at 10% level.

TABLE 7.15 Source-wise changes in gross income between streams (% change)

Source of income	Year	HUN1 (Anantapur/Kurnool)				HUN2 (Prakasam)				HUN1–HUN2
		UP-MID	UP-DOWN	MID-DOWN	WS-CONT	UP-MID	UP-DOWN	MID-DOWN	WS-CONT	
Agriculture	Normal	(−19.7)−(−15.4)	(−19.7)−(36)[a]	(−15.4)−(36)	(3.0)−(282.8)[a]	(184.5)−(243)[b]	(184.5)−(116.5)	(243)−(116.5)	(145.3)−(105.4)	(11.7)−(137.7)[a]
	Drought	(−34.7)−(−15.1)	(−34.7)−(39)[a]	(−15.1)−(39)[c]	(1.4)−(21.2)	(27.1)−(−27.4)	(27.1)−(15.1)	(−27.4)−(15.1)[c]	(11.2)−(39.8)	(−1.4)−(14.1)[c]
Livestock	Normal	(−193.2)−(123.2)	(−193.2)−(3.2)	(123.2)−(3.2)	(131.4)−(39.6)	(129.9)−(101.6)	(129.9)−(94)	(101.6)−(94)	(109.6)−(117.6)	(96.5)−(110.9)
	Drought	(−14.7)−(22.6)	(−14.7)−(124.5)	(22.6)−(124.5)	(38.6)−(−29.3)	(57.1)−(157.1)	(57.1)−(79.9)	(157.1)−(79.9)	(82.4)−(56.8)	(22.8)−(76.4)

Labor	Normal	(289.4)—(193.1)[b]	(289.4)—(164.6)[a]	(193.1)—(164.6)	(195.1)—(319.3)[a]	(164.1)—(193.8)[c]	(164.1)—(184)	(193.8)—(184)	(176.6)—(172.3)	(215.5)—(175.8)
	Drought	(51.4)—(87.8)	(51.4)—(112.2)[b]	(87.8)—(112.2)	(90.4)—(102.9)	(90.5)—(68.4)	(90.5)—(34.8)[b]	(68.4)—(34.8)	(72.6)—(68.8)	(92.6)—(71.8)[c]
Others	Normal	(143.8)—(454.9)	(143.8)—(400.5)	(454.9)—(400.5)	(371.6)—(1290.3)	(122)—(313.1)	(122)—(227.5)	(313.1)—(227.5)	(190)—(173.3)	(481.8)—(187.7)
	Drought	(163.9)—(401.5)	(163.9)—(277.6)[b]	(401.5)—(277.6)	(286.8)—(189.4)	(262.9)—(161.1)	(262.9)—(254.2)	(161.1)—(254.2)	(235)—(373.7)	(260)—(250.5)
Total	Normal	(−5.6)—(2.9)	(−5.6)—(58.1)	(2.9)—(58.1)	(22.2)—(331.7)[a]	(162.5)—(231.3)	(162.5)—(128.5)	(231.3)—(128.5)	(154.2)—(120.9)	(36.1)—(148.1)[a]
	Drought	(−11.1)—(34.4)[c]	(−11.1)—(88.7)[b]	(34.4)—(88.7)[b]	(44.6)—(37)	(86.7)—(37.1)	(86.7)—(37.6)	(37.1)—(37.6)	(49.5)—(78.4)	(43.4)—(53.3)

[a] Level of significance at 1% level.
[b] Level of significance at 5% level.
[c] Level of significance at 10% level.

Similarly, except in the downstream village of HUN1 and the midstream village of HUN2, the SMF HHs have gained more during the drought year in both HUNs. However, the differences between the farm sizes are not statistically significant.

Agriculture is the major source of income for the HHs in the sample villages and contributes 77% of income in HUN1 and 68% in HUN2 during the normal year (after the watershed), although its share in the total income has declined (except in upstream and midstream villages of HUN2; Figure 7.15). The percentage change in income from agriculture has increased in all locations, except in the upstream and midstream villages of HUN1, and these changes are also statistically significant (Table 7.14). In HUN2, the midstream village recorded a higher increase in income than the upstream village; the differences between SMF and LMF HHs are not significant as far as the changes in the share of agricultural income are considered.

The share of livestock as a source of income, although marginal in both HUNs, has increased in HUN1, except in the downstream and control villages (Figure 7.15), and has declined in HUN2. However, the percentage change in livestock income is positive in all villages, and the differences are statistically significant in all watershed villages of HUN2 and in the midstream and control villages of HUN1 (Table 7.14). The increase in livestock income appears to be greater in the upstream and less in the downstream villages in both HUNs, although the stream-wise differences are not statistically significant (Table 7.15). Similarly, the differences between the LMF and SMF HH livestock incomes are also not statistically significant.

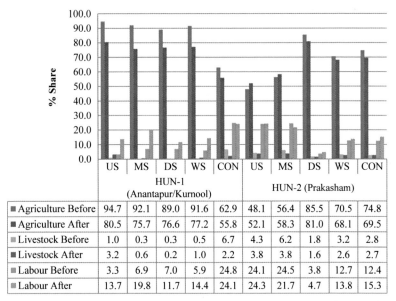

	US	MS	DS	WS	CON	US	MS	DS	WS	CON
		HUN-1 (Anantapur/Kurnool)					HUN-2 (Prakasham)			
■ Agriculture Before	94.7	92.1	89.0	91.6	62.9	48.1	56.4	85.5	70.5	74.8
■ Agriculture After	80.5	75.7	76.6	77.2	55.8	52.1	58.3	81.0	68.1	69.5
■ Livestock Before	1.0	0.3	0.3	0.5	6.7	4.3	6.2	1.8	3.2	2.8
■ Livestock After	3.2	0.6	0.2	1.0	2.2	3.8	3.8	1.6	2.6	2.7
■ Labour Before	3.3	6.9	7.0	5.9	24.8	24.1	24.5	3.8	12.7	12.4
■ Labour After	13.7	19.8	11.7	14.4	24.1	24.3	21.7	4.7	13.8	15.3

FIGURE 7.15 Source-wise changes in the share of household income: normal year. US, upstream; MS, midstream; DS, downstream; CON, control.

The share of labor income has gone up substantially in HUN1, where it has more than doubled over the period of watershed intervention; this increase is marginal in HUN2 (Figure 7.15). The percentage change in labor income is substantial and significant in all villages including the control villages, which have reported a similar growth in labor income.

The growth in labor income is more in the upstream village and least in the downstream village in HUN1 during the normal year, and the reliance on labor as a source of income is more prominent during the drought year than during the normal years in both HUNs (except the control village of HUN1 and midstream village of HUN2), and more so, in HUN1 (Figure 7.16). Between normal and drought years, with respect to the percentage changes in labor income, the differences are found to be statistically significant in HUN2, whereas in HUN1, the differences are found to be significant only in upstream and control villages (Table 7.15).

While the percentage change in labor income is more among the HHs in the control village than their counterparts in the watershed villages of HUN1, the reverse was observed in HUN2 during both years. However, the difference with respect to changes in labor income between watershed and control villages is significant only in HUN1 during the normal year.

When the percentage changes in labor income of the SMF and LMF HHs are compared within their respective villages, the LMF HHs are observed to have experienced more changes in labor income than their SMF counterparts, except in the downstream village during the normal year in HUN1. Further, when all watershed villages were combined in HUN2, the SMF HHs appeared to have benefited more due to an increase in their labor income compared with the LMF HHs. The changes in labor income over the years are mainly from the advent of the employment guarantee program rather than the watershed

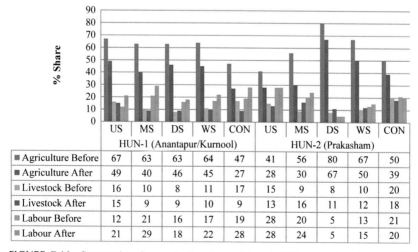

	US	MS	DS	WS	CON	US	MS	DS	WS	CON
	\multicolumn{5}{}{HUN-1 (Anantapur/Kurnool)}					
■ Agriculture Before	67	63	63	64	47	41	56	80	67	50
■ Agriculture After	49	40	46	45	27	28	30	67	50	39
▨ Livestock Before	16	10	8	11	17	15	9	8	10	20
■ Livestock After	15	9	9	10	9	13	16	11	12	18
▨ Labour Before	12	21	16	17	19	28	20	5	13	21
▨ Labour After	21	29	18	22	28	28	24	5	15	20

FIGURE 7.16 Source-wise changes in the share of household income: drought year. US, upstream; MS, midstream; DS, downstream; CON, control.

intervention per se. For employment, the benefits are equally good, if not more, even in control villages.

7.4.3.3 Debt-savings Ratio

The debt-savings ratio indicates the financial viability and credit worthiness of the HHs. In general, the debt-savings ratio in agriculture is on the rise due to the declining viability of agriculture (declining savings) and improved access to institutional credit (supply of debt) for crop loans, etc. Access and utilization of crop loans is reflected in the differences in the debt-savings ratios of normal and drought years (Figure 7.17).

The debt-savings ratio is observed to be better during the drought year, which could be because the farmers do not utilize their crop loans because of poor crops or no prospects of a crop. In both HUNs, the watershed villages have a lower debt-savings ratio compared with the control villages. Similarly, the downstream villages have a lower debt-savings ratio compared with the midstream and upstream villages in both HUNs. The lower debt-savings ratio reported in the upstream village of HUN1 is mainly due to poor access to credit in this village because of its location and social composition (it is a remote tribal village).

7.4.4 Social Capital

Two indicators of social capital are used to assess the watershed impacts, such as group membership and migration.

7.4.4.1 Group Membership

Group membership is measured in terms of the average number of persons in a village having membership in groups, and the percentage of HHs having

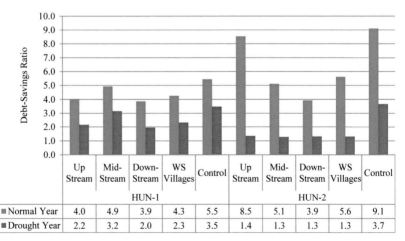

	Up Stream	Mid-Stream	Down-Stream	WS Villages	Control	Up Stream	Mid-Stream	Down-Stream	WS Villages	Control
			HUN-1					HUN-2		
■ Normal Year	4.0	4.9	3.9	4.3	5.5	8.5	5.1	3.9	5.6	9.1
■ Drought Year	2.2	3.2	2.0	2.3	3.5	1.4	1.3	1.3	1.3	3.7

FIGURE 7.17 Debt-savings ratio: normal and drought years.

membership in active groups—while the former represents the supply side, the latter reflects the demand side of the social capital. WSD is expected to increase the group memberships, as the formation of user groups is part of the watershed initiatives.

Group membership has increased substantially and significantly in all villages except one (Table 7.16)—the upstream village in HUN1, which, according to the villagers, did not record a significant rise due to official apathy. The increase in membership is greater in the control villages compared with the watershed villages in both HUNs, which could be due to the low base in the control villages; the situation is not much different even during the drought year.

Between the streams, the increase in membership is more in the downstream village in HUN1, while it is the reverse in HUN2. This could be due to the prevailing socioeconomic conditions in the locations. The presence of active dairy cooperatives in the midstream village of HUN2 is reflected in the substantial increase in the membership of groups here. In case of membership in active groups, the increase is not significant in any of the villages during any year—the watershed intervention does not seem to have any specific impact either on groups or membership.

7.4.4.2 Migration

Migration is an indicator of poor economic conditions in the villages. Watershed interventions are expected to reduce migration, since they are expected to generate employment directly (through works) and indirectly (through increased crop production). The extent of migration is measured in terms of the number of persons migrating, number of days of migration, and income from migration. During the normal year, only the income from migration has gone up significantly in the watershed villages in both HUNs (Table 7.17). However, no significant increase is observed in the control villages indicating that the watershed interventions do not have any positive impact on migration. Further, it is seen in both HUNs that the upstream villages have experienced greater impact in the number of days of migration and income from migration; while in HUN2, even the downstream village has reported a significant increase in migration income. The extent of changes in migration is stronger during the drought year, where the watershed villages reported an increase in the number of days as well as income from migration in both HUNs—the control village in HUN1 has reported higher growth in migration (persons and income). This does seem to give an edge to the watershed interventions during the drought year.

7.4.5 Human Capital

The age composition of the population provides the basis of assessment for human capital. There are variations in age and gender composition across the streams and between the HUNs as well as watershed and control villages

TABLE 7.16 Changes in group membership across the streams

Particulars	Location of the watersheds				
HUN (district)	Upstream	Midstream	Downstream	Watershed villages	Control village
Normal year					
HUN1 (Anantapur/Kurnool)					
% Change in average number of persons having membership	114	417[a]	525[a]	390[a]	1433[a]
Membership in a functional group (% of HHs)	17	63	59	51	70
HUN2 (Prakasam)					
% Change in average number of persons having membership	1340[a]	2450[a]	475[a]	1027[a]	2700[a]
Membership in a functional group (% of HHs)	59	78	68	67	54
Drought year					
HUN1 (Anantapur/Kurnool)					
% Change in average number of persons having membership	271[a]	216[a]	333[a]	273[a]	1400[a]
Membership in a functional group (% of HHs)	27	44	66	49	74
HUN2 (Prakasam)					
% Change in average number of persons having membership	2100[a]	1900[a]	219[a]	705[a]	1050[a]
Membership in a functional group (% of HHs)	46	88	68	64	40

[a]Level of significance at 1% level.

TABLE 7.17 Changes in migration across streams

HUN (district)	% Change	Upstream	Midstream	Downstream	Watershed villages	Control village
				Location of the watersheds		
Normal year						
HUN1 (Anantapur/Kurnool)	Persons	21	22	33	23	3
	Days	46[b]	34	-27	24	-3
	Income	137[a]	67[a]	-49	61[a]	38
HUN2 (Prakasam)	Persons	17	-50	22	8	+ve
	Days	46[c]	-99	427[c]	68	+ve
	income	257[c]	-97	371[c]	165[c]	+ve
Drought year						
HUN1 (Anantapur/Kurnool)	Persons	-3	50[c]	3	18	59
	Days	28	43[b]	11	31[a]	152[b]
	Income	70[a]	137[a]	99[c]	105[a]	556[a]
HUN2 (Prakasam)	Persons	160[b]	+ve	-11	57[b]	-33
	Days	483[b]	+ve	-16	114[c]	-25
	income	336	+ve	60	148[a]	26

[a] Level of significance at 1% level.
[b] Level of significance at 5% level.
[c] Level of significance at 10% level.

(Table 7.18). The proportion of the working population is greater in the watershed villages when compared with the control village in HUN1, while these differences are marginal in HUN2. In both HUNs, the working population, including women, in upstream villages is larger.

Skill training, along with awareness building through exposure visits, is part of watershed interventions; however, skill improvement in watershed villages is observed only in HUN2, whereas in HUN1 skill improvement is be greater in the control village (Table 7.19).

On the other hand, the role of women in financial, family, and farm management has increased substantially in the watershed villages when compared with the control villages. It is further observed that the role of women in management decisions has increased more in HUN1 than in HUN2.

Thus, we see that the assessment of watershed impacts using the five capitals framework has provided a comprehensive view of the nature of impacts. However, the impacts are neither from nor specific to any indicator or capital. Furthermore, there is no concrete evidence to suggest that the changes in the five capitals over the period could be attributed to watershed interventions. Although there are positive and significant changes in various indicators of the five capitals and farmers have gained directly from land improvement (natural capital) activities under the WSD, these improvements have not translated to any substantial improvement in crop yield (financial capital). This is mainly due to the reason that WSD interventions have not

TABLE 7.18 Age-wise distribution (%) of population (% of workers)

		Up to 15 years		16–60 years		>60 years	
HUN	Location	Total	Female	Total	Female	Total	Female
HUN1 (Anantapur/ Kurnool)	US	29 (8)	28 (9)	68 (91)	67 (93)	4 (50)	5 (50)
	MS	24 (2)	23 (0)	67 (90)	70 (85)	10 (66)	7 (43)
	DS	27 (0)	26 (0)	68 (85)	70 (86)	5 (75)	4 (56)
	WS	26 (3)	25 (2)	67 (88)	69 (87)	7 (67)	5 (48)
	Control	27 (2)	31 (3)	66 (82)	60 (85)	7 (44)	9 (36)
HUN2 (Prakasam)	US	28 (0)	31(0)	68 (87)	66 (91)	4 (56)	4 (14)
	MS	27 (0)	24 (0)	62 (88)	68 (83)	11 (68)	8 (56)
	DS	23 (0)	21 (0)	63 (76)	67 (74)	14 (69)	12 (54)
	WS	27 (0)	26 (0)	65 (84)	67 (84)	8 (66)	7 (45)
	Control	21 (0)	21 (0)	66 (85)	67 (89)	12 (77)	12 (70)

Figures in parentheses indicate % of workers. US, upstream; MS, midstream; DS, downstream; WS, watershed.

TABLE 7.19 Changes in skills and gender roles

	Location	Skills acquired (% of HHs)	Increasing role for women in decision making (% of HHs)		
			Financial management	Farm management	Family management
HUN1	US	2	31	31	63
	MS	4	42	34	54
	DS	5	35	32	47
	Control	10	16	16	18
HUN2	US	10	17	20	20
	MS	5	28	27	28
	DS	3	15	13	13
	Control	2	16	14	16

US, upstream; MS, midstream; DS, downstream; WS, watershed.

resulted in any significant increase in area under irrigation (natural capital) despite the changes in irrigation assets—bore wells and motors replacing open wells (physical capital). Further, although this is a common phenomenon seen across the rainfed regions, it may be deduced that WSD interventions have further aggravated groundwater exploitation. This has not resulted in any direct financial benefit to the farmers from crop production in a sustainable way, as the gains through cropping pattern shifts toward horticultural crops in some watershed villages were short term in nature and could not be sustained due to the degradation of groundwater resources.

Next, although the household income has increased substantially and significantly over the years, the share of agricultural income has declined in most cases, while income from wage labor has gone up substantially. The increase in income from wage labor is mainly due to the employment-guarantee programs rather than from the watershed interventions. The improvements in agricultural and livestock incomes that could be directly linked to WSD have shown only marginal changes.

Further, the impact of WSD on social capital is little in terms of membership in groups. In contrast, income from migration has gone up in the watershed villages, which indicates a negative impact.

In human capital, skill improvement in watershed villages is evident only in HUN2. On the other hand, female participation in management decisions has improved substantially in the watershed villages.

Between HUNs, the impacts are marginally better in HUN2 (Prakasam) due to better rainfall conditions. The watershed impacts at scale revealed that the downstream villages performed better in most indicators—the downstream villages appear to be more stable during the drought year when compared with the normal year. Further, the farm-size analysis of all indicators revealed that the impacts are not significantly different for these groups in most cases, and there is no clear evidence that WSD interventions have benefited the SMFs or LMFs differentially.

The limited evidence on the positive impact of WSD across the streams could be attributed to the time lag between the interventions and the impact assessment. In the absence of baseline data, the impacts were assessed on the basis of before and after situations as well as with and without situations. This kind of assessment has limitations due to the memory lapse of the respondents. In the event of longer time lag, the impacts are often not sustained long enough due to the lack of, or poor, maintenance of the WSD structures. While this points toward lack of sustainability of watershed impacts, capturing the direct impacts and attributing them to WSD interventions becomes difficult. This gives rise to the need for alternative impact indicators on which WSD interventions have a bearing. One such indicator is the resilience of the households in terms of their ecological and livelihood attributes in the context of watershed interventions.

7.5 IMPACT OF WSD ON RESILIENCE

To have an aggregate assessment, resilience of the households is measured in terms of the perceived number of droughts a household can survive with their existing five capitals. Households reporting survival of two or more droughts are considered highly resilient, while all others have low resilience. Households possessing all five capitals and having enough to survive drought are called strong capital households and those with less than five capitals or not having enough to survive are called weak capital households. Based on the resilience and capitals, the sample households are grouped under four categories: high resilience with strong capitals (all five capitals strong), high resilience with weak capitals (less than five capitals and weak), low resilience and strong capitals (all five capitals strong), and low resilience and weak capitals (less than five capitals and weak).

High resilience is observed in the watershed villages when compared with the control villages (Table 7.20)—the proportion of households reporting high resilience in the watershed villages is more than double that of control villages. The high resilience (75% of the households) observed in HUN2 could be from better rainfall conditions, as the impact of watershed is greater in the better rainfall regions [1,2]. A higher proportion of households is observed to be highly resilient from upstream to downstream locations. One interesting observation is that a higher proportion of households in the upstream village of HUN2 reported high resilience when compared with the midstream and downstream villages. This needs to be probed further in the context of access to five capitals. The status of the five capitals at the household level appears to

TABLE 7.20 Access to five capitals and level of resilience

		Level of resilience			
		HUN1 (Anantapur/ Kurnool)		HUN2 (Prakasam)	
Location of watersheds	Status of capitals	High	Low	High	Low
Upstream	Strong	13 (30)	31 (70)	48 (66)	25 (34)
	Weak	0 (0)	1(100)	9 (90)	1(10)
Midstream	Strong	25 (35)	46 (65)	38 (79)	10 (21)
	Weak	5 (50)	5 (50)	0 (0)	1 (100)
Downstream	Strong	36 (48)	39 (52)	37 (84)	7 (16)
	Weak	3 (38)	5 (62)	5 (100)	0 (0)
WSD villages	Strong	74 (39)	116 (61)	123 (75)	42 (25)
	Weak	8 (42)	11 (58)	14 (88)	2 (12)
Control village	Strong	7 (18)	31 (82)	10 (29)	25 (71)
	Weak	2 (50)	2 (50)	3 (43)	4 (57)

be influencing resilience, although the number of households reporting weak capitals is limited in the sample villages (<10%).

Resilience across socioeconomic groups clearly indicates that the LMF HHs are more resilient compared with the SMF HHs in both HUNs (Table 7.21). Similar observations could be made even in social groups—the Scheduled Caste (SC) households are least resilient, while the other caste households are most resilient in both HUNs. The logical reason for this is the difference between access to the five capitals across the HUNs, streams, and socioeconomic groups. There appears to be a clear linkage between the resilience and possession of five capitals (Tables 7.22–7.26). The exception seems to be the midstream village in HUN2, which has lower resilience despite better access to physical and financial capitals when compared with the upstream village. Only in financial capital are the socioeconomic groups doing better, i.e., the majority of them possess strong financial capital. This is mainly due to the wage labor component of the financial capital (also observed in the earlier analysis; Table 7.23). Since the SMF HHs depend more on wage labor, they are able to get more income. On the other hand, the LMF HHs receive more income from agriculture.

The access or possession of social capital is rather weak across the villages, i.e., the majority of the households reported weak social capital (Table 7.25). This could be because the status of migration has not improved after the

TABLE 7.21 Resilience across socioeconomic groups

| | Economic groups | | | | Social groups | | | | | |
| | High resilience | | Low resilience | | High resilience | | | Low resilience | | |
Location	SMF	LMF	SMF	LMF	SC	BC	OC	SC	BC	OC
HUN1	61	77	39	23	53	67	81	47	33	19
US	50	58	50	42	50	54	NA	50	46	NA
MS	49	78	51	22	20	59	72	80	41	28
DS	73	96	27	4	56	78	89	44	22	11
Control	63	78	37	22	68	78	60	32	22	40
HUN2	78	89	22	11	67	75	83	33	25	17
US	83	100	17	0	84	76	92	16	24	8
MS	87	100	13	0	67	92	81	33	8	19
DS	92	100	8	0	NA	87	92	NA	13	8
Control	32	71	68	29	25	31	64	75	69	36
Grand total	71	80	29	20	59	70	82	41	30	18

US, upstream; MS, midstream; DS, downstream.

TABLE 7.22 Access to natural capital across socioeconomic groups

| | SMF | | LMF | | SC | | BC | | OC | |
	WCa	SCa	WCa	SCa	WCa	SCa	WCa	SCa	WCa	SCa
HUN1	87	13	71	29	89	11	88	12	65	35
US	79	21	67	33	100	0	66	34	NA	NA
MS	93	7	88	12	100	0	94	6	84	16
DS	80	20	34	66	44	56	90	10	49	51
Control	100	0	100	0	100	0	100	0	100	0
HUN2	56	44	26	74	70	30	70	30	35	65
US	62	38	0	100	53	47	73	27	8	92
MS	45	55	0	100	100	0	58	42	34	66
DS	28	72	0	100	NA	NA	54	46	20	80
Control	96	4	64	36	100	0	94	6	77	33
Grand total	69	31	61	39	81	19	80	20	47	53

WCa, weak capital; SCa, strong capital; US, upstream; MS, midstream; DS, downstream.

TABLE 7.23 Access to physical capital across socioeconomic groups

	SMF		LMF		SC		BC		OC	
	WCa	SCa	WCa	SCa	WCa	SCa	WCa	SCa	WCa	SCa
HUN1	**97**	**3**	**88**	**12**	**95**	**5**	**98**	**2**	**83**	**17**
US	93	7	97	3	100	0	96	4	NA	NA
MS	100	0	94	6	100	0	100	0	92	8
DS	95	5	63	37	77	23	96	4	76	24
Control	100	0	100	0	100	0	100	0	100	0
HUN2	**75**	**25**	**49**	**51**	**83**	**17**	**86**	**14**	**56**	**44**
US	81	19	25	75	74	26	89	11	34	66
MS	59	41	0	100	100	0	67	33	49	51
DS	61	39	40	60	NA	NA	87	13	51	49
Control	100	0	79	21	100	0	100	0	87	13
Grand total	**84**	**16**	**79**	**21**	**90**	**10**	**92**	**8**	**67**	**33**

WCa, weak capital; SCa, strong capital; US, upstream; MS, midstream; DS, downstream.

TABLE 7.24 Access to financial capital across socioeconomic groups

	SMF		LMF		SC		BC		OC	
	WCa	SCa	WCa	SCa	WCa	SCa	WCa	SCa	WCa	SCa
HUN1	**33**	**67**	**30**	**70**	**44**	**56**	**29**	**79**	**24**	**76**
US	65	35	54	66	0	0	66	34	NA	NA
MS	41	59	25	75	70	30	31	69	32	68
DS	20	80	15	85	44	56	13	87	18	82
Control	27	73	21	79	36	64	61	39	20	80
HUN2	**26**	**74**	**35**	**65**	**37**	**63**	**25**	**75**	**31**	**69**
US	24	76	13	87	21	79	23	77	25	75
MS	20	80	0	100	67	33	16	84	19	81
DS	20	80	10	90	NA	NA	20	80	23	77
Control	50	50	71	29	63	37	44	56	64	36
Grand total	**28**	**72**	**31**	**69**	**41**	**59**	**27**	**73**	**29**	**71**

WCa, weak capital; SCa, strong capital; US, upstream; MS, midstream; DS, downstream.

TABLE 7.25 Access to social capital across socioeconomic groups

	SMF		LMF		SC		BC		OC	
	WCa	SCa	WCa	SCa	WCa	SCa	WCa	SCa	WCa	SCa
HUN1	83	17	67	33	88	12	77	27	67	33
US	93	7	93	7	100	0	85	15	NR	NR
MS	87	13	68	32	100	0	88	12	60	40
DS	79	21	44	56	100	0	76	24	51	49
Control	73	27	52	48	77	23	39	61	80	20
HUN2	82	18	52	48	83	17	83	17	67	33
US	87	13	25	75	79	21	89	11	50	50
MS	82	18	67	33	66	34	83	17	85	15
DS	69	31	40	60	NA	NA	NA	NA	NA	NA
Control	85	15	71	29	100	0	80	20	59	41
Grand total	83	17	63	37	86	14	94	6	68	32

WCa, weak capital; SCa, strong capital; US, upstream; MS, midstream; DS, downstream.

TABLE 7.26 Access to human capital across socioeconomic groups

	SMF		LMF		SC		BC		OC	
	WCa	SCa	WCa	SCa	WCa	SCa	WCa	SCa	WCa	SCa
HUN1	36	64	23	77	40	60	29	71	25	75
US	43	57	29	71	100	0	35	65	NA	NA
MS	43	57	31	69	50	50	37	63	36	74
DS	34	66	4	96	55	45	24	76	14	86
Control	21	79	22	78	23	77	11	89	60	40
HUN2	26	74	17	83	23	77	5	95	25	75
US	17	83	0	100	5	95	21	79	8	92
MS	22	78	33	67	33	67	17	83	30	70
DS	26	74	30	70	NA	NA	20	80	31	69
Control	54	66	14	86	62	38	63	27	18	82
Grand total	30	70	21	79	33	67	27	63	25	75

WCa, weak capital; SCa, strong capital; US, upstream; MS, midstream; DS, downstream.

advent of watershed and skill development is also limited. With social capital the difference between watershed and control villages is observed only in HUN1, while the differences are marginal in HUN2. Downstream villages, LMF HHs, and Other Caste (OC) HHs have better access to social capital. That is, in addition to natural and physical capitals, the households in these locations are better equipped with social capital as well.

With access to human capital, the control village is better off when compared with the watershed villages in HUN1, while the reverse is true in HUN2. Further, while in HUN1 the upstream village has poorer access to human capital compared with the midstream and downstream villages, in HUN2 the upstream village is better off. This could be because of the socioeconomic conditions of the households in the upstream village of HUN1. In the socioeconomic groups, the LMF and OC households, in general, have better access to human capital, except in the midstream and downstream villages of HUN2 (Table 7.26)—in these villages, a higher proportion of SMF and Backward Caste (BC) households have reported better access to human capital when compared with their counterparts.

Thus, the assessment of watershed impacts on resilience provided clarity and demonstrated a logical pattern. The households appear to have related their five capital assets to resilience better than to watershed impacts. This could be because resilience is the perceived data, rather than factual data, regarding the changes in yield rates, income, etc. Often the changes due to watershed (improved soil moisture, reduced runoff, and increased physical assets) would not have translated into material benefits such as irrigation, yields, and incomes. However, the households might perceive that they can withstand natural adversities better when compared with earlier periods; these perceived benefits could be attributed to the WSD interventions, as evident from the comparisons between watershed and control villages. These differences, however, need to be tested for statistical significance and attribution to watershed intervention. In the following section, we consider the analysis of factors influencing resilience with the help of multiple regression.

7.6 FACTORS INFLUENCING RESILIENCE

Resilience is defined as the capacity of a household to survive consecutive years of drought. To capture the differential level of resilience capacity of the households, the variable indicating different levels of resilience is coded from 0 to 2 (where 0 reflects vulnerability and 1 and 2 indicate weak and high resilience, respectively). While households that have expressed their inability to survive even a single drought are called vulnerable and assigned a code of 0, households reporting survival of only one drought are considered to be weak resilient (coded as 1), and those reporting survival of two or more droughts are termed as high resilient (coded as 2).

The ordered probit model was used to examine the predicted probabilities of resilience. The use of an ordered probit model is justified by the nature of the dependent variable, which is an ordinal variable—it is neither continuous, nor normally distributed; hence, the use of an OLS^2 model would lead to biased estimates of the variables. The ordered probit technique of multiple regression was, therefore, chosen for this analysis to avoid this measurement problem with the dependent variable.

All the independent variables used in the ordered probit model are grouped under the five capitals, i.e., natural, physical, financial, social, and human. We have also included variables to address the attribution problems, such as watershed, stream, and HUN. The rationale behind each of these variables and their measurement is presented in the following subsections.

7.6.1 Natural Capital

The study uses two variables to capture the household's access to natural capital: land ownership and share of CPRs in availability of fodder.

7.6.1.1 Land Ownership

This is the household's land ownership in acres (1 acre = 0.417 ha). It covers the total area of all parcels owned by the household, excluding the area that is rented. Farm size is hypothesized to be positively associated with resilience capacity. Larger farm size is associated with greater wealth and increased availability of capital, and it increases the risk-bearing capacity of the households possessing more land. Because of sufficient production capacity and income, households having large landholdings are expected to have greater flexibility to engage in new activities, which in turn improve their resilience capacity.

7.6.2 Physical Capital

Physical capital assets, such as wells and livestock, are considered in this analysis to assess their impact on the household's resilience capacity.

7.6.2.1 Livestock (Density of TLU^3)

This variable indicates the density of livestock units (TLU per acre) that a household owns. Livestock is generally considered to be an asset that could be used either in the production process or exchanged for cash or other productive assets. Therefore, resilience capacity is expected to increase with the livestock population.

---

2. Standard methods of multiple regressions like the OLS assume that all variables are measured on an interval scale.
3. As households in the area own different types of livestock, all types of livestock are converted into a common unit of measurement, i.e., TLU. The TLU is arrived at by converting small ruminants to big ruminants on a 3:1 ratio.

7.6.2.2 Density of Bore Wells (Number Per Acre)

This variable is used to capture the availability of coping mechanisms at the household level. As availability of wells indicates increased water supply for the households, more bore wells per acre are expected to enhance the households' resilience capacity.

7.6.3 Financial Capital

Household income from various sources (crop production, livestock, services, etc.), saving, and debt are considered under financial capital.

7.6.3.1 Income Diversity Index

Income is regarded as one of the most important factors that influence a household's resilience capacity. Considering this aspect, an income diversity index (IDI) was constructed using the various sources of income on which the households generally rely for their livelihoods, i.e., 1 minus the sum of the squares of the proportion of income from each source:

$$\text{Index of income Diversity} = 1 - \sum_{i}^{n} P^2$$

Here, P_i is the proportion of household's income from the ith source.

IDI varies from 0 to 1, indicating a maximum value of 1 when there is extreme diversity, and a minimum value of 0 when there is no diversity or there is perfect homogeneity.

7.6.4 Social Capital

Social networks like membership in groups, family relationships, friends, and administrative and political connections are included as indicators of social capital.

7.6.4.1 Membership in Groups

The percentage of people in a household having membership in different types of groups that exist at the village level is used to capture the social capital at the household level. Membership in groups and associations broadens people's access to and influence over other institutions, which in turn is expected to support coordinated strategies for attaining livelihood goals and is thus hypothesized to have a positive impact on resilience.

7.6.4.2 Social Category

Caste continues to shape the social position, access to knowledge, and opportunities in life of individuals in rural India; therefore, it is a commonly used indicator for social status in much of South Asia. Considering its importance in the socioeconomic and political life of rural India, it is expected that the relative strength of a household's social group[4] in the community will influence his/her level of resilience. In this study, a positive relationship is hypothesized between the relative strength of the household's social group and its level of resilience.

7.6.5 Human Capital

Human capital is related to a number of variables including the gender, age, dependency ratio, health, and education level of the households in each village.

7.6.5.1 Gender of the Household's Head

With regard to the gender of the household's head, it is assumed that male-headed households are more resilient when compared with female-headed households; female-headed households are less likely to have labor and disposable income (they have specific characteristics and face competing demand on the time of the household head), which may adversely affect their resilience capacity and are thus expected to have a lower level of resilience. Therefore, it is hypothesized that the variable would have a positive sign (+), indicating higher level of resilience for men than women.

7.6.5.2 Age of the Household's Head

Age is an important variable for explaining variation in resilience capacity. Its effect on a person's level of resilience may be negative or positive, while experience comes with age, and is equated with older household heads, which is expected to have a positive effect on their resilience capacity, younger household heads may have longer planning horizons and, hence, may be more likely to invest resources in a new undertaking such as watershed-related activities. The role of age is thus more ambiguous, because age as a proxy for experience may be offset by greater reluctance to try new things, including new technologies or government-sponsored programs. The net effect on the level of resilience, therefore, cannot be determined a priori [3].

7.6.5.3 Dependency Ratio

Dependency ratio is calculated as the ratio of the number of family members in the labor force/number of workers to the household size (number of persons living together or sharing a common kitchen for at least a year). Apart from reflecting the labor availability at the household level, it also measures the burden on the members of the labor force within the household. Households having a

---

4. Using the conventional caste classification systems in India, households are classified into four groups: SC-1, ST-2, BC-3, and OC-4 (ST, Scheduled Tribes).

greater dependency ratio are less likely to have labor and disposable income at their disposal and are thus expected to have a lower level of resilience.

7.6.5.4 Health

At the household level, the head of a household is considered central to the decision-making process. All important decisions related to choosing an appropriate livelihood strategy for the family or in case of participation in any collective activity are generally channeled through this person. Therefore, the resilience capacity of a household to a large extent depends on the status of the health of its head and is thus hypothesized to have a positive impact on resilience. The health-related variable included in the model is an ordinal variable representing the frequencies of illness pertaining to the head of the household (Chronic: 1; 15 days in a month: 2; 6 months in a year: 3; and no sickness: 4).

7.6.5.5 Education

For the purpose of this study, the average level of education of the entire household is considered. This variable is measured as the average number of years of schooling completed by all adult members of the household. Education has a number of positive externalities: in addition to enhancing a household's ability to acquire and process information, it may also be correlated with the level of information access, which is hypothesized to be a key for better resilience. Moreover, more educated households, as opposed to less educated households, are expected to have privileged access to markets/alternative job opportunities, and thus have higher exit options, which will help them in improving their resilience capacity.

7.6.6 Results and Discussion (Ordered Probit Model)

The following discussion is based on the main findings of the ordered probit analysis. (Table 7.27). Econometric estimates for the two models are statistically significant ($p < 0.01$), with chi-squared statistics of 103.7 and 98.6, respectively. No suggestion of substantive multicollinearity between explanatory variables was found in any of the models. The signs of the estimated coefficients are informative for the probabilities associated with the ordered rankings of the level of resilience. A positive coefficient implies an increased chance that a household with a higher score on the independent variable will be observed in the higher category. A negative coefficient indicates a chance that a household with a higher score on the independent variable will be observed in a lower category.

The estimates clearly indicate that watershed intervention has a positive and significant impact on resilience (Table 7.27). The positive and significant association between stream dummy and resilience indicates that resilience increases as one moves from upstream to downstream locations; that is, downstream villages

TABLE 7.27 Regression results for factors influencing household resilience

Variable	Specification I			Specification II		
	Coef.	Std. Err.	z	Coef.	Std. Err.	z
HUN dummy	0.483[a]	0.144	3.350	0.362[a]	0.141	2.560
Stream dummy	0.280[a]	0.062	4.520	—	—	—
Watershed_dummy	—	—	—	0.590[a]	0.149	3.950
Social category dummy	0.023	0.061	0.380	0.079	0.059	1.330
Gender of household's head dummy	—	—	—	0.039	0.213	0.190
Age of household's head (years)	0.006	0.005	1.350	0.008	0.005	1.550
Average family education (years)	0.055[b]	0.020	2.700	0.066[a]	0.020	3.250
Dependency ratio	−0.002	0.002	−0.800	−0.003	0.002	−1.090
Health status of household head (dummy)	0.148[b]	0.085	1.750	0.134	0.085	1.580
Total land (farm size in acres)	0.036[b]	0.018	2.030	0.029[c]	0.017	1.680
Membership in groups (% of people in HH)	0.000	0.003	0.080	0.002	0.003	0.450
TLU	0.069[b]	0.039	1.760	0.056	0.039	1.430
Share of CPRs in fodder (%)	−0.004	0.003	−1.540	−0.005[c]	0.003	−1.800
Density of bore wells (no. per acre)	0.551[b]	0.253	2.180	0.610[a]	0.252	2.430
IDI	−0.730[b]	0.301	−2.430	−0.757[a]	0.301	−2.520

[a]*Level of significance at 1% level.*
[b]*Level of significance at 5% level.*
[c]*Level of significance at 10% level.*

have higher resilience when compared with midstream and upstream villages. The HUN dummy has also revealed a positive impact on resilience, indicating that households in HUN2 are more resilient when compared with HUN1. Of these three variables, watershed has the highest impact on resilience, and watershed intervention enhances the resilience by 59%. Households in HUN2 are 36−48% more resilient than the households in HUN1; household resilience goes up by 28% as one moves from upstream to midstream and downstream locations.

We see that most of the indicators of the five capitals have a positive impact on resilience. Natural capital—land or farm size—has a positive impact on resilience; that is, large farmers are more resilient than small farmers. However, the other natural capital indicator, CPR, has a negative impact on resilience. This could be because the households that are less resilient (SMF) are more dependent on CPRs.

Both the physical capital indicators, livestock and density of bore wells, have a positive impact on resilience. Hence, livestock and bore wells help improve resilience. However, the increasing density of bore wells is not a sustainable way to improve resilience in these regions in the absence of appropriate replenishing mechanisms.

Only one indicator of financial capital turned out to be significant—the IDI revealed a positive association with household resilience; that is, the higher the number of economic activities that a household is involved in the less resilient it is. This could be because less resilient households tend to participate in a number of activities to make a living.

Human capital indicators of education and health of the household have revealed a positive impact on resilience—educated and healthy households are more resilient.

On the contrary, none of the social capital indicators turned out to be significant, indicating that social capital such as group membership does not improve resilience. Although this sounds unreasonable, its nonsignificance might be because most of the households have group membership and, hence, the variation across the households is low.

On the whole, it appears that resilience as an indicator of watershed impacts provides a reasonable assessment. In the absence of baseline information and time lag between implementation and evaluation, it would be hard to expect farmers to recall the precise tangible impacts. Also, the impacts may be marginal to capture in the low rainfall regions such as the present one.

7.6.7 Assessing Watershed Impacts at Scale: An Integrated Approach

The watershed is a technical intervention that needs to be designed taking hydrogeological and biophysical aspects into account; these aspects are rarely taken into consideration at the design stage. Impacts are also assessed in isolation of hydrogeological and the biophysical aspects. As a result, impacts are attributed purely to watershed interventions. This adds to the attribution problems that are statistical in nature and already prevalent. Placing the watersheds in the technical context helps understand the impacts better, i.e., the positive or negative impacts may not necessarily be due to watershed interventions per se or the implementation problems. For instance, variations in rainfall between watersheds could influence the impact substantially. Apart from rainfall, other biophysical attributes such as soils, land use, and irrigation vary between the two sample HUNs (Table 7.28; see also Chapter 6).

TABLE 7.28 Biophysical attributes of the sample HUNs

Details	HUN1	HUN2
Total geographical area (ha)	14323	9498
% of forest	22	13
% of cultivable wasteland	16	40
% fallow	09	NA
% of dry land	45	17
% of wet land	8	30
Slope	$1-2°_0$	$2-3°$
% of black soil	09	26
% of red soil	69	40
% of mixed soil	17	2
% of sandy loam	4	20
Annual rainfall (mm)	641	702
Major irrigated crops	Paddy, vegetables, groundnut, maize, sunflower, and bajra	Paddy, cotton, sunflower, and bajra, and vegetables
Major unirrigated crops	Groundnut, red gram, bajra, castor, and sorghum	Cotton, sunflower, bajra, red gram, and green chili

From the hydrogeological perspective, the aquifer geometry in the study areas can be categorized into three zones: (1) moderate to deep weathering and fracturing zone suitable for artificial recharge measures by water-spreading methods, (2) areas with deep fractures suitable for artificial recharge methods by injection methods, and (3) areas with a very shallow basement not suitable for any intervention (for technical details see Chapter 3). Based on the drainage order, mini-percolation and percolation tanks can be proposed on the first- to third-order streams, and when the area gets into plane topography check dams need to be considered (Table 7.29). The interventions required for a watershed tapping only the weathered zone are different from watersheds tapping the fractured zone, while groundwater is tapped from deep aquifers in both types of watersheds. In these areas, injection wells are more suitable for recharge. In the upstream locations of HUN1 (Vajralavanka–Maruvavanka) and in the control village (Karadikonda), the aquifer is shallow in nature. In these areas, groundwater recharge is affected through water-spreading methods such as check dams, percolation tanks, and farm ponds. The trends

TABLE 7.29 Biophysical aspects and WSD interventions

Hydrogeological feature	Suitable for interventions	Status of hydrogeological features		Present interventions	
		HUN1	HUN2	HUN1	HUN2
Very shallow basement	Suitable for on-farm interventions only	Upstream and midstream	Lower portion of upstream locations, beginning of midstream locations and upper portion of downstream locations	More focus on check dams, although on-farm interventions are also evident	More focus on check dams, although on-farm interventions are also evident
Moderate to deep weathering and fracturing (shallow aquifer)	Artificial recharge (check dams, percolation tanks, farm ponds, etc.)	Upstream and control villages	Upper portion of upstream locations and lower portion of downstream locations	Check dams are provided everywhere with varying density; no percolation ponds	Check dams are provided everywhere with varying density; as well as percolation ponds
Deep fracture/deep weathering zone	Artificial recharge (injection wells and check dams)	Southern edge of mid and downstream; small portion in the middle of midstream locations	Middle portion of midstream locations and lower portion of downstream locations	No injection wells	One injection well provided by APFAMGS in downstream locations of HUN2

(Continued)

TABLE 7.29 Biophysical aspects and WSD interventions—cont'd

Hydrogeological feature	Suitable for interventions	Status of hydrogeological features		Present interventions	
		HUN1	HUN2	HUN1	HUN2
Forest fringe areas	Needs integrated treatment with croplands and alignment with existing water bodies	Mostly upstream locations	All around the HUN	Trenches in upstream locations; not integrated with crop lands	Trenches are made around most parts of forest area; no integrated cropland and existing water bodies
Conversion of wastelands to crops (land distribution)	Changes in runoff and recharge	All areas	All areas	Not considered in WSD design (unexpected); land pressure and investment support by government to these programs	Not considered in WSD design (unexpected); land pressure and investment support by government to these programs
Changes in cropping pattern	To be planned according to groundwater availability	Midstream	Midstream and downstream	Horticultural crops are promoted and sustained so far	Horticultural crops are promoted that could not be supported even with WSD
Mechanized plowing	Changes in runoff and recharge	Shifted to mechanized plowing	Shifted to mechanized plowing	WSD design did not consider these changes	WSD design did not consider these changes

of water level fluctuation are the key to assessing the groundwater recharge, which directly depicts the changes in groundwater storage in a given area. Furthermore, the shift in cropping pattern from irrigated crops to horticulture in the upstream of HUN1 as well as in HUN2 also impacted the groundwater storage and its sustainability (Table 7.29).

The existing interventions do not match the required or the most appropriate interventions for these two HUNs (Table 7.29). Stream interventions such as check dams are the main focus of the WSD in these regions (Table 7.30), although farm treatments are also evident. As depicted in Chapters 3 and 6, the concentration of check dams is more in the upstream regions, which are not very effective in improving the groundwater recharge.

Our observation in the upstream village of HUN1 (S. Rangapuram) clearly indicated that despite the construction of check dams, there is no improvement in the groundwater situation. Attempts to install bore wells have failed and the normal rainfall supports only swallow wells. Although swallow wells (only about two to three) overflow during excess rainfall years, the hydrogeology is not conducive for deep water storage. Hence, the construction of check dams has not helped beyond marginal improvement of swallow wells. During the below normal rainfall years, the village even faces drinking water shortage. This is mainly because of the limited groundwater potential in this village due to its very shallow basement. Hence, attempts to improve the groundwater situation through on-stream interventions are

TABLE 7.30 Density of water-harvesting structures in sample HUNs and watershed villages

	No. at the HUN level	No. in the sample villages	Area irrigated (sample villages)	No. of farmers covered (sample villages)
Maruvavanka and Vajralavanka (Kurnool/Anantapur): HUN1				
Check dams	128 (122)	39 (47)	180	80
Percolation tanks	0	0	0	0
tanks	13	0	0	0
Peethuruvagu (Prakasam): HUN2				
Check dams	11 (863)	4 (375)	120	95
Percolation tanks	23	7 (214)	140	66
tanks	14	2	232	NA

Figures in parentheses are number of hectares of geographical area per structures.

efficient investment decisions, and on-farm interventions with judicious land use planning are expected to be more beneficial as well as sustainable. This is strongly reflected in the declining income from agriculture and the resilience of the households (the proportion of households reporting high resilience is small).

In HUN2, which is characterized by a moderate shallow basement in the midstream and downstream locations, the situation is better as the check dam interventions have helped improve the groundwater, although the aquifer storage potential is limited to moderate shallowness. Unfortunately, in the absence of such information, horticultural (sweet lime) crops were promoted on a large scale. While these crops were sustained as long as rainfall was normal and above normal, they could not be protected during the below normal rainfall years; hence, due to two consecutive below normal rainfall years in recent years, most of the horticultural crop has dried up and the farmers have incurred losses. Thus, it appears that promotion of horticulture (land use) in these conditions is not the right approach as these aquifers fill up and deplete faster in good and bad rainfall years. On the other hand, water-intensive crops even in good rainfall years may not be sustained, as rainfall fluctuations are quite normal in these regions.

Further, in this HUN, the midstream village has reported poor performance when compared with upstream as well as downstream villages. This has resulted in a decline in agricultural income and lower proportion of households reporting high resilience during the drought year. The midstream village has revealed the lowest impact when compared with the upstream and downstream villages in this HUN. Although this goes against the given wisdom, the poor performance could be explained in terms of the hydrogeological attributes of the streams.

Similarly, the land use changes, such as distribution of wastelands, mechanization, etc., have reduced the runoff in recent years. While water bodies were filled during normal years, after the advent of these changes they are filled only during excess rainfall years. In some locations (upstream locations of HUN1), the interventions in the forest fringe areas have reduced the inflows into the water bodies, adversely affecting the cropping pattern. The farmers feel that filling water bodies is more beneficial than groundwater recharge. According to them, once the upstream water body is filled, it not only facilitates growing of paddy, but also helps in the recharge of downstream groundwater on a wider scale. Infiltration rates are quite high in the upstream tank bed (Chapter 6), and the recharge also includes return flows from paddy cultivation. Thus, it is necessary to integrate WSD interventions in the forest fringes with the water bodies and land use in the downstream locations. This could be another reason for the poor performance of the midstream villages.

Thus, we see that to assess the possibilities for reallocation of water between streams, storing, and using more water in places where water productivity is relatively high, there is a need for a holistic approach. We have done this by estimating the production function for paddy, which is the only

TABLE 7.31 Water productivity: paddy (multiple regression analysis)

Year	Upstream	Downstream
Normal year	0.02*	0.02*
Drought year	0.01*	0.01*

Note: * indicates 1 percent level of significance.
Dependent variable: Paddy yield per acre in quintals.
Independent variables: Average education, owned area, bore wells, IDI, variety, pesticides (no. of sprays), pumping hours per day, hired labor, big ruminants density, fertilizer use, Farm Yard Manure (FYM) use, etc.

common crop with enough observations in upstream and downstream locations. As far as paddy is concerned, water productivities are similar, indicating that there is no economic rationale for reallocation of water (Table 7.31). Given that the area under paddy is on the decline and the shift toward high-value horticultural crops is not very successful, water use and productivities of other crops such as groundnut and cotton would more useful. However, in the absence of reasonable sample size for a common crop across the streams we could not complete that exercise. It may be noted that paddy continues to be the most preferred crop for the farmers in these regions, as the watershed interventions have not really helped in moving toward an alternative land use that is sustainable. This could be due to the inappropriateness of the intervention to the hydrogeological and biophysical attributes of the region. At the same time, changes in land use are short-sighted, as they did not consider the hydrogeology of the locations.

7.7 CONCLUSIONS

This chapter attempts to assess the impacts of watershed interventions at the scale of a HUN. Treating the watersheds within a HUN provides scope for capturing the externalities. We used two indicators for impact assessment: the standard approach for measuring the impacts on various socioeconomic indicators, where the SRL framework of five capitals has been adopted, and the resilience of the households in the context of watershed interventions. Of these two broad indicators, resilience has provided clear evidence of impact when compared with the five capitals approach. In the five capitals, the impacts are subdued and do not provide any clear evidence of the impacts in terms of statistical significance. On the other hand, resilience is positively associated with the location (stream) and the watershed in the HUN—HUN2 with better rainfall is more resilient than HUN1. Further, downstream locations are more resilient than upstream and midstream locations, and watershed villages are more resilient than non-watershed (control) villages. This supports the formulated hypothesis.

However, there are deviations to this logical pattern: the extremely poor performance of the upstream village in HUN1 despite it being a model watershed (acclaimed as a best-implemented watershed) and the unexpected poor performance of the midstream village in HUN2 when compared with the upstream village, despite the shifts to high-value horticultural crops.

The explanation for these deviations lies in the hydrogeology of the locations: the hydrogeology of the upstream village in HUN1 (very shallow basin) does not suit any on-stream interventions for groundwater recharge. As a result, despite well-constructed and maintained check dams, the village could not benefit from groundwater recharge and continues to depend on shallow wells; the situation worsens during below normal rainfall years.

In the midstream in HUN2, the land use pattern is not in line with the groundwater potential. This village is characterized by a moderately shallow basin with limited groundwater potential. Due to the nature of the aquifer, groundwater swells and depletes faster during good and bad rainfall years. In the absence of this hydrological information, horticultural crops were promoted. When the demand for water was surpassed, the potential wells started failing and the horticultural crops started drying up; that is, groundwater was exploited beyond its potential (sustainable yields). As long as there is balance between demand and supply, the shift toward water-intensive crops would sustain, which is observed in parts of HUN1 (horticultural crops are still being sustained).

These two cases clearly demonstrate the role and importance of hydrogeological and land use practices in explaining and understanding the watershed impacts. In the absence of such information, the impacts are often attributed to the quality of watersheds (implementation) or at the most to rainfall variations (if any). This vindicates our basic premise of considering the biophysical and hydrogeological aspects while assessing the impacts. Such integration and assessment becomes convenient and comprehensive when watersheds are placed in the context of an HUN.

On the whole, the assessment of impacts validates the hypothesis that meso-watersheds could generate differential benefits at scale (upstream/downstream). It is clear that there is some mismatch of perceptions of the outcome of WSD in terms of benefits and costs to other parties that appear to relate to hydrogeological and biophysical characteristics of the location. In the following, some important concerns and challenges that need policy attention while implementing the IWMP watersheds are listed:

- Hydrology-based approach for placing IWMP watersheds within the hydrological coordinates would help in understanding the upstream/downstream linkages better.
- Technical inputs need to be used for assessing surface and groundwater hydrology and their linkages in the context of biophysical attributes, while designing the watersheds. Contributions from this project have clearly shown the usefulness of these models in designing IWMP watersheds (see Chapters 3—6).

- In this context, there is a need for differential allocations within and between watersheds as against the present blanket, which is fixed per hectare allocations. However, the components to which these funds should be allocated need to be location specific, depending on the agro-climatic and hydrogeological factors.
- Financing of WSD needs to be changed to asset-based planning, instead of the one-time, program-based approach, to ensure the sustainability of the watershed structures. This could be done following a life cycle cost approach where capital (asset) management is part of project costing; that is, watersheds should be provided with asset management funds on a continuous basis. This would help to enhance the benefits from the watershed-linked livelihood and income-support policies such as dairy and other allied activities.
- This calls for permanent institutional arrangements that have constitutional validity as well as linkages to manage the watersheds on a continuous basis. This would have double impact on managing the structures in a more systematic manner and using the fund flows efficiently (see Chapter 10).

REFERENCES

[1] Deshpande RS, Ratna Reddy V. Differential Impact of Watershed Based Technology: Some Analytical Issues. *Indian J agric Econ* 1991;vol. 46.(No. 3):261−9. July−September.

[2] Joshi PK, Jha AK, Wani SP, Joshi Laxmi, Shiyani RL. *Meta-analysis to assess impact of watershed program and people's participation.* Colombo, Sri Lanka: IWMI, Research Report 8; 2004.

[3] Lapar M, Lucila A, Pandey S. Adoption of soil conservation: the case of Philippine uplands. *Agric Econ* 1999;**21**:241−56.

[4] Reddy V, Ratna Gopinath Reddy M, Galab S, Soussan J, Springate Baganski O. Participatory Watershed Development in India: Can it Sustain Rural Livelihoods? *Dev Change* 2004;**35**(2):297−326. April.

[5] Reddy V, Ratna Gopinath Reddy M, Malla Reddy YV, Soussan John. "Sustaining Rural Livelihoods in Fragile Environments: Resource Endowments and Policy Interventions − A Study in the Context of Participatory Watershed Development in Andhra Pradesh". *Indian J Agric Econ* 2008;vol. 63.(No. 2). April−June, 2008.

[6] Reddy V, Ratna Gopinath Reddy M, Soussan John. *Political Economy of Watershed Management: Policies, Institutions, Implementation and Livelihoods.* Jaipur: Rawat Publishers; 2010.

[7] Tompkins EL, Boyd E, Nicholson-Cole SA, Weatherhead K, Arnell WN, Adger WN. *Linking adaptation research and practice. A report submitted to DEFRA as part of the climate change impacts and adaptation cross-regional research programme.* Norwich: Tyndall Centre for Climate Change Research, University of East Anglia; 2005.

[8] UKCIP. *Climate adaptation: risk, uncertainty and decision-making.* Oxford: UKCIP Technical Report, UKCIP; 2003. http://www.ipcc.ch/pub/syrgloss.pdf.

[9] Inter Governamental Panel on Climate Change, Third Assessment report (IPCC, TAR) "Climate Change 2001: Working Group I: The Scientific Basis" 2001. http://www.grida.no/publications/other/ipcc_tar/?src=/climate/ipcc_tar/.

Chapter 8

Evaluating the Determinants of Perceived Drought Resilience: An Empirical Analysis of Farmers' Survival Capabilities in Drought-Prone Regions of South India

Ram Ranjan *, Deepa Pradhan *, V. Ratna Reddy [§] and Geoffrey J. Syme [¶]

*Department of Environment and Geography, Faculty of Science, Macquarie University, Sydney, Australia, [§]Livelihoods and Natural Resource Management Institute, Hyderabad, India, [¶]Edith Cowan University, Perth, Australia

Chapter Outline

8.1 INTRODUCTION

Global climate change is contributing to more frequent and more intense droughts. Several studies point to the disastrous consequences of prolonged droughts on farming [1−3]. Developing countries, such as India, are particularly vulnerable to climate change [4], and within such economies, the farming communities in rainfed regions are especially vulnerable to prolonged droughts. In such conditions, the threat of repeated droughts poses significant challenges to farmers' survival capabilities. This is underlined by the increasing incidence of suicide by farmers in rural India [5,6]—the number of suicides between 2001and 2005 was a little less than 100,000 [7].

Integrated Assessment of Scale Impacts of Watershed Intervention
http://dx.doi.org/10.1016/B978-0-12-800067-0.00008-6.

In response to rainfed conditions and their associated food insecurity and poverty, watershed development (WSD) programs have been implemented in several parts of Asia and Africa to try to provide livelihood support to farmers by augmenting their natural resource base through better management of soil, water (surface runoff and groundwater recharge), and forestry resources [8]. In India, WSD programs have been ongoing since 1970. Several studies have evaluated their impact [9−11], and various methods have been adopted for assessing them. For instance, Hope [8] used a propensity score matching (PSM) method to compare the benefits to watershed-treated areas with respect to the untreated areas and found that WSD projects in Madhya Pradesh did not make farmers any better off compared with those in the non-WSD regions. It was also observed that the main benefits of WSD programs are short-term in these regions [8]. A study, using the sustainable livelihoods framework in sample watersheds of Andhra Pradesh, showed that WSD is a necessary condition for enhancing livelihoods but not sufficient to ameliorate poverty in rainfed conditions [11]. Joshi et al. [10], in their meta-analysis, reviewed a number of impact studies using different methods, and concluded that even though the benefits of watershed intervention have been modest overall, the long-term implications could be significant. The nature and intensity of WSD impacts could depend on a number of factors such as rainfall, quality of implementation, time lag between implementation and impact assessment, methods of assessment, and so on. These problems could be reduced by considering the long-term impact of WSD interventions on the resilience of farmers in the face of challenges such as prolonged drought.

Enhancing resilience is the implicit goal of WSD in farming regions, even though it has not received due attention. For example, soil and water conservation efforts at the farm level directly influence sustainability and therefore presumably resilience, especially when climate change increases the frequency of droughts. Also, the approach of using resilience as an impact indicator addresses some of the difficulties, such as time lag and quality of implementation, to a large extent. Yet, quantifying and measuring resilience poses its own challenges.

There has also been an increasing trend toward using the five capital types—social, physical, financial, natural, and human—for assessing farmers' livelihood impacts. Focusing on the five capitals not only provides the true impacts of WSD interventions on farmers' drought survival (DS) capabilities, but it also helps in understanding how exactly the WSD programs augment the key capitals possessed by the farmers. Farmers' survival capabilities in the wake of repeated drought can be determined by these five types of capital.

For instance, social capital and social norms could be a crucial factor in fostering community lifestyles that are resilient ecologically, economically, and in terms of individual well-being (human capital). Social capital not only lowers the transaction costs but also induces confidence in investment by building trust [12].

Similarly, studies point to the importance of human capital in DS among households in Bangladesh, Ethiopia, and Malawi [13]. Both biological (nutrition and health) aspects as well as educational and skills aspects of human capital can be important.

Natural capital resources such as common pool resources could also be vital to resilience building among farmers [14]. Diamond [15] identified mismanagement of natural resources as one of the reasons for the collapse of the Mayan and Easter Island civilizations.

While the previous studies point to the relevance of the five capitals, what still remains to be assessed is how these capitals combine to provide resilience for farmers. More important, from an empirical perspective, which of these capitals is most relevant for the DS of farmers? Further, based on their endowments of these capitals, can any inference be made about the most resilient or the most vulnerable community types? While financial income may be very important for survival, farmers with no income may also survive if they have good social networks, common pool resource access, and physical and natural assets such as big and small ruminants and agricultural land. Human capital such as health could play a key role as well, given that most governmental interventions working toward mitigating drought severity require farmers to do manual labor in return for daily minimum wages. For instance, the Mahatma Gandhi National Rural Employment Guarantee Act (MGNREGA) in India guarantees 100 days of employment in the manual unskilled sector to every rural household [16]. However, these programs may not cater to the most vulnerable sections of society that perennially suffer from low overall health status.

Drought resilience, despite its attractiveness as a policy tool, is not an easily quantifiable variable. Attempts have been made lately, however, to do just that [17–19]. Ranjan and Athalye [17] defined and measured resilience as the ability of a farmer to survive a certain number of consecutive droughts, in the context of groundwater-intensive farming. In the developing countries, this has been redefined as the capability to survive a certain number of repeated droughts while maintaining minimum consumption levels [20,21].

Keeping in mind these special characteristics of farmers' survival determinants, in this chapter we simply explore what determines households' survival over a set of consecutive droughts and then use the household characteristics and their endowments of the five capitals to draw inferences. Rather than deriving an implicit measure of resilience, we directly question the farmers regarding their capability to survive a certain number of repeated droughts. While these are called DS estimates and hence contain their own biases, it is more likely that farmers' responses are based on their evaluation of publicly and privately known endowments of measurable and immeasurable capitals as well as other factors such as their psychological resilience and determination to survive.

Additionally, we explore whether the WSD programs have had any positive effect on improving the drought resilience of the farmers in the

WSD-implemented areas. Finally, using PSM, we compare drought resilience (or DS) outcomes for farmers in the WSD-treated regions to those in the untreated regions.

This study is based on a survey of 522 households in three districts of South India—Anantapur, Kurnool, and Prakasam—the districts of Anantapur and Kurnool fall under a different hydrological unit (HUN) as compared with Prakasam. In this study, the HUN in Anantapur and Kurnool districts is referred to as HUN1 and the HUN in the Prakasam District is referred to as HUN2. These districts have been characterized by repeated droughts in the past and by depleted groundwater reservoirs; they are also home to farming households with sustained low or negative annual crop incomes.

Water scarcity in the study areas threatens farmers' livelihoods directly— by reducing crop yields and income—as well as indirectly by adversely affecting the human and natural capitals relied on to supplement their meager incomes. For instance, water scarcity depletes the common pool resources such as grazing lands and water reservoirs including ponds, tanks (lakes), and wells, as well as forestry. Farmers' health could decline from a reduction in the quantity and quality of drinking water as well as from reduction in consumption, which is based on common property resources (CPRs). In addition, excessive reliance on CPRs such as forestry leads to their rapid degradation and depletion and subsequent soil erosion, which has a negative feedback effect on crop output. This could intensify the poverty traps in which farmers are already caught.

In this chapter, we evaluate the perceived capability of farmers to survive repeated droughts based on their capital endowments, which comprise human, natural, social, physical, and financial capitals (refer to Table 8.1 for a description of the variables selected to represent these five capitals).

If farmers expect to survive a higher number of repeated droughts in the future, they are considered to be more resilient than farmers who expect to survive fewer drought years. The farmers were provided with a description of a scenario in which repeated droughts manifested in the future, and based on their ability to draw from the pool of their five capital types, they were asked to provide an estimate of the number of years of repeated droughts they expected to survive. Survival is interpreted as the capability of sustaining a reasonable level of livelihood without having to undergo catastrophic outcomes such as loss of human lives within the household.

The WSD programs comprise construction of concrete check dams at various upstream and downstream locations to facilitate higher surface water infiltration underground [22]. Additionally, farmers are also encouraged and supported to carry out soil conservation and afforestation efforts under this scheme. To assess the effectiveness of the watershed intervention programs, the farmers were also questioned about their ability to survive (repeated droughts) in the absence of WSD interventions.

TABLE 8.1 Definition of variables

Variable	Description	Unit
Resilience		
DSWSD	DS with WSD intervention	1–4 years
DSWWSD	DS without WSD intervention	1–4 years
Physical capital		
avgsrumnt	Average number of small ruminants in the household	Number
avgbrumnt	Average number of big ruminants in the household	Number
Natural capital		
totland	Total landholdings per household	Acres
avercpr	Income from CPRs averaged over the last 5 years	Rupees
HUN2	Dummy = 1 for Prakasam District, and 0 for Anantapur and Kurnool district	Scalar
streamcode	Category 1 = upstream, 2 = midstream, and 3 = downstream	Scalar
Social capital		
avergrpmem	Average group membership over the last 5 years	Number
ocaste	Dummy = 1 for other castes, and 0 for backward and scheduled castes and tribes	Scalar
Human capital		
totedumem	Total number of educated members in the household	Number
averernmem	Average number of earning members in the household	Number
avdepend	Average number of dependents in the household	Number
avhlth	Average health of the household members over the last 5 years	1–3 (3 = poor health)
avskills	Average skill level of the household over the last 5 years	1–3 (3 = poor skill)
Financial capital		
ncincm	Net crop income	Rupees

(Continued)

TABLE 8.1 Definition of variables—cont'd

Variable	Description	Unit
farmlaborincome	Income earned per year working on agricultural farms	Rupees
farmlabordays	No. of days spent working as farm labor	Days/year
non-farm income	Income earned per year working as nonfarm labor	Rupees
non-farm days	No. of days spent working as nonfarm labor	Days/year
nregsincome	Income earned through MGNREGA program per year	Rupees
nregsdays	No. of days spent on MGNREGA program	Days/year
migrationincome	Income earned through migration remittances	Rupees
migrationdays	No. of days emigrated	Days/year
Expenses		
exfood	Annual expenditure on food	Rupees
exedu	Annual expenditure on education	Rupees
exhlth	Annual expenditure on health	Rupees
exalchl	Annual expenditure on alcohol	Rupees
exwells	Annual expenditure on wells	Rupees
extractr	Annual expenditure on tractors	Rupees
exothrimplement	Annual expenditure on other implements	Rupees
exlnddev	Annual expenditure on land development	Rupees
exentmnt	Annual expenditure on entertainment	Rupees

8.2 METHODOLOGY AND FINDINGS

The sample villages were classified into upstream, midstream, and downstream villages, based on their physical location with respect to the underground stream movement (see Chapter 1 for details). Sample respondents were selected from upstream, midstream, and downstream villages of these hydrological systems from the three districts to assess the effect of stream-related locational advantages to the farmers, which may allow for better groundwater access and water harvesting. For instance, during plenty or normal rainfall years, downstream farmers are expected to benefit from WSD structures built upstream; however, during drought years, upstream farmers would benefit more if they would harvest most of the water before it could reach downstream. Two control villages with no WSD interventions (also referred to as untreated villages) were also surveyed, including one from each HUN. Figure 8.1 shows the DS responses of the households within the HUNs.

DS responses with WSD interventions (Figure 8.1) are regressed on several explanatory variables comprising the five capitals as well as household characteristics including the location, social stratification, and other behavioral variables of the farmers. A description of the variables is provided in Table 8.1. The DS responses are a proxy and measure the farmers' capabilities to sustain their livelihoods when faced with long-term drought.

In the literature, the five capitals have been identified to be important toward enhancing the survival capabilities of the farmers. For instance, physical capital, such as big and small ruminants, can insure against crop failure as farmers can either use their produce to survive or sell them off to convert into cash. Therefore, it may be expected that farmers with greater number of ruminants are more resilient to repeated droughts. Similarly,

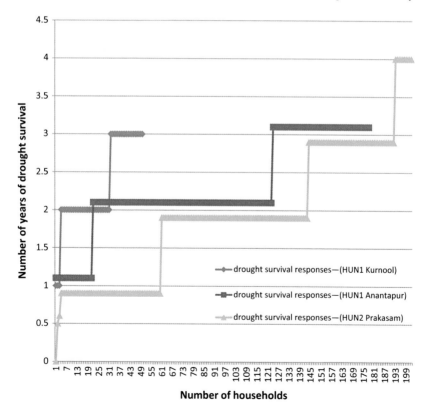

FIGURE 8.1   DS responses in the HUNs. Note: Some DS responses for 1, 2, and 3 years were marginally increased or lowered to visually distinguish them in the graph.

natural capital—farm location (in a certain HUN or location within the watershed such as upstream, midstream, or downstream), land size, and level of access to common pool resources—could also affect the drought resilience level of a given household. For example, farmers in HUN1 (comprising the districts of Anantapur and Kurnool), in general, are less privileged compared with farmers in HUN2 (comprising the district of Prakasam) in terms of common pool resources. Therefore, farmers in HUN2 may be expected to give a better drought resilience response. Indicators of social capital used in this study include the average group membership of households and social stratification (caste). Group membership could be an indicator of the social network of a household, which could come in handy in times of hardship. Similarly, farmers belonging to a higher caste may have an advantage because of their ability to corner resources or because of better social networks.

Human capital—number of educated household members, number of earning members, number of dependents, health status, and skill level of the household members—may also significantly affect drought resilience. Educated farmers may have better access to nonfarm employment opportunities because of their larger networks or better skill sets. Similarly, variables representing financial capital such as income from various sources including farming, farm labor, MGNREGA, and migration have a direct bearing on farmers' survival capability during lean years.

In addition to the capital endowments of the farmers, the choices made by them, as well as their personal circumstances, could also affect their DS capabilities. For instance, household expenditure on food, education, agricultural implements, etc., can determine the availability of cash during drought years. Therefore, drought resilience of farmers in the study areas is defined as a function of the five capital types along with their expenditure patterns:

$$droughtresilience = fn(physicalcaptial, naturalcaptial, socialcaptial,$$
$$humancaptial, financialcaptial, expences)$$

The following equation is used for estimating the DS response:

$$(1)\ DSWSD = \alpha_0 + \alpha_1 \cdot avgsrumnt + \alpha_2 \cdot avgbrumnt + \alpha_3 \cdot totland + \alpha_4 \cdot avercpr$$
$$+ \ \alpha_5 \cdot HU - 2 + \alpha_6 \cdot streamcode + \alpha_7 \cdot avergrpmem + \alpha_8$$
$$\cdot ocaste + \alpha_9 \cdot totedumem + \alpha_{10} \cdot averernmem + \alpha_{11} \cdot avdepend$$
$$+ \alpha_{12} \cdot avhlth + \alpha_{13} \cdot avskills + \alpha_{14} \cdot ncincm + \alpha_{15}$$
$$\cdot farmlaborincome + \alpha_{16} \cdot nonfarmincome + \alpha_{17} \cdot nregsincome$$
$$+ \alpha_{18} \cdot migrationincome + \alpha_{19} \cdot exfood + \alpha_{20} \cdot exedu + \alpha_{21}$$
$$\cdot exhlth + \alpha_{22} \cdot exalchl + \alpha_{23} \cdot exwells + \alpha_{24} \cdot extractr$$
$$+ \alpha_{25} \cdot exothrimplement + \alpha_{26} \cdot exlndev$$

Table 8.2 presents the regression results, based on which a number of observations can be made. The variables that significantly affected DS responses

TABLE 8.2 Regression results for equation 1: explaining DS with watershed interventions

Variables	Coefficients		Std. error	t value	Pr(>|t|)
Intercept	**1.896966**	***	0.221	8.600	0.000
Physical capital					
Average number of small ruminants	0.001894		0.002	0.880	0.377
Average number of big ruminants	−0.027557		0.021	−1.320	0.186
Natural capital					
Total acres of land	0.004095		0.014	0.300	0.768
Average CPR (2005−2006 to 2010−2011; Rs.)	−0.000004		0.000	−0.390	0.700
HUN2	−0.113480		0.087	−1.310	0.192
Watershed intervention stream level (Upstream is base)					
Midstream	−0.024659		0.095	−0.260	0.794
Downstream	**−0.213432**	*	0.104	−2.060	0.040
Social capital					
Average group membership (2005−2006 to 2010−2011)	0.033511		0.060	0.560	0.577
Higher caste	**0.338508**	***	0.089	3.810	0.000
Human capital					
Total number of educated members in household	**0.094320**	**	0.033	2.850	0.005
Average number of earning members (2005−2006 to 2010−2011)	0.023599		0.037	0.650	0.519
Average number of dependents (2005−2006 to 2010−2011)	−0.054982		0.035	−1.560	0.119
Average health status (2005−2006 to 2010−2011; lower the value the better the health status)	**0.115165**	.	0.070	1.650	0.099

(Continued)

TABLE 8.2 Regression results for equation 1: explaining DS with watershed interventions—cont'd

Variables	Coefficients		Std. error	t value	Pr(>|t|)
Average skill level (2005–2006 to 2010–2011; lower the value the better the skill level)	−0.184575	***	0.049	−3.790	0.000
Financial capital					
Net crop income (Rs.)	−0.000001		0.000	−0.770	0.440
Farm labor income (2010–2011; Rs.)	0.000003		0.000	0.770	0.442
Non-farm labor income (2010–2011; Rs.)	−0.000001		0.000	−1.380	0.167
Income from MGNREGA (2010–2011; Rs.)	0.000017		0.000	1.300	0.196
Income from migration (2010–2011; Rs.)	0.000003		0.000	0.830	0.406
Expenses					
Expenses on food	0.000003		0.000	0.700	0.483
Expenses on education	−0.000007	**	0.000	−2.930	0.004
Expenses on health	0.000001		0.000	0.620	0.538
Expenses on alcohol	0.000012	.	0.000	1.710	0.087
Expenses on wells	0.000019	**	0.000	2.930	0.004
Expenses on tractors	0.000019	*	0.000	2.130	0.034
Expenses on other implements	0.000017		0.000	1.500	0.134
Expenses on land development	0.000020		0.000	1.350	0.177

$N = 429$; R-squared $= 0.2629$; adjusted R-squared $= 0.2133$; F statistic p value $= 0.0000$. Significant codes: 0; ***, 0.001; **, 0.01; *, 0.05; ., 0.1.

include stream; caste; total number of educated members within the family; health status of the household; skill level of the household; and household expenditure on education, alcohol consumption, wells, and tractors. With human capital, it was found that the total number of educated members in a household (*totedumem*) helps with DS and the regression coefficient is significant. Furthermore, the average number of dependents (*avdepend*) reduces the DS (although the regression coefficient is not significant); a higher number of earning members (*averernmem*) has a positive, although not significant, impact on DS; and households with higher skills perceived that they could survive a higher number of consecutive drought years. Note that value 1 measures higher skills and value 3 measures lower skills.

Results imply that better health is not important for DS. Since a value of 1 stands for good health and 3 for poor health, a positive sign implies that farmers with poor health may have a higher subjective perception of DS. We will explore further the possible reasons behind this anomaly later.

The results also showed that expenditure on education (*exedu* variable) makes households more vulnerable to repeated droughts (this result is consistently found for all regression analyses performed in the study), and expenditure on food is positively related with DS responses (regression co-efficient insignificant).

With physical capital, the results show that ownership of big ruminants (*avgbrumnt* variable) has a negative impact on DS, while the ownership of small ruminants (*avgsrumnt*) has a positive impact on DS (however, the regression coefficients are not significant). Households often stock up on ruminants as a drought risk hedging strategy, and it is possible that sustained droughts make it harder for households with ruminants to survive. We also see that expenditure on wells, tractors (regression coefficient significant), implements, and land development (regression coefficient not significant) is positively associated with DS.

Location-wise, downstream (regression coefficient significant) and midstream (regression coefficient insignificant) households are observed to have lower DS compared with upstream households. The fact that perceived DS is lower for downstream farmers is surprising—one reason for this result could be that the downstream farmers have higher water availability during normal rainfall years, whereas during drought years all the limited water is appropriated by the upstream farmers thus making their downstream counterparts more vulnerable. This could possibly explain their perceived lower responses to DS during drought years.

Farmers accumulate social capital through a variety of activities. In the survey, we mainly questioned them regarding their participation in different village level groups to get an idea of their social capital. Although not significant, the results show that DS has a positive relationship with average group membership (*avergrpmem*). Similarly, the caste of the household also affects their social capital as social networking efforts could be based on caste alliances. Higher caste farmers are observed to have a significantly higher DS response. This may be an indicator of increased inequality as higher caste households are able to appropriate a major share of the water resources with WSD interventions. Figure 8.2 depicts the landholdings across lower and higher caste categories. It is clear from the figure that among the sampled population, there is a larger number of farmers with very small landholdings. If the size of land is any indicator of political or social influence, one would expect the small farmers to be disadvantaged with respect to water allocations.

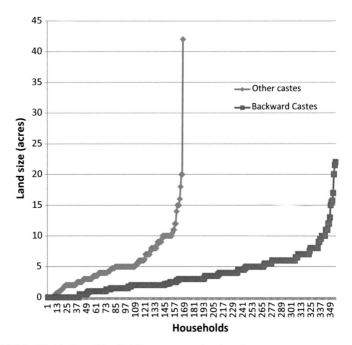

FIGURE 8.2   Distribution of landholdings across caste categories.

It is important to note that net crop income (measured by the variable *ncincm*) does not add much to DS. Figure 8.3 shows the net crop income for the surveyed households. It is interesting to note that out of the 522 households surveyed, there were only 100 households with a positive net crop income. We further see that MGNREGA-based income (*nregsincome* variable) has a positive impact on DS (although the coefficient is not significant). Finally, HUN2 (Prakasam District) was found to have a positive DS (although the regression coefficient is not significant).

Next, we estimate the determinants of stated DS ability without WSD, which is regressed similarly using the following equation:

$$(2)\ DSWWSD = \alpha_0 + \alpha_1 \cdot avgsrumnt + \alpha_2 \cdot avgbrumnt + \alpha_3 \cdot totland$$
$$+ \alpha_4 \cdot avercpr + \alpha_5 \cdot HU - 2 + \alpha_6 \cdot streamcode$$
$$+ \alpha_7 \cdot avergrpmem + \alpha_8 \cdot ocaste + \alpha_9 \cdot totedumem$$
$$+ \alpha_{10} \cdot averernmem + \alpha_{11} \cdot avdepend + \alpha_{12} \cdot avhlth$$
$$+ \alpha_{13} \cdot avskills + \alpha_{14} \cdot ncincm + \alpha_{15} \cdot farmlaborincome$$
$$+ \alpha_{16} \cdot nonfarmincome + \alpha_{17} \cdot nregsincome$$
$$+ \alpha_{18} \cdot migrationincome + \alpha_{19} \cdot exfood + \alpha_{20} \cdot exedu$$
$$+ \alpha_{21} \cdot exhlth + \alpha_{22} \cdot exalchl + \alpha_{23} \cdot exwells + \alpha_{24} \cdot extractr$$
$$+ \alpha_{25} \cdot exothrimplement + \alpha_{26} \cdot exlndev$$

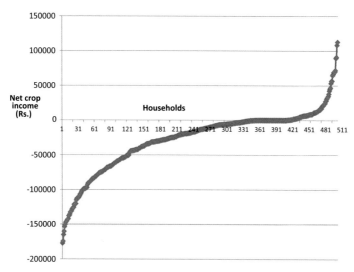

FIGURE 8.3   Distribution of net crop income across households.

Note that the explanatory variable data remain unchanged; only the DS responses of the same farmers are elicited based on their capability to survive consecutive droughts (without WSD).

Some important observations can be made based on the results presented in Table 8.3. The variables found to significantly affect DS responses in absence of WSD include the following: average number of big ruminants, location within the stream, average number of dependents, and educational expenses.

With human capital, we see that health does not play an important role in DS without WSD, just as was the case with WSD. However, the total number of educated members does not help in DS without WSD, unlike the results with WSD, in which the total number of educated members did play a role with DS. Similarly, without WSD intervention, the households with skills felt less confident to survive droughts as compared with WSD intervention. This implies that having skills has become more important in the presence of WSD intervention, indicating that WSD intervention may have created opportunities for employment. The drought resilience of households was also found to have a significant negative relationship with the average number of dependents in the situation without WSD.

Further, the results indicate that households with big ruminants (which are a part of the physical capital) stated a higher DS, in contrast to the results with WSD, where we found the opposite result, although the regression coefficient was not significant. Thus, it is possible that WSD projects have not been very helpful to farmers with large ruminants. This may be due to the decrease in fodder availability as a result of restrictions that might have been placed with respect to accessing the CPRs. Location-wise, it is observed that downstream

TABLE 8.3 Regression results for equation 2: explaining DS without watershed interventions

Variables	Coefficients		Std. error	t value	Pr(>|t|)
Intercept	**0.866689**	***	0.138	6.260	0.000
Physical capital					
Average number of small ruminants	0.000966		0.001	0.720	0.473
Average number of big ruminants	**0.023114**	.	0.013	1.770	0.077
Natural capital					
Total acres of land	0.005803		0.009	0.670	0.505
Average CPR (2005−2006 to 2010−2011; Rs.)	−0.000009		0.000	−1.530	0.126
HUN2	−0.043048		0.055	−0.790	0.430
Watershed intervention stream level (Upstream is base)					
Midstream	−0.083563		0.059	−1.410	0.160
Downstream	**−0.182098**	**	0.065	−2.800	0.005
Social capital					
Average group membership (2005−2006 to 2010−2011)	0.009404		0.038	0.250	0.803
Higher caste	0.025427		0.056	0.460	0.649
Human capital					
Total number of educated members in household	0.020406		0.021	0.980	0.327
Average number of earning members (2005−2006 to 2010−2011)	−0.010431		0.023	−0.450	0.650
Average number of dependents (2005−2006 to 2010−2011)	**−0.043322**	.	0.022	−1.960	0.051
Average health status (2005−06 to 2010−2011; lower the value the better the health status)	0.062343		0.044	1.430	0.155
Average skill level (2005−2006 to 2010−2011; lower the value the better the skill level)	0.007231		0.031	0.240	0.813

TABLE 8.3 Regression results for equation 2: explaining DS without watershed interventions—cont'd

Variables	Coefficients		Std. error	*t* value	*Pr*(>|*t*|)
Financial capital					
Net crop income (Rs.)	−0.000001		0.000	−1.440	0.151
Farm labor income (2010−2011; Rs.)	0.000002		0.000	0.940	0.349
Nonfarm labor income (2010−2011; Rs.)	0.000000		0.000	0.440	0.657
Income from MGNREGA (2010−2011; Rs.)	−0.000013		0.000	−1.520	0.130
Income from migration (2010−2011; Rs.)	0.000000		0.000	−0.060	0.955
Expenses					
Expenses on food	0.000004		0.000	1.390	0.166
Expenses on education	**−0.000004**	**	0.000	−2.650	0.008
Expenses on health	0.000000		0.000	0.110	0.909
Expenses on alcohol	0.000002		0.000	0.440	0.660
Expenses on wells	0.000001		0.000	0.360	0.720
Expenses on tractors	0.000000		0.000	0.070	0.948
Expenses on other implements	−0.000001		0.000	−0.100	0.923
Expenses on land development	−0.000003		0.000	−0.320	0.747

$N = 429$; R-squared $= 0.0968$; adjusted R-squared $= 0.0360$; F statistic p value $= 0.0323$. Significant codes: 0; ***, 0.001; **, 0.01; *, 0.05; ., 0.1.

households are still worse off compared with the upstream households without WSD intervention.

Another observation of significance is that it appears that the effects of social stratification have been magnified as a result of WSD intervention: the "Other Castes category" (which includes all forward castes) was no more significantly resilient than the Backward Castes in the situation without WSD intervention. This implies that WSD intervention may have contributed to an increase in social inequality. Further, none of the capital types helps with DS in the without WSD situation; whereas, with WSD intervention, MGNREGA-based income was found to be positive but with a minimal impact (regression coefficient not significant).

Next, we explore the determinants of some key income sources, which may affect DS. First, migration income is examined. The following equation is used for explaining migration income:

$$(3)\ migrationcome = \alpha_0 + \alpha_1 \cdot avgsrumnt + \alpha_2 \cdot avgbrumnt + \alpha_3 \cdot totland + \alpha_4 \cdot avercpr$$
$$+ \alpha_5 \cdot HU - 2 + \alpha_6 \cdot streamcode + \alpha_7 \cdot avergrpmem + \alpha_8 \cdot ocaste$$
$$+ \alpha_9 \cdot totedumem + \alpha_{10} \cdot averernmem + \alpha_{11} \cdot avdepend + \alpha_{12}$$
$$\cdot avhlth + \alpha_{13} \cdot avskills + \alpha_{14} \cdot ncincm + \alpha_{15} \cdot farmlabrdays$$
$$+ \alpha_{16} \cdot nonfarmdays + \alpha_{17} \cdot nregsdays + \alpha_{18} \cdot migrationdays$$
$$+ \alpha_{19} \cdot exfood + \alpha_{20} \cdot exedu + \alpha_{21} \cdot exhlth + \alpha_{22} \cdot exalchl + \alpha_{23}$$
$$\cdot exwells + \alpha_{24} \cdot extractr + \alpha_{25} \cdot exothrimplement + \alpha_{26} \cdot exlndev$$

From the results presented in Table 8.4, we can make the following observations. The variables that were found to significantly affect the migration income of the surveyed households in the study area are caste, number of days of migration, number of MGNREGA days, and expenses on alcohol and other implements. We see that the higher castes have significantly higher migration income, which perhaps implies the ability to tap into their social networks in the cities. Migration income is obviously influenced most by the number of days of migration income. Similarly, those who spend more days on MGNREGA programs are also more likely to migrate for job opportunities (coefficient significant). Likewise, expenditure on land development is seen to have a positive impact on the migration income (although the regression coefficient is not significant). Another interesting point to note is that migration income is positively related to alcohol consumption, which perhaps is due to the higher availability of cash.

An important category of income is the income derived through participation in the MGNREGA scheme; that is, being able to derive a higher MGNREGA-based income, which adds to the resilience of the farmers. We therefore test for variables that contribute to higher MGNREGA-based income. The initial hypothesis is that the health of the farmers should be key to determining higher MGNREGA-based income as these schemes are labor intensive:

$$(4)\ nregsioncome = \alpha_0 + \alpha_1 \cdot avgsrumnt + \alpha_2 \cdot avgbrumnt + \alpha_3 \cdot totland + \alpha_4 \cdot avercpr$$
$$+ \alpha_5 \cdot HU - 2 + \alpha_6 \cdot streamcode + \alpha_7 \cdot avergrpmem + \alpha_8 \cdot ocaste$$
$$+ \alpha_9 \cdot totedumem + \alpha_{10} \cdot averernmem + \alpha_{11} \cdot avdepend$$
$$+ \alpha_{12} \cdot avhlth + \alpha_{13} \cdot avskills + \alpha_{14} \cdot ncincm + \alpha_{15} \cdot farmlabrdays$$
$$+ \alpha_{16} \cdot nonfarmdays + \alpha_{17} \cdot nregsdays + \alpha_{18} \cdot migrationdays$$
$$+ \alpha_{19} \cdot exfood + \alpha_{20} \cdot exedu + \alpha_{21} \cdot exhlth + \alpha_{22} \cdot exalchl$$
$$+ \alpha_{23} \cdot exwells + \alpha_{24} \cdot extractr + \alpha_{25} \cdot exothrimplement$$
$$+ \alpha_{26} \cdot exlndev$$

Some key observations can be made here based on the results presented in Table 8.5. The significant variables explaining this category of income include: HUN, health status, skill level, and the number of MGNREGA days.

TABLE 8.4 Regression results for equation 3: explaining migration income

Variables	Coefficients		Std. error	t value	Pr(>|t|)
Intercept	158.062800		1233.204	0.130	0.898
Physical capital					
Average number of small ruminants	−5.338779		12.041	−0.440	0.658
Average number of big ruminants	−117.074700		116.620	−1.000	0.316
Natural capital					
Total acres of land	−83.463270		78.165	−1.070	0.286
Average CPR (2005−2006 to 2010−2011; Rs.)	−0.042897		0.052	−0.820	0.410
HUN2	77.764810		496.841	0.160	0.876
Watershed intervention stream level (Upstream is base)					
Midstream	−210.546100		531.566	−0.400	0.692
Downstream	−52.286970		584.659	−0.090	0.929
Social capital					
Average group membership (2005−2006 to 2010−2011)	−144.252500		336.784	−0.430	0.669
Higher caste	**1032.291000**	*	502.559	2.050	0.041
Human capital					
Total number of educated members in household	70.811100		187.067	0.380	0.705
Average number of earning members (2005−2006 to 2010−2011)	195.068100		205.586	0.950	0.343
Average number of dependents (2005−2006 to 2010−2011)	77.404060		201.184	0.380	0.701

(Continued)

TABLE 8.4 Regression results for equation 3: explaining migration income—cont'd

Variables	Coefficients		Std. error	t value	Pr(>|t|)
Average health status (2005–2006 to 2010–2011; lower the value the better the health status)	−118.792000		395.218	−0.300	0.764
Average skill level (2005–2006 to 2010–2011; lower the value the better the skill level)	−233.169000		272.943	−0.850	0.393
Financial capital					
Net crop income (Rs.)	−0.001669		0.005	−0.340	0.731
Farm labor days (2010–2011)	1.321711		2.093	0.630	0.528
Non-farm labor days (2010–2011)	−1.166567		1.513	−0.770	0.441
MGNREGA days (2010–2011)	**15.614830**	*	6.537	2.390	0.017
Migration days (2010–2011)	**143.603200**	***	2.954	48.610	0.000
Expenses					
Expenses on food	−0.041848		0.026	−1.600	0.110
Expenses on education	0.008326		0.014	0.610	0.542
Expenses on health	−0.000328		0.006	−0.060	0.954
Expenses on alcohol	**0.130600**	**	0.040	3.250	0.001
Expenses on wells	0.009327		0.035	0.260	0.792
Expenses on tractors	0.002710		0.050	0.050	0.956
Expenses on other implements	**0.129732**	*	0.065	2.000	0.046
Expenses on land development	0.114870		0.083	1.390	0.166

$N = 427$; R-squared $= 0.8953$; adjusted R-squared $= 0.8882$; F statistic p value $= 0.0000$. Significant codes: 0; ***, 0.001; **, 0.01; *, 0.05; ., 0.1.

TABLE 8.5 Regression results for equation 4: explaining MGNREGA-based income

Variables	Coefficients		Std. error	t value	$Pr(>|t|)$
Intercept	−143.570600		167.291	−0.860	0.391
Physical capital					
Average number of small ruminants	2.231949		1.633	1.370	0.173
Average number of big ruminants	1.096242		15.820	0.070	0.945
Natural capital					
Total acres of land	2.142285		10.604	0.200	0.840
Average CPR (2005−2006 to 2010−2011; Rs.)	−0.005899		0.007	−0.840	0.404
HUN2	**−167.564100**	*	67.399	−2.490	0.013
Watershed intervention stream level (Upstream is base)					
Midstream	34.590450		72.110	0.480	0.632
Downstream	−9.521398		79.312	−0.120	0.905
Social capital					
Average group membership (2005−2006 to 2010−2011)	−27.438170		45.687	−0.600	0.548
Higher caste	10.706990		68.175	0.160	0.875
Human capital					
Total number of educated members in household	18.074130		25.377	0.710	0.477
Average number of earning members (2005−2006 to 2010−2011)	25.568700		27.889	0.920	0.360
Average number of dependents (2005−2006 to 2010−2011)	−26.361950		27.292	−0.970	0.335
Average health status (2005−2006 to 2010−2011; lower the value the better the health status)	**−128.733900**	*	53.614	−2.400	0.017

(Continued)

TABLE 8.5 Regression results for equation 4: explaining MGNREGA-based income—cont'd

Variables	Coefficients		Std. error	t value	Pr(>|t|)
Average skill level (2005–2006 to 2010–2011; lower the value the better the skill level)	**117.688600**	**	37.026	3.180	0.002
Financial capital					
Net crop income (Rs.)	−0.000265		0.001	−0.400	0.687
Farm labor days (2010–2011)	−0.398205		0.284	−1.400	0.161
Nonfarm labor days (2010–2011)	0.138780		0.205	0.680	0.499
MGNREGA days (2010–2011)	**85.771350**	***	0.887	96.720	0.000
Migration days (2010–2011)	0.292853		0.401	0.730	0.465
Expenses					
Expenses on food	0.002472		0.004	0.700	0.486
Expenses on education	0.001402		0.002	0.760	0.449
Expenses on health	0.000805		0.001	1.040	0.301
Expenses on alcohol	−0.003570		0.005	−0.660	0.513
Expenses on wells	0.000317		0.005	0.070	0.947
Expenses on tractors	0.001594		0.007	0.240	0.813
Expenses on other implements	−0.006550		0.009	−0.740	0.457
Expenses on land development	−0.010206		0.011	−0.910	0.363

$N = 427$; R-squared $= 0.9743$; adjusted R-squared $= 0.9726$; F statistic p value $= 0.0000$. Significant codes: 0, ***, 0.001; **, 0.01; *, 0.05; ., 0.1.

MGNREGA-based income e declines as health declines. This implies that less healthy households are less reliant on MGNREGA-based income for survival. This is a problematic finding as it implies that government assistance programs are not reaching the most vulnerable households. However, it is important to note that the MGNREGA scheme is not a welfare program by design.

The results further show that MGNREGA-based income increases with decrease in the skill level. This is logical as MGNREGA offers unskilled labor work and households with skills may have more options. It is also observed that the households in HUN2 are less reliant on the MGNREGA-based income, which could be due to their higher relative prosperity.

Since health could be affected by income, there is a possibility for the presence of endogeneity in the relationship between income and health. To account for this, we use an instrumental variable approach where health is first predicted based on the following equation:

$$(5)\ \widehat{avh\ lth} = \alpha_1 + \alpha_2 \cdot avgsrumnt + + \alpha_3 \cdot avgbrumnt + \alpha_4 \cdot avercpr +$$
$$+ \alpha_5 \cdot totedumem + \alpha_6 \cdot averernmem + \alpha_7 \cdot eavdepend$$
$$+ \alpha_8 \cdot exfood + \alpha_9 \cdot exedu + \alpha_{10} \cdot exhlt + \alpha_{11} \cdot exalchl$$
$$+ \alpha_{12} \cdot exentmnt$$

Next, the MGNREGA-based income is regressed on the predicted average health and other key variables using the following equation:

$$(6)\ nregsioncome = \widehat{\alpha_0} + \alpha_1 \cdot HU - 2 + \alpha_2 \cdot streamcode + \alpha_3 \cdot avergrpmem$$
$$+ \alpha_4 \cdot ocaste + \alpha_5 \cdot \widehat{vhl\ th} + \alpha_6 \cdot avskills + \alpha_7 \cdot ncincm$$
$$+ \alpha_8 \cdot farmlabordays + \alpha_9 \cdot nonfarmlabordays + \alpha_{10} \cdot nregdays$$
$$+ \alpha_{11} \cdot migrationdays + \alpha_{12} \cdot exwells + \alpha_{13} \cdot extractr$$
$$+ \alpha_{14} \cdot exothrimplement + \alpha_{15} \cdot exlnddev$$

It is observed that even when MGNREGA-based income is estimated using an IV method (Table 8.6) to account for endogeneity in health, the results do not change. Further, healthier farmers are found to have higher reliance on MGNREGA-based income (although the regression coefficient is not significant). It was also found that the farmers in HUN1 (Kurnool and Anantapur districts) are more likely to rely on MGNREGA-based income for their survival.

Farmers are found to rely significantly on CPRs for their survival during the drought years; this is especially true in the case of marginal farmers. For this reason, we tested for household characteristics that may suggest higher reliance on CPR income using the following equation:

$$(7)\ cprincome = \alpha_0 + \alpha_1 \cdot avgsrumnt + \alpha_2 \cdot avgbrumnt + \alpha_3 \cdot totland + \alpha_4 \cdot HU - 2$$
$$+ \alpha_5 \cdot streamcode + \alpha_6 \cdot avergrpmem + \alpha_7 \cdot ocaste + \alpha_8 \cdot totedumem$$
$$+ \alpha_9 \cdot averernmem + \alpha_{10} \cdot avdepend + \alpha_{11} \cdot avhlth + \alpha_{12} \cdot avskills$$
$$+ \alpha_{13} \cdot ncincm + \alpha_{14} \cdot farmlabordays + \alpha_{15} \cdot nonfarmlabourdays$$
$$+ \alpha_{16} \cdot nregsdays + \alpha_{17} \cdot migrationdays + \alpha_{18} \cdot exfood + \alpha_{19} \cdot exedu$$
$$+ \alpha_{20} \cdot exhlth + \alpha_{21} \cdot exalchl + \alpha_{22} \cdot exwells + \alpha_{23} \cdot extractr$$
$$+ \alpha_{24} \cdot exothrimplement + \alpha_{26} \cdot exlndev$$

TABLE 8.6 Instrumental variable regression for equations 5 and 6

Variables	Coefficients		Std. Error	t value	$Pr(>|t|)$
Intercept	61.450630		186.722	0.330	0.742
Natural capital					
HUN2	**−186.75150**	**	62.200	−3.000	0.003
Watershed intervention stream level (Upstream is base)					
Midstream	29.813310		73.942	0.400	0.687
Downstream	1.243100		73.787	0.020	0.987
Social capital					
Average group membership (2005−2006 to 2010−2011)	−5.624891		45.365	−0.120	0.901
Higher caste	23.480450		65.066	0.360	0.718
Human capital					
Average health status (2005−2006 to 2010−2011; lower the value the better the health status)	−202.02		137.541	−1.470	0.145
Average skill level (2005−2006 to 2010−2011; lower the value the better the skill level)	**115.499600**	**	38.324	3.010	0.003
Financial capital					
Net crop income (Rs.)	−0.000558		0.000	−1.240	0.216
Farm labor days (2010−2011)	−0.373763		0.245	−1.530	0.127
Nonfarm labor days (2010−2011)	0.219243		0.187	1.170	0.240
MGNREGA days (2010−2011)	**85.603440**	***	0.843	101.580	0.000
Migration days (2010−2011)	0.422538		0.589	0.720	0.473

TABLE 8.6 Instrumental variable regression for equations 5 and 6—cont'd

Variables	Coefficients	Std. Error	t value	Pr(>|t|)
Expenses				
Expenses on wells	0.000636	0.003	0.250	0.799
Expenses on tractors	0.003454	0.005	0.670	0.504
Expenses on other implements	−0.005033	0.007	−0.740	0.460
Expenses on land development	−0.007847	0.009	−0.850	0.394

$N = 427$; R-squared = 0.9736; Chi-squared p value = 0.0000. Significant codes: 0; ***, 0.001; **, 0.01; *, 0.05; ., 0.1.
Instrumented variables: avhlth Instrument variables: hun2 2.streamcode 3.streamcode avergrpmem ocaste avskills ncincm farmlabordays nonfarmdays nregsdays migrationdays exwells extractr exothrimplmnt exlndev avercpr exfood exedu exhlth exalchl exentmnt averernmem avdepend avgsrumnt avgbrumnt

Based on the results presented in Table 8.7, we see the variables found to significantly affect CPR income include: average number of small and big ruminants; stream location; total number of educated members in a household; total number of earning members in a household; health status; number of MGNREGA days; number of migration days; and expenditure on education, wells, and tractors.

We also see that the average number of dependents increases CPR reliance (although regression coefficient is not significant). Furthermore, reliance on CPR decreases with the increase in the total number of educated members in the households—the variable "average earning members" is positively related with CPR income. This simply means that most vulnerable households still rely on CPR and would need more hands to gather food from the CPRs; also, lower health implies lower reliance on CPR income, highlighting the labor-intensive nature of CPR resource harvesting.

The results show that CPR-based income significantly increases for farmers with big and small ruminants (for obvious reasons), while it decreases with land size (although regression coefficient is insignificant) and well ownership (regression coefficient is significant). It is observed that downstream farmers have lower CPR reliance compared with the upstream farmers. On the other hand, in the earlier regression on drought resilience, downstream farmers were found to be more vulnerable compared with the upstream farmers.

Low water availability reduces DS capability directly and CPR levels, and hence, farmers' reliance on CPR income. Also, CPR-based income increases

TABLE 8.7 Regression results for CPR income estimation

Variables	Coefficients		Std. error	t value	Pr(>|t|)
Intercept	3471.551000	**	1172.478	2.960	0.003
Physical capital					
Average number of small ruminants	86.549040	***	10.734	8.060	0.000
Average number of big ruminants	954.010300	***	101.430	9.410	0.000
Natural capital					
Total acres of land	−77.582270		75.026	−1.030	0.302
HUN2	443.233800		477.009	0.930	0.353
Watershed intervention stream level (Upstream is base)					
Midstream	434.552100		510.436	0.850	0.395
Downstream	−2567.570000	***	547.066	−4.690	0.000
Social capital					
Average group membership (2005−2006 to 2010−2011)	−433.483900		322.963	−1.340	0.180
Higher caste	−317.805600		482.758	−0.660	0.511
Human capital					
Total number of educated members in household	−647.251200	***	176.857	−3.660	0.000
Average number of earning members (2005−2006 to 2010−2011)	359.405200	.	196.774	1.830	0.069
Average number of dependents (2005−2006 to 2010−2011)	222.169700		193.042	1.150	0.250
Average health status (2005−2006 to 2010−2011; lower the value the better the health status)	−1188.301000	**	375.177	−3.170	0.002

TABLE 8.7 Regression results for CPR income estimation—cont'd

Variables	Coefficients		Std. error	t value	Pr(>|t|)
Average skill level (2005–2006 to 2010–2011; lower the value the better the skill level)	−156.252400		262.215	−0.600	0.552
Financial capital	0.004346		0.005	0.930	0.350
Net crop income (Rs.)					
Farm labor days (2010–2011)	−2.127691		2.009	−1.060	0.290
Nonfarm labor days (2010–2011)	1.584101		1.452	1.090	0.276
MGNREG days (2010–2011)	**12.246950**	.	6.253	1.960	0.051
Migration days (2010–2011)	**7.291788**	**	2.816	2.590	0.010
Expenses					
Expenses on food	0.003810		0.025	0.150	0.879
Expenses on education	**0.076739**	***	0.013	6.130	0.000
Expenses on health	0.000639		0.006	0.120	0.908
Expenses on alcohol	0.001334		0.039	0.030	0.972
Expenses on wells	**−0.073530**	*	0.034	−2.170	0.030
Expenses on tractors	**0.105131**	*	0.047	2.220	0.027
Expenses on other implements	0.051038		0.062	0.820	0.413
Expenses on land development	0.089247		0.079	1.120	0.261

$N = 427$; R-squared $= 0.5273$; adjusted R-squared $= 0.4966$; F statistic p value $= 0.0000$. Significant codes: 0; ***, 0.001; **, 0.01; *, 0.05; ., 0.1.

for MGNREGA-reliant farmers. Based on earlier results, this simply highlights the fact that healthier households are not only more likely to participate in MGNREG schemes but are also likely to have higher CPR reliance; we see that migrant families also show higher CPR income.

It is observed that CPR-based reliance in HUN2 is higher (although not significant). The above results indicate that HUN2 is a relatively well-off district, although higher reliance on CPR income in this area is simply due to higher forested area.

So far, we have used the conventional regression analysis to infer results based on correlation between variables. One of the drawbacks of such an approach is its inability to establish causality between the independent and dependent variables. Next, we use semiparametric methods to separate the effects of watershed interventions from other variables on enhancing perceived DS.

8.3 TESTING FOR THE IMPACT OF WATERSHED INTERVENTIONS ON DROUGHT RESILIENCE

To evaluate any positive or negative influence of watershed interventions on farmers' livelihoods, we need to compare the DS responses of the households within the regions with watershed interventions (or treated regions) with those of neighboring regions (or control regions) where such programs have yet to manifest. The PSM method has been extensively used in situations where the effect of a treatment on a parameter of interest needs to be assessed by separating the influence of any other factors (for review, see [23] and references therein). Hope [8] explored the impacts of WSD programs in the state of Madhya Pradesh, India, using PSM and found that farmers planting both kharif and rabi crops fared badly in WSD regions compared with farmers in untreated regions. The study also revealed that the main impacts of the WSD projects were mostly short-term and confined to wage labor.

By matching individuals with similar characteristics within the treated category to those in the control category or region, the PSM method evaluates the overall difference in the parameter of interest that could be solely ascribed to a particular treatment. Since it is unlikely that an exact match will be found between any two respondents in the control and treated areas, use of logit and probit methods has been recommended (see [24,25]). Once the propensity scores have been created, various matching algorithm options exist to compare the outcomes. When the sample size is large, the selection of matching algorithms is expected to have little influence on the final outcomes. For a small sample size, results may vary depending on the particular method selected [26]. In this study, we primarily use the "radius caliper" method—after briefly comparing the outcomes using the "nearest neighbor matching" option. Using STATA (which is a statistical software), the effect of treatment (WSD in this study) is derived as the average treatment effect on the treated (ATT).

We derived the impact of watershed interventions in the two HUNs and the results are as follows:

1. For HUN2 the three sampled villages are Ondutla, Thaticherla, and Penchikalapadu, and the control village is Alsandalapalli. There were 204 observations from the treated villages within the HUN2 and 46 from the control village.

2. For the HUN1 region, the treated villages are S. Rangapuram, Utakallu, and Basinepalle, and the control village is Karidikonda. The number of household observations for the treated villages was 227 and for the control village was 45.

To assess DS benefits from WSD programs, the respondents were questioned about their DS capabilities with respect to financial as well as human capitals; that is, we asked them how many years of repeated droughts they could survive given their current level of financial (or human) capital.

To assess the impact of human capital, we elicited separate DS responses in terms of their average health and the average level of skills. For DS with respect to financial capital, we surveyed whether the households thought they had enough financial wealth to survive one or more consecutive drought years. The same question also was repeated with respect to human capital (health and the level of skills). For instance, a household may project 3 years of consecutive DS capability based on its financial wealth, but its human capital levels may lead to a lower DS projection—it is possible that a household may perceive a higher DS for human capital than for financial capital or vice versa.

Table 8.8 depicts the outcomes of the propensity matching exercise conducted for the two HUNs separately. First, the nearest neighbor matching option was used in PSM, which identifies and matches a farmer from the treated group with the closest neighbor (or propensity score) from the control group. The propensity score was calculated as the probability of being in the treated group based on the logistic regression of the dependent variable (which is 1 for treated and 0 for untreated) on several independent variables that

TABLE 8.8 ATT generated using the nearest neighbor matching option ($n = 1$)

	Treated	Control	Difference	Std. error	t statistic
HUN2 (Prakasam)					
DSWFC	2.27	1.93	0.34	0.2226	1.53
DSWHLTHC	2.41	2.08	0.3333	0.28	1.19
DSWSKLLC	1.04	1.09	−0.0533	0.27	−0.20
HUN1 (Anantapur and Kurnool)					
DSWFC	2.32	2.09	0.22	0.25	0.88
DSWHLTHC	2.30	2.36	−0.063	0.27	−0.23
DSWSKLLC	1.34	1.402	−0.058	0.346	−0.17

DSWFC, number of years of DS with financial capital; DSWHLTHC, number of years of DS with health capital; and DSWSKLLC, number of years of DS with skill capital.

included household and other socioeconomic characteristics. More specifically, the independent variables were as follows: *farmlaborincome, nonfarmincome, nregsincome, ncincm, avercpr, agergrpmem, exedu, exfood, avhlth, avskills, totland, totedumem, avgbrum, avgsrum,* and *ocaste*.

For assessing the resilience related to financial capital, the variable *dswfc* was considered as the dependent variable, while for resilience related to human capital, *dswhlthc* and *dswskllsc* were chosen as the dependent variables. As is evident from Table 8.8, for HUN2, measures of resilience such as financial capital and health capital show an improvement in the watershed-treated regions. It was found that skill-related resilience responses were marginally lower in the watershed regions. However, the *t* statistics are not significant for any of this difference when using the nearest neighbor matching method. Furthermore, resilience related to financial capital is higher by a factor of 0.34 in the WSD regions. This means that farmers in the WSD regions have seen an increase in their DS capabilities by almost four months (one season) compared with similar farmers in the control villages. For HUN1, however, the change has been insignificant and even negative (although statistically insignificant) as is evident from the health and skill-related resilience outcomes.

One challenge we faced was with the survey of a smaller number of households captured under the control villages. To make up for the paucity of data, we also used the radius caliper option in the PSM method—we picked the radius caliper option of 0.1. The outcomes obtained are depicted in Table 8.9.

Differences in financial and human capital resilience were found to increase in the watershed regions of HUN2 as compared with the nearest neighbor matching and they are significant as well. The farmers seemed to have gained an additional six months' worth of resilience in the treated regions. However, we find that HUN1 shows no improvement in stated resilience measures even while using the radius caliper method. Table 8.9 also depicts the *t* statistics values and the revised standard errors after performing bootstrapping in STATA using 500 replications.

We also checked for other important indicators of resilience such as changes in crop income and farm labor income to see whether they have been influenced by WSD interventions. In HUN2, as well as in HUN1, the net crop income is lower in WSD districts (as compared with the control regions in the same districts). The variable *farmlaborincome* has also undergone a negative change in the treated areas in the two districts, although the difference is not statistically significant.

When we combine the two regions and analyze the entire region for any changes in these resilience indicators, the results still hold in terms of health and financial resilience improvement in the treated areas: the net crop income seems to have dropped by roughly Rs.12,000 in the treated areas. Tables 8.10—8.12 depict the outcomes of the Rosenbaum Test performed to

TABLE 8.9 PSM generated using the radius caliper matching option ($r = 0.1$)

	Treated	Control	Difference	Std. error	t statistic	Std. error bootstrap	t Statistic bootstrap
HUN1 (Prakasam)							
DSWFC	2.27	1.79	0.479	0.165	**2.9**	0.1819	**2.64**
DSWHLTHC	2.41	1.9	0.5122	0.21	**2.41**	0.252	**2.03**
DSWSKLLC	1.04	0.855	0.185	0.20	0.89	0.22	0.85
ncincm	−25982	−7929	−18052	10561	**−1.71**	8688	**−2.08**
farmlaborincome	8843	9488	−645	2660	−0.24	2824	−0.23
HUN2 (Anantapur and Kurnool)							
DSWFC	2.32	2.19	0.124	0.168	0.74	0.165	0.75
DSWHLTHC	2.30	2.26	0.35	0.172	0.21	0.16	0.21
DSWSKLLC	1.34	1.61	−0.266	0.225	−1.18	0.225	−1.19
ncincm	−39930	−22680	−17250	8330	**−2.07**	8615	**−2**
farmlaborincome	7971	8995	−1023	1938	−0.53	1783	−0.57
HUN1 and HUN2 combined							
DSWFC	2.311	1.954	0.3564	0.095	**3.75**	0.094	**3.78**
DSWHLTHC	2.32	2.11	0.208	0.104	**1.99**	0.100	**2.07**
DSWSKLLC	1.20	1.14	0.052	0.123	0.43	0.098	0.53
ncincm	−31920	−19550	−12369	5919	**−2.09**	4634	**−2.67**
farmlaborincome	8669	8819	−149	1252	−0.12	1237	−0.12

The last two columns present standard errors and t statistic values after performing bootstrapping with 500 replications in STATA.

TABLE 8.10 Rosenbaum's Rbounds estimation for *DSWFC* for HUN1 and HUN2 combined using the radius caliper matching method

Gamma	Sig+	Sig−	That+	That−	CI+	CI−
1	0	0	0.499	0.499109	0.3914	0.5394
1.5	1.1e−12	0	0.134	0.560015	0.090092	0.58171
2	6.0e−07	0	0.0849	0.5857	0.0606	0.6132
2.5	0.000573	0	0.0633	0.6091	0.033	0.66
3	0.02492	0	0.0424	0.63552	3.4e−07	0.9178

TABLE 8.11 Rosenbaum's Rbounds estimation for *DSWHLTH* for HUN1 and HUN2 combined using radius caliper matching method

Gamma	Sig+	Sig−	That+	That−	CI+	CI−
1	0.000096	0.000096	0.3714	0.3714	0.353	0.383
1.5	0.38576	1.2e−13	0.3113	0.3923	−0.0957	0.407
2	0.9835	0	−0.1	0.413	−0.117	0.469
2.5	0.999	0	−0.1160	0.4431	−0.133	0.852
3	1	0	−0.127	0.84	−0.152	0.865

TABLE 8.12 Rosenbaum's Rbounds estimation for *ncincm* for HUN1 and HUN2 combined using radius caliper matching method

Gamma	Sig+	Sig−	That+	That−	CI+	CI−
1	0.0030	0.0030	−5853.4	−5853	−10669	−1631
1.1	0.000173	0.027	−7822	−3973	−1273	85
1.2	7.0e−06	0.1207	−9685	−2426	−14611	1565

test the null hypothesis that treated areas are no different from the untreated areas as the level of hidden bias from unobserved variables increases (gamma measures the log odds ratio resulting from bias through unobserved variables) [27]. It is evident that the resilience related to financial capital is robust even when the log odds ratio increases (or decreases) by a factor of 3.

On the other hand, for resilience related to human capital, the robustness is not high at all. The same holds true for the sensitivity of the differences related to net crop income in the treated and untreated areas. Furthermore, the negative change in the treated areas is not robust from the effect of excluded variables bias in our model.

8.4 CONCLUSION

We performed parametric and semiparametric analyses of farmers' perceived DS responses to assess the role of the five types of capitals as well as the households' characteristics in making farmers resilient to repeated droughts. We tested for drought resilience with and without WSD intervention as well as identified variables that influenced farmers' nonagricultural incomes such as MGNREGA employment, CPR reliance, and migration income. The results show that farmers with a significant source of nonagricultural income could either come from vulnerable or resilient categories.

The findings from the above analysis present a clear picture of household aspects that either make them more resilient or more vulnerable. First, the role of human capital, such as health and education, in influencing drought resilience becomes crucial. Healthy individuals are not only observed to show higher participation in the MGNREG scheme but also have higher CPR income; yet, healthy individuals did not give higher DS responses.

We found that having a higher number of educated members in the household also helps with DS. However, households that spend more on education perceived a marginally lower DS. This highlights the trade-offs between accumulating higher human capital (which could provide long-term resilience) at the cost of reducing current or short-term resilience.

One interesting observation is the increase in inequality, both among social groups as well as geographically, with WSD interventions. It was found that downstream users of groundwater seemed to have become worse off with WSD interventions with respect to drought resilience, while social status, characterized by caste groups, also led to differing drought resilience perceptions. This could imply either historical inequality or dominance over natural resources by the higher caste cohorts.

It was further seen that ownership of physical assets such as equipment also helps with drought resilience; better educated and more landed households have higher drought resilience with WSD interventions. From a policy perspective, this points to the need for complementing WSD schemes with interventions that redistribute the gains from groundwater augmentation among the more disadvantaged cohorts of the society.

It turns out that farmers relying more on MGNREGA-based income also have higher reliance on CPRs as well as a higher share of food expenditure in their total income. Further, healthier farmers are found to be more reliant on MGNREGA programs.

When we tested for the impact of WSD programs on enhancing the DS of farmers in the treated villages vis-à-vis the same in the control villages, it turned out that farmers in HUN2 had significantly benefited from WSD intervention programs, and their DS capabilities were higher by almost six months for some forms of resilience (such as financial and human capital related resilience). This, however, is not the case for farmers in HUN1 who do not show any improvement, which could be because of the differences in the hydrogeological and biophysical aspects between these two HUNs. Further investigation is required to delve into the causes that make HUN1 villages unresponsive to WSD treatment. Further, a decline in the net crop income for the entire region with WSD intervention also does not reflect positively on the WSD programs.

REFERENCES

[1] IPCC. (2012). Managing the Risks of Extreme Events and Disasters to Advance Climate Change Adaptation, Special Report of the Intergovernmental Panel on Climate Change, http://www.ipcc-wg2.gov/SREX/images/uploads/SREX-All_FINAL.pdf

[2] Dai A. Characteristics and Trends in Various Forms of the Palmer Drought Severity Index During 1990–2008. J Geophysical Res 2011;vol. 116.:D12115.1–26. http://dx.doi.org/10.1029/2010JD015541.

[3] Romn J. Desertification: The Next Dust Bowl. Nature 2011;478:450–1. http://dx.doi.org/10.1038/478450a.

[4] Brenkert AL, Malone EL. Modeling Vulnerability and Resilience to Climate Change: A Case Study of India and Indian States. Climatic Change 2005;71(1–2):57–102.

[5] Das A. Farmers' Suicide in India: Implications for Public Mental Health. Int J Soc Psychiatry 2011;57:21.

[6] NYT. On India's Farms, A Plague of Suicide; 2006. URL, http://www.nytimes.com/2006/09/19/world/asia/19india.html.

[7] Mishra S. Risks, Farmers' Suicides and Agrarian Crisis in India: Is there a Way Out? Mumbai: Indira Gandhi Institute of Development Research; 2007.

[8] Hope RA. Evaluating Social Impacts of Watershed Development in India. World Dev 2007;vol. 35(8):1436–49.

[9] Kerr JG, Pangare G, Pangare VL. Watershed Development Projects in India—An Evaluation Research Report 127. Washington DC: International Food Policy Research Institute; 2002.

[10] Joshi PK, Jha AK, Wani SP, Joshi L, Shiyani RL. Meta-analysis to Assess Impact of Watershed Program and People's Participation, Comprehensive Assessment Report 8. Colombo: IWIMI; 2005.

[11] Reddy VR, Reddy MG, Galab S, Soussan J, Springate-Baginski O. Participatory Watershed Development in India: Can it Sustain Rural Livelihood? Dev Change 2004;vol. 35(No. 2):297–326.

[12] Pretty J. Social Capital and Collective Co-management of Resources. Science12 2003;Vol. 302(No. 5652):1912–4.

[13] Yamauchi F, Yohannes Y, Quisumbing A. Natural Disasters, Self-Insurance, and Human Capital Investment: Evidence from Bangladesh, Ethiopia and Malawi, IFPRI Discussion Paper, 00881; 2009. url, http://www.ifpri.org/sites/default/files/publications/ifpridp00881.pdf.

[14] Beck T, Naismith C. Building on Poor People's Capacities: The Case of Common Property Resources in India and West Africa. World Dev 2001;29(1):119−33.

[15] Diamond JM. Collapse: How Societies Choose to Fail or Succeed. New York: Viking Books; 2005.

[16] Government of India. MGNREGA SAMEEKSHA: An Anthology of Research Studies on the Mahatma Gandhi National Rural Employment Guarantee Act, 2005. New Delhi: Government of India, Ministry of Rural Development; 2012.

[17] Ranjan R, Athalye S. Drought Resilience in Agriculture: The Role of Technological Options, Land Use Dynamics and Risk Perception,. Nat Res Model 2009;vol. 22(No. 3): 437−62.

[18] Ranjan R. The Role of Credit in Enhancing Drought Resilience in Agriculture. J Environ Econ Policy 2013a;vol. 2(3):303−27.

[19] Ranjan R. Combining Social Capital and Technology for Drought Resilience in Agriculture, Natural Resource Modeling, published online; Aug 9, 2013. url, http://onlinelibrary.wiley.com/doi/10.1111/nrm.12021/pdf.

[20] Ranjan R. Technology Adoption for Long-term Drought Resilience. Water Res Plann Manag 2014;vol. 140(No. 3). March 1, 2014.

[21] Ranjan R. Mathematical Modeling of Drought Resilience in Agriculture. Nat Res Model 2012;vol. 26(2):237−58.

[22] Government of India. Guidelines for Watershed Development. New Delhi: Government of India, Ministry of Rural Development; 1994.

[23] Heinrich C, Maffioli A, Vazqez G. A Primer for Applying PropensityScore Matching, Impact-Evaluation Guidelines, Technical Notes No. IDB-TN-161; 2010. url, http://idbdocs.iadb.org/WSDocs/getdocument.aspx?docnum=35320229.

[24] Rosenbaum PR, Rubin DB. The Central Role of Propensity Score in Observational Studies for Casual Effects. Biometrika 1983;vol. 70(1):41−55.

[25] Khandker SR, Koolwal GB, Samad HA. Handbook on Impact Evaluation. Washington DC: The World Bank; 2010.

[26] Caliendo M, Kopeinig S. Some Practical Guidance for the Implementation of Propensity Score Matching. J Econ Surveys 2008;vol. 22(1):31−72.

[27] Rosenbaum PR. Observational Studies. 2nd ed. New York: Springer Series in Statistics; 2002.

Chapter 9

Modeling Livelihood Indicators and Household Resilience using Bayesian Networks

Wendy Merritt *, Brendan Patch *, V. Ratna Reddy [§], Sanjit Kumar Rout [§] and Geoffrey J. Syme [¶]

Fenner School of Environment and Society, The Australian National University, Canberra, ACT, Australia, [§]Livelihoods and Natural Resource Management Institute, Hyderabad, India, [¶]Edith Cowan University, Perth, Australia

Chapter Outline

9.1 INTRODUCTION

Effective management of natural resources requires an understanding of the aims, interactions, and implications of different policies or activities on complex (and interlinked) biophysical, economic, and social processes. Predicting outcomes of policy choices can be highly complex where trade-offs between sectors (or objectives) or where impacts across time and/or space need to be assessed. Integrated assessment (IA) methodologies have therefore

Integrated Assessment of Scale Impacts of Watershed Intervention
http://dx.doi.org/10.1016/B978-0-12-800067-0.00009-8
287

increasingly received attention as a useful approach for developing and understanding complex environmental and societal issues as well as for supporting decision-making processes. IA is a problem-focused activity in which research is linked to policy in an adaptive and iterative process, and emphasis is placed on developing frameworks and processes that encourage stakeholder involvement and consideration of complex interactions and interdependencies between the human and natural environment [1].

A component of IA widely recognized as necessary to understanding and evaluating the nature of trade-offs is the development of a model (often called an IA model). Models are simplified representations of real systems, which can be used for a number of roles ranging from explicitly documenting system understanding, knowledge gaps, and key assumptions to facilitating social learning, communication, and stakeholder participation to the assessment of trade-offs of alternate management options.

The Bayesian network (BN) modeling approach is used in this chapter to model relationships between household categories (geographic, economic, and social) and the stocks of the livelihood capitals (e.g., social capital) and between stocks and household capacity to survive consecutive drought years. Numerous modeling approaches have been used in IA studies. Five modeling approaches to integrating knowledge that have been widely used in the scientific literature, reviewed by Kelly et al. [2], accommodate multiple issues, values, scales, and uncertainty considerations as well as facilitate stakeholder engagement including system dynamics (SD), BNs, coupled component models, agent-based models (ABM), and knowledge-based models (also referred to as expert systems) [2]. BNs can be used for prediction, system understanding, and social learning purposes as well as to support decision making under uncertainty. For prediction, BNs offer advantages over coupled component models when qualitative information will be explicitly used in model specification and parameterization or where decision making under uncertainty occurs. Furthermore, BNs are appropriate for system understanding or social learning where it is sufficient or desired to consider the aggregated effects of individuals on a system and where the representation of dynamic processes or feedback loops is not considered critical [2]. On the other hand, the ABM and SD modeling approaches are more appropriate if the representation of individual behavior or dynamic process and feedback are a necessary model feature.

The BNs outlined in this chapter are the central component of the integrated model used in Chapter 12 to assess the impacts of biophysical and social policy scenarios on water resources, capital strength, and resilience. This chapter is organized as follows: Section 9.2 provides an overview of the theory behind BNs and their application to environmental and societal issues similar to those associated with Watershed Development (WSD) as well as previous applications using the sustainable livelihoods framework. Section 9.3 briefly describes the capital submodels (Section 9.3.2) and the integration of these submodels to model household resilience (Section 9.3.3). A more in-depth discussion of the social capital BN is used in Section 9.4 to demonstrate how the submodels can be analyzed and used to explore the influence of WSD on capital stocks in relation to other geographical and social variables.

9.2 BNs

This section is an overview of the theory behind BNs, their application, and comparison to previous applications using the sustainable livelihoods framework.

9.2.1 Introduction

The preceding chapters have explored in detail the impact of scale on the implementation of WSD. When dealing with such complex problems, it can be useful to extract important information and display it in a simplified model so that it can be used to clearly display an understanding of how the system operates, therefore providing an avenue for discussion on the system where views and opinions can be shared and evaluated. Given their graphical nature, BNs are an ideal choice of model for improving system understanding through the elicitation and combination of stakeholder knowledge [3] or analysis of qualitative and/or quantitative datasets [4].

This section introduces the theoretical background of BNs, the typical model development process in environmental science applications, and how BN models can be used and evaluated. For a thorough discussion of BN modeling and guidelines to support the development and evaluation of environmental systems, readers are referred to Chen and Pollino [5].

Mathematically, BNs are a probabilistic modeling approach combining a causal graphical structure or network (often referred to as an influence diagram) with conditional probability density functions, which specify the strength and nature of the relationships between variables (or nodes) in the network. The first step in developing a BN is to define the purpose and context of the model, including the definition of objectives for the model and the end users. After this, a conceptual model representing system understanding is often developed to inform the structure of the BN [5]. Once the structure of the model is defined, model variables (or nodes) are assigned "states" and parameterized using qualitative or quantitative information. Once populated, this model can be evaluated using sensitivity analysis and cross-validation, as well as by testing model scenarios and behavior with experts and other stakeholders. Thus, development of a BN model is typically an iterative procedure that lends itself to collaboration and engagement between model developers, "experts," and community members at all stages of model development.

BNs are structured using a directed acyclic graph—a graph in which variables (or nodes) are connected by unidirectional arrows (arcs). In its basic form, this structure does not allow representation of feedback loops.[1] Given

---

1. Some software packages allow representation of dynamics through the representation of time slices in separate networks. These are referred to as dynamic Bayesian networks.

two variables, X and Y, there is said to be a direct connection between the two when the state of X directly influences the probability density over the states of Y or vice versa. In the graph structure, this is represented by an arc starting in one of the variables (called the "parent"), and ending in the other node (called the "child"). The flow of influence between variables is determined using Bayes' Theorem, which is stated as follows:

$$P(X|Y) = \frac{P(Y|X)P(X)}{P(Y)},$$

$P(X)$ is the probability that the event X occurs (e.g., probability that a household has a functioning well) and $P(Y)$ is the probability that the event Y occurs (e.g., probability that a household owns land). These are called the "marginal probabilities of X and Y." If it is known that Y has occurred and this influences the probability that X will occur, then the new probability that X will occur becomes $P(X|Y)$ (e.g., the probability that a household has a functioning well given that the household owns land). This is called the "conditional probability of X given Y." Marginal probabilities are used to describe variables that have no parents, while conditional probabilities are defined for each child node. For example, when X and Y are not directly connected, they may still influence each other through some other variable, or set of variables, Z. Figure 9.1 displays the four ways that Bayes' Theorem may be used to create indirect connections between an input variable X and an output variable Y [6].

There exist algorithms that can be used to determine graph structures from data [7]. Such algorithms can be useful for modeling poorly understood systems or when there is a large amount of available data. However, these algorithms do not typically perform well in natural resource management, and environmental science applications, where processes are complex and highly stochastic, and where there is limited data from which to learn the model structure [5]. Consequently, the structure of the network is typically defined using expert opinion in these fields of study.

BNs are often structured in a hierarchical manner; in Section 9.3, we define a general hierarchy of household class variables, explanatory variables (where applicable), stocks of capital assets, the extent to which the stocks of assets

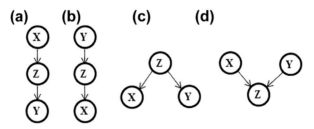

FIGURE 9.1   The four possible indirect connections from X to Y via Z: (a) an indirect causal effect, (b) an indirect evidential effect, (c) a common cause, and (d) a common effect (*Source*: [6] p.70).

support the survival of consecutive drought years, strength of each capital, and resilience to consecutive droughts.

The graph structure used in BNs is very useful. In the absence of the graph structure, the joint probability density function for n variables of interest is specified as follows:

$$P(X_1, X_2, ..., X_n).$$

This describes the probability of the different combinations of states across all the variables. Assuming that our system could be modeled using n variables, each with at least two possible states, specification of the joint probability density function would require at least 2^n parameters (four parameters in the above example and 32 parameters if there were five variables each with two states). However, once the graph structure has been specified, the assumption of conditional independence allows the entire joint probability distribution over all the variables to be quantified using only parameters, which specify the relationship of each variable with its immediate neighbors.

For example, the joint distribution is normally factorized as follows:

$$P(X, Y, Z) = P(X) \, P(Z|X) \, P(Y|X, Z).$$

Using the graph structure in Figure 9.1a, however, it is possible to factorize the joint distribution more simply as follows:

$$P(X, Y, Z) = P(X) \, P(Z|X) \, P(Y|Z).$$

This factorization has fewer parameters because $P(Y|Z)$ is a simpler object compared to $P(Y|X, Z)$.

Each child variable (i.e., Y) in a BN has an associated conditional probability table (CPT) that describes the probability of being in a state, given a combination of values of all parent states (say X_1, ..., X_n), which is written as $P(Y|X_1, ..., X_n)$.

BNs can be used to perform statistical inference in a number of ways. For example, analysis on the model can be performed evidentially (Figure 9.1b) by setting the value of the child node and analyzing how this influences the distribution over the causes (i.e., the states of the parent node). Evidential (or diagnostic) reasoning is useful when the study is attempting to identify the most likely cause of a particular state of the child node. Such reasoning is used commonly in medical diagnostics [8]. Alternatively, causal (or predictive) reasoning (Figure 9.1a) can be performed by setting the state of the parent node(s) and studying how this influences the distribution over the child node. Causal reasoning is useful when the effect of alternative states of the parent nodes on a system is under study, for example, the effect of WSD on the livelihoods of households. It is possible to use evidential and causal reasoning together to investigate intercausal relationships.

Figure 9.2 shows the CPT underlying the functional well node in a network consisting only of the nodes *Watershed development* (as parent) and

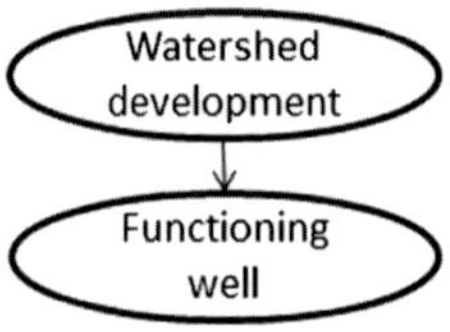

Watershed development → Functioning well	State of WSD Applicability	Probability of Functioning Well Ownership Conditional on State of WSD Applicability	
		No	Yes
	No	0.85	0.15
	Yes	0.61	0.39

FIGURE 9.2   A simple two node network with binary states.

Functioning well (as child). Each of these nodes may only take on the states "yes" or "no." In Figure 9.2, an entry corresponds to the probability of the state of the column variable conditional on the row variable being set so that each row now defines a conditional probability density function in one variable.

Conditional probability tables can be parameterized using observed data, probabilistic or empirical equations, outputs from model simulations, or expert elicitation. The capital strength and resilience BNs described in Section 9.3 have mostly been parameterized using the resilience survey dataset described in Chapter 7 with expert elicitation used to define some variables. For data-based BNs, algorithms that can "learn" these tables are readily available within software packages such as Netica, Hugin Expert, Analytica, and BUGS. When observed data used, parameters are commonly estimated using either maximum likelihood estimates (MLE) or maximum a posteriori estimates (MAP).

In the MLE approach, model parameters are chosen to maximize the probability of the model replicating the data. On the other hand, MAP estimates place a belief on the values of parameters before any data are presented, and then data are used to change this belief. For example, if a coin is flipped and it came up heads on both flips the MLE estimate of the probability of a head on the next flip would be one. It could, however, be argued that the coin is likely to be fair, perhaps due to a belief that most coins are fair, and 10 artificial flips (five heads and five tails) could be assigned to this belief so that the MAP estimate of the probability of heads on the next flip is closer to 7/12. Thus, MAP estimates are typically more reliable than MLE estimates, especially in low data environments [6, p. 751]. In this study, MAP estimates have been used.[2]

Once the parameters have been specified, the behavior and performance of the model can be evaluated quantitatively (sensitivity analyses, cross-validation, and accuracy assessments) or qualitatively (evaluation of model

---

2. States in the variables in BNs can be categorical, Boolean, discrete, or continuous. However, a common requirement of the software packages used to implement BN models is that they require variables to be discrete to "learn" the CPT of a variable. This requirement necessitates a step between graph specification and probability density specification, where all variables are discretized. This study has attempted to ensure that there is sufficient data for accurate parameterization in all states and, wherever possible, align breakpoints with thresholds that have some interpretation in terms of the model objectives.

outputs by experts). Sensitivity analysis can be used to ascertain variables that have the most influence on the outcome variable, evaluate model plausibility, and identify sensitive variables that need further quantification. In the analyses presented in Section 9.4, results from sensitivity analyses refer to the mutual information (MI) statistic, which indicates the variance in the examined variable that is explained by changes in the input nodes.[3]

Comparison of model predictions with observations not used in the estimation procedure is a common method of testing model validity. To do this, the dataset is partitioned into a "training set" and a "testing set." The model parameters can be learnt using the training set and then the outputs using these parameters are compared with the testing set. It is possible to repeat this procedure multiple times with different partitions of the dataset to gain an overall view of model performance (see [6, p. 706] for more details). Further, review by experts should involve a structured assessment of the purpose and structure of the model as well as the model relationships. In this project, sensitivity analyses were used in conjunction with expert review of model behavior and comparison with results from other analytical techniques.

9.2.2 Integration of Societal and Environmental Aspects of Water Management using BNs

Over the last 20 years, the utility of BNs in modeling environmental problems has become widely recognized, particularly in areas where there is high uncertainty, a need to integrate across issues or disciplines, a desire to develop models in a participatory process, or where qualitative (or a mix of qualitative and quantitative) information is to be used in the model parameterization process [4,5,9−12].

Calder et al. [13] developed pilot BNs to investigate the biophysical and societal impacts of interventions undertaken as part of the Jala Samvardhane Yojana Sangha and Sujala WSD projects in India. The aims of the pilot study were to develop common understanding between stakeholders on the causal linkages between factors that are critical to the success of the projects and also to identify the potential for BNs to improve tactical decision making over space and time. To achieve the latter, the nodes in the BN represent disaggregated impacts of project interventions, which can be populated with data routinely collected from WSD projects. Next, efforts were made to keep the models simple enough such that a "numerate graduate" could use them with minimal training. The authors concluded that the first indications were that BNs were useful in analyzing the data collected as part of the WSD projects and, with further testing and development on real-world situations, could help organizations implement and evaluate WSD projects to make better use of collected information.

---

3. MI is a relative measure. A value of 0 indicates no influence while a value of 1 indicates a perfect causal relationship.

BNs have been used on a couple of occasions to implement the sustainable livelihoods framework to develop an understanding of the impact of different economic, social, and technical interventions on people's livelihoods. In an early operationalization of the sustainable livelihoods framework within an analytical model, Newton et al. [14] developed a BN to model the impacts of commercializing, non-timber forest products on rural livelihoods. The analytical framework constitutes capital assets before commercialization, which are influenced by a range of environmental, political, and socioeconomic factors, and the change in the availability of capital assets as a result of commercialization, which subsequently impacts livelihoods. About 66 factors that influence the success of commercialization of non-timber forest products were identified, and experts assigned a score to each factor for each of the 19 case studies of commercialization from Mexico and Bolivia. While only a limited number of case studies were used to develop model relationships, the authors argued that the tool allowed diagnosis of the causes of success and failure of the case studies and could potentially be used to explore the potential impacts of policy options and other interventions on livelihoods.

Of most relevance to the mesoscale project was the study by Kemp—Benedict et al. [15], which implemented the sustainable livelihoods framework within a BN model to explore the links between water-related intervention and livelihood outcomes. The authors selected the BN approach to explicitly account for fundamental uncertainties and challenges in the analysis of the impacts of interventions.

The model structure is shown in Figure 9.3. In this figure, all physical and natural assets other than water infrastructure and natural water availability are encompassed by physical and natural assets, respectively. All asset types can influence water productivity and are directly linked to livelihood outcomes. Livelihood outcomes are also affected by production, which is determined by water availability and water productivity. The assets available to households and the strategies they employ affect the level of outcomes achieved from changes in production. The authors explicitly considered the influence of institutions within the framework, and noted that the way in which they function can be

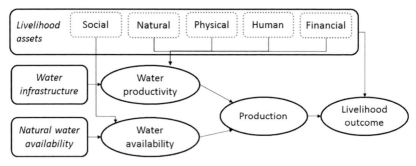

FIGURE 9.3   Structure of the Kemp—Benedict et al. [15] sustainable livelihoods BN.

difficult to represent. Hence, Kemp−Benedict et al. [15] considered it more important to represent the effect that institutions have on livelihoods through changes in assets and capabilities than to explicitly include institutional variables. For example, in an application of the framework to a dataset from the Si Saket province in Northeast Thailand, institutions as a variable was replaced by variables such as wealth and education level of the household head.

The authors used the model to identify which type of water infrastructure helped poor households most to achieve positive livelihood outcomes (e.g., through increase in income/profit from aquaculture, livestock, or crop production). The model indicated quite weak response to interventions. The authors concluded that this was a feature of BNs and stated that "fuzzy relationships between probabilistic variables lead to relatively weak responses." They argued that this is an advantage of the approach as it reflects the reality that communities commonly respond weakly to interventions in the short term and, more commonly, interventions may affect the relative chances of achieving a positive outcome in any given year which, over time, may lead to noticeably improved livelihoods. The study highlighted that the range of livelihood strategies used by households to protect against shocks like drought can make designing and evaluating the impact of individual policy interventions difficult [15].

9.3 CAPITAL STRENGTH AND RESILIENCE BNs

This section is an overview of the development process and structure of the BN submodels.

9.3.1 Model Structure and Development Process

BN submodels have been developed for each of the five capitals based on the resilience survey data detailed in Chapter 7. These submodels are linked to a measure of resilience, which is defined in this study as the capacity of a household to survive consecutive drought years (Figure 9.4a). Stocks of each capital during 2010−2011 are related to household class variables and, if applicable, additional explanatory variables. The level of each capital stock is linked to the variables describing the number of years for which those levels of stock would support household survival of droughts. These "*Drought support*" variables are linked to a final outcome variable (e.g., *Natural capital*), which describes the strength of each household's capital (Figure 9.4b). Figure 9.5 summarizes the technical steps involved in the model development process.

9.3.2 Capital Strength Submodels

The capital strength submodels have been developed to demonstrate the impacts of WSD on livelihood capitals, both within and between the study villages. Household class variables reflect treatment (*Watershed development*),

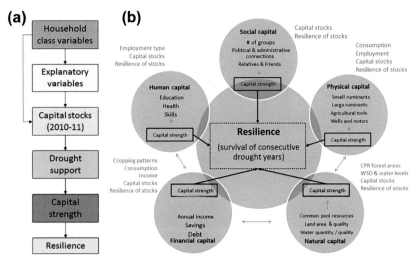

FIGURE 9.4 Capital strength and resilience BNs: (a) submodel BNs for each type of capital are linked to resilience or the capacity of households to survive consecutive drought years. (b) Structure of the BN models.

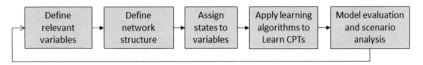

FIGURE 9.5 Model development process.

geographic (*Hydrological unit* and *Location*), and socioeconomic categories (*Economic category* [farm size] and *Social category* [caste]). Table 9.1 provides a description of the variables used in each BN).

The capital stocks represented in the financial submodel are the quantity of savings, debt, and income. The model also represents the types of crop grown and their respective sale prices and quantities as well as the income streams for survey households (Figure 9.6). The parent variables for *Savings* are *Income* and *Consumption*; *Debt* is determined by *Consumption*, *Income*, *Investment*, and *Years of crop loss*. *Income* is the child node (variable) of *Wage (Average)*, *Household earners*, and *Crop revenue*.

The human capital stocks represented in Figure 9.7 are skills, education, and health. The *Skills* variable is the child node of the *Economic category*, *Social category*, *Hydrological unit*, and *Watershed development* variables. *Drought support (Education)* is related to both *Literate household members* (which is determined by *Economic category*, *Watershed development*, and *Hydrological unit*) and *Education expenditure*, which is linked to the *Economic category* and *Location*. *Drought support (health)* is linked to the *Health* and *Health expenditure* variables. These variables are, in turn, linked to the variables *Economic category*, *Watershed development*, and *Hydrological unit*. *Health* is also linked to *Social category* and is a parent of *Health expenditure*.

TABLE 9.1 Variables in the BNs

Name	Description	States
Household class variables		
Watershed development	Whether or not the household's village has received WSD	No, yes
Hydrological unit	The hydrological unit within which the household's village is located	Anantapur/Kurnool, Prakasam
Location	The location within the hydrological unit within which the household's village is located	Downstream, midstream, upstream
Economic category	The hydrological unit within which the household's village is located	Landless, small marginal, medium large
Social category	The hydrological unit within which the household's village is located	Scheduled Caste, Scheduled Tribe, Backward Caste, Other Caste
Climate	Characterization of rainfall years (see Chapter 4)	Deficit, normal, excess
Resilience and capital strength variables		
Drought support (<indicator>)	How many consecutive drought years the household's stocks for a particular indicator (e.g., income) would support household capacity to survive. These variables are the blue nodes in Figures 9.6 to 9.10.	Not applicable,[a] no drought,[a] one drought, two droughts, three droughts
Resilience	How many consecutive droughts a household could survive given the strength of five capitals	No drought, one drought, two droughts, three droughts
Capital strength (e.g., Financial capital)	These variables are populated using the rule that capital is strong if any of the capital indicator stocks could last two or more consecutive drought years	Weak, strong

(Continued)

TABLE 9.1 Variables in the BNs—cont'd

Name	Description	States
Variables common to two or more capital BN sub-models		
Land area (acres)	Total area of land owned by a household (in acres)	0 acres, 0–5 acres, 5–10 acres, >10 acres
Rainfed area (acres)	Area of rainfed land owned by a household (in acres)	0 acres, 0–5 acres, 5–10 acres, >10 acres
Investment (land)	Whether or not households invest in land improvement	Not applicable, 0, 0–2000, ≥2000
Functioning well	Whether or not the household has a functioning well	No, yes
Land quality	Self-assessed quality of land	Not applicable, poor to medium, good
Irrigated area (acres)	Area of irrigated land owned by a household (in acres)	0 acres, 0–5 acres, >5 acres
Literate household members	Number of literate household members in 2010–2011	0, 1, 2–3, ≥4
Skills	Qualitative description of the level of skills of household members in 2010–2011	Low, medium, high
Household earners	Number of household earners in 2010–2011	0–1, 2–3, ≥4
Household dependents	Number of household dependents in 2010–2011	0–1, 2–3, ≥4
Investment (irrigation)	Annual investment in irrigation	No, yes
Access to CPR forests	Access to CPR forests	No, yes
Value (Rs.) of fodder	Annual value (in Rs.) of fodder obtained from CPR forests	0, <3000, 3000–5000, >5000
Investment (tools)	Annual investment in tools (in Rs.)	0, <8000, >8000
Crop area (kharif)	Area under cropping in the kharif season (in acres)	0, <2.5, 2.5–5, 5–7.5, >7.5

Variables unique to financial capital submodel		
Investment	Annual investment in irrigation, tools, and land development (in Rs.)	0, <10,000, >10,000
Crop revenue	Annual crop revenue (in Rs.)	0, <25,000, 25,000 to 50,000, >50,000
Rainfed crop type	Primary crop type on rainfed land	Not applicable, groundnut system, other
Irrigated crop type	Primary crop type on irrigated land	Not applicable, paddy, vegetables, other
Rainfed land productivity	Maximum output per acre (in Rs.) household achieves from rainfed land	0, 0–4800, 4800–9600, 9600–18,000, >18,000
Irrigated land productivity	Maximum output per acre (in Rs.) household achieves from irrigated land	0, 0 –20,000, >20,000
Years of crop loss	Number of years of crop loss during 2007–2011 (inclusive)	0, 1, >1
Consumption	Total annual consumption of household (in Rs.)	0–40000, 40,000–80,000, >80,000
Income	Total annual income of household (in Rs.)	0–50,000, 50,000–80,000, >80,000
Debt	Total debt of household (in Rs.)	0–40,000, 40,000–80,000, >80,000
Savings	Total savings of household (in Rs.)	0, 0–10,000, 10,000–20,000, 20,000–40,000, >40,000
Drought income	Additional income household receives from work during drought year (in Rs.)	0, 0–10,000, >10,000
Variables unique to human capital sub-model		
Work days (farm)	Number of days in 2010–2011 that the household worked on farms	0, 0–200, 200–400, ≥ 400

(Continued)

TABLE 9.1 Variables in the BNs—cont'd

Name	Description	States
Work days (non-farm)	Number of days in 2010–2011 that the household worked in nonfarming employment (excluding NREGS).	0, 0–200, 200–400, ≥ 400
Work days (NREGS)	Whether or not members of the household have been used through the NREGS (yes/no).	No, yes
Education expenditure	Average annual expenditure on education (in Rs.)	0, 0–25,000, >25000
Health expenditure	Average annual expenditure on household health (in Rs.)	0–5000, 500010,000, 10,000–15,000, >15,000
Health	Qualitative description of the health of household members in 2010–2011	Good, satisfactory, bad
Variables unique to natural capital submodel		
Water stocks	Self-assessed adequacy of water stocks	Not applicable, more than adequate, adequate, inadequate
Value (Rs.) of fuel	Annual value (in Rs.) of fuel obtained from CPR forests	0, <2000, 2000–4000, >4000

Indirect value	Whether or not indirect value is obtained from CPR forests (e.g., mitigation against erosion)	No, yes
Direct CPR value (Rs.)	Annual value (in Rs.) obtained from direct use of CPR.	0, <2000, 2000−4000, >4000
Value from non-timber	Whether or not direct value is gained from non-timber uses of CPR forests.	No, yes
Potential access to groundwater (%)	The maximum percentage of bores at the village level that would be operational accessed under different climates	<5, 5−20, 20−65, >65
Variables unique to physical capital submodel		
Agricultural tools	Quantity of agricultural tools owned by household	0, 1−4, >4
Big ruminants	Quantity of buffaloes, cows and young stock owned by household	0, 1−2, 3−4, >4
Small ruminants	Quantity of sheep and goats owned by household	0, 1, 2, >2
Variables unique to social capital submodel		
Groups	Number of groups to which the household members belong	0, 1, ≥2
Members	Number of household group members	0, 1, ≥2
Administrative connections	Presence of administrative connections	No, yes
Political connections	Presence of political connections	No, yes

NREGS, National Rural Employment Guarantee Scheme.
^a*Not relevant to all drought support (<indicator>) variables.*

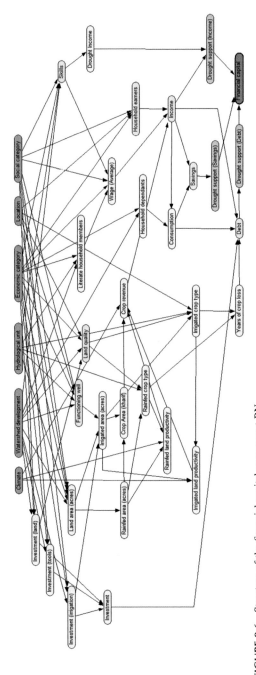

FIGURE 9.6 Structure of the financial capital component BN.

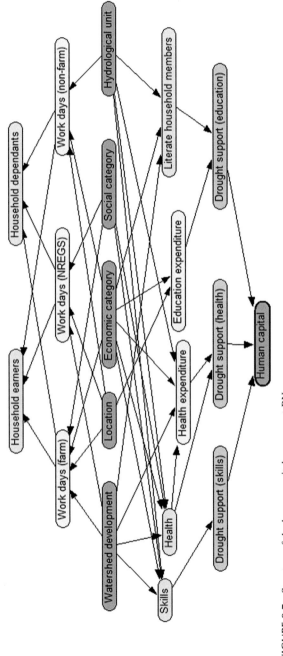

FIGURE 9.7 Structure of the human capital component BN.

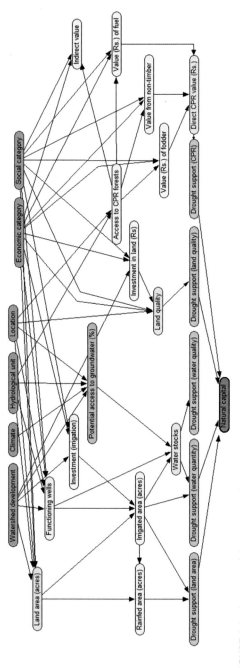

FIGURE 9.8 Structure of the natural capital component BN.

The natural capital stocks represented include the quantity and quality of land and water resources that households own or have access to as well as the direct value of common pool resources (CPRs) accessed by each household (Figure 9.8). The direct value gained from CPRs comprises the value gained from the collection of fodder, fuel, and non-timber products. All these variables have the following parent variables: *Access to CPR forests, Economic category*, and *Social category*. Also represented in the network, with the same parent variables, is indirect value gained from CPR forests (e.g., protection against soil erosion). The *Location, Economic category, Social category*, and *Investment in land* variables are used to determine *Land quality*. The adequacy of the *Water stocks* variable is determined by the available groundwater *(Groundwater MCM)* and *Irrigated area (acres)*. The *Potential access to groundwater* (%) variable has been populated using expert elicitation by the authors of Chapters 3 and 6 and is described as the maximum percentage of wells within a village area from which water could be extracted under dry, normal, and wet climate years. It has been related to the geographic variables *Location* and *Hydrological unit* as well as *Climate* and *Watershed development*.

Climate is classified as below average (dry), average (normal), or above average (wet) climate years. *Irrigated area (acres)* is related to the total land area, whether or not households own functioning wells, and the level of investment in irrigation. *Functioning Wells* is the child variable of *Location, Hydrological unit, Watershed development, Economic category*, and *Social category*. These variables are also the parents of the total household land area, and *Land area (acres)*. Given the near identical response by the households for the quantity and quality of water resources, the *Natural capital* variable is not linked to the *Drought support (Water quality)* variable.

The physical capital stocks represented in Figure 9.9 are the quantity of agricultural tools, numbers of ruminants (big and small) owned by households, and ownership of a functional well. *Functional well* is the child variable of *Watershed development, Hydrological unit, Social category, Location*, and *Economic category*. The quantities of both small and big ruminants are determined by *Value (Rs.) of fodder, Social category, Economic category*, and *Crop area (kharif)*. The parent variables of *Agricultural tools* are *Investment (tools), Rainfed area (acres)*, and *Irrigated area (acres)*. These variables are defined consistently with the financial and natural capital submodels.

The stocks represented in the social capital submodel are membership in groups, administrative connections, and political connections (Figure 9.10). The *Social category* and *Economic category* variables are directly linked to the drought support gained from relatives and friends in nearby villages, i.e., the *Drought support (relatives & friends)* variable. The variable representing the number of groups that households participate in, *Groups*, is the child variable of *Social category, Economic category*, and *Watershed development*. The parents of both the *Administrative connections* and *Political connections* variables are *Economic category, Watershed development, Hydrological unit*, and *Groups*. The *Groups* variable is linked as a surrogate

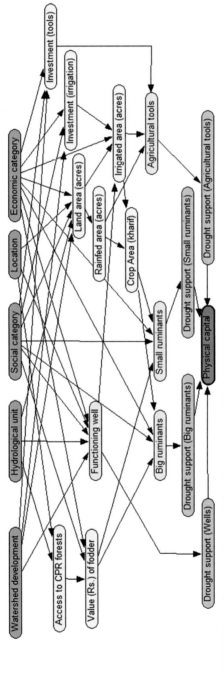

FIGURE 9.9 Structure of the physical capital component BN.

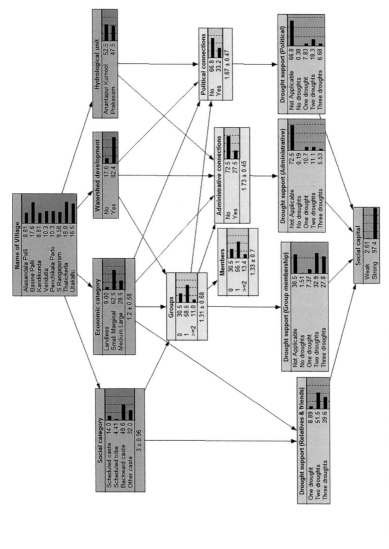

FIGURE 9.10 Structure of the social capital component BN.

for possible effectiveness of group participation in gaining connections with people of influence.

9.3.3 Capital Strength and Resilience to Droughts

In accordance with the analyses in Chapter 7, capital strength is defined as strong if any of the household resilience indicators have a value of two or more droughts. The strength of each type of capital is linked to the number of consecutive drought years that the household could survive (*Resilience*; Figure 9.11).

9.4 ANALYZING SOCIAL CAPITAL USING THE BN SUBMODEL

This section uses the social capital submodel to demonstrate how BN models can be used to analyze social datasets. The social capital submodel in Figure 9.10 shows the "unconditioned" probability distributions derived using the resilience survey dataset. For example, based on the dataset, 30.5% of the households reported that no members of the household were part of a group; 58.5% reported membership of one group; and the remaining 11% reported membership of at least two groups.

9.4.1 Analyzing Social Data using BNs

Figure 9.12 shows the strong effect that both group membership *(Groups)* and farm size *(Economic category)* have on the likelihood of having administrative and political connections. In the figure, each bar for administrative and political connections corresponds to setting the parent variables to 100% of a variable state (e.g., no or yes for the *Watershed development* variable), leaving all other parameters untouched. We see that group membership is not essential for administrative or political connections as there is still some likelihood of connections ($\sim$20% or more) with no group membership. However, the likelihood of connections increases to >50% when the number of groups that households participate in is two or more. Further, landless households are much less likely to report administrative or political connections compared to the medium-to-large farmers, and compared to the influence of the group membership and farm size variables, there is less difference in the likelihood of connections between the two hydrological units and villages that have or have not received WSD.

Interestingly, villages that have received WSD reported slightly lower ($\sim$5%) administrative connections compared to the control villages. This is explored further in Figure 9.13 to look at differences between households in WSD and control villages by the level of group membership. We see that for households that had no group membership, the reported administrative

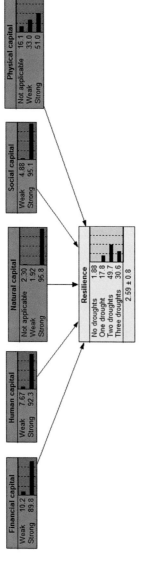

FIGURE 9.11 Relating the five capitals to resilience (i.e., how many consecutive drought years households could survive).

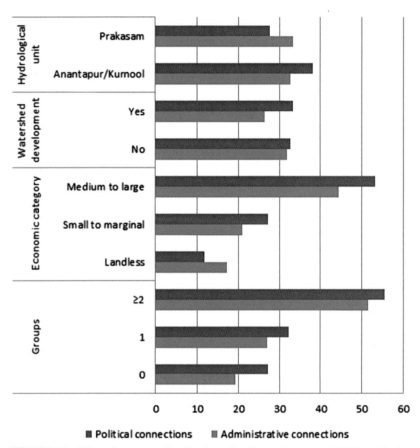

FIGURE 9.12   Effect of group membership, farm size *(Economic category)*, WSD, and hydrological unit have on the likelihood of having administrative and political connections.

connections are higher in the WSD villages (double that of the control villages). Conversely, for both categories where households participate in groups, the reported administrative connections are higher in the control villages than in the WSD villages. Furthermore, of the 92 surveyed households that live in the control villages, only 20 households reported no membership in groups while 10 reported membership of multiple groups. Since this is a relatively small sample for developing model relationships, care should be taken while interpreting the results.

The *Drought support (group membership)* variable is most sensitive to changes in the *Groups* variable (MI = 0.89) compared with changes in *Administrative connections* (MI = 0.02). Across the levels of group membership, whether or not households have administrative connections does not greatly change the likelihood of *Drought support (group membership)* being in the "three droughts" state (Figure 9.14). However, the presence of

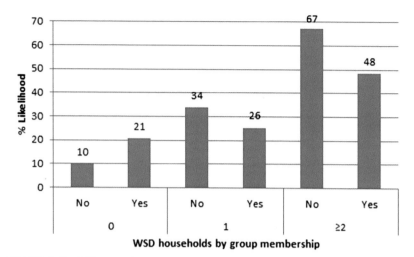

FIGURE 9.13 Differences in administrative connections between households in WSD and control villages (by the level of group membership).

administrative connections seems to have a positive effect on lower states of drought support—the likelihood of *Drought support (group membership)* being in the "two droughts" state increases from 46 to 52% when households participating in one group also have administrative connections, while for households participating in multiple groups, the likelihood increases from 38 to 50% (Figure 9.14).

With most households reporting drought support of two or more droughts for at least one of the four indicators of social capital (511 of 522 households), the strength of social capital across the sample population is mostly strong (97.4% in Figure 9.11). It is observed that the *Social capital* variable is most sensitive to *Drought support (relatives & friends)* (MI = 0.098), followed by *Drought support (group members)* (MI = 0.037), *Drought support (political)* (MI = 0.011), and *Drought support (administrative)* (MI = 0.007). When households reported low drought support for their connections with relatives and friends—that is *Drought support (relatives & friends)* is set to "one drought"—the likelihood of social capital being weak is 30%, compared with 5.2, 3.4, and 2.9% when *Drought support (group membership)*, *Drought support (political)*, and *Drought support (administrative)*, respectively, are set to less than two droughts. The sources for the differences between the social capital variables in the BN and Chapter 7 are discussed in Section 9.4.4.

9.4.2 Scenario Analysis

A small percentage (7.4%) of small or marginal farmer households reported a low level of drought support ("one drought") related to their connections with

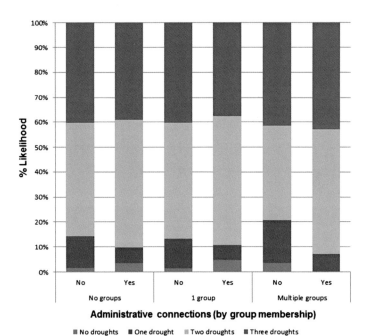

FIGURE 9.14 Relationship between drought support (group membership) and the levels of group membership and presence of administrative connections.

relatives and friends. For these households, the strength of their social capital is relatively poor, with a 28% likelihood of being "weak" compared with the households that report drought support for two or more consecutive drought years (Figure 9.15a). To demonstrate how BNs can be used for scenario analysis, Figure 9.15b shows the probability distribution for *Social capital* if small or marginal farmer households with low *Drought support (relatives and friends)* that are not members of groups are targeted by a successful program to involve all these households in groups. In Figure 9.15a, we see that 28% of the small or marginal farmer households are not involved in groups. By reducing this to 0% in Figure 9.15b, and assuming the same proportional split between membership in one group and two or more groups, the likelihood of *Social capital* being weak drops from 28 to 8.4% due to the large increase in *Drought support (group membership)*.

9.4.3 Data Issues

For some variables, there are insufficient data to populate conditional probability tables using the BN learning algorithms. Despite the 522 households surveyed, not all combinations of the states for the household variables are well-sampled; this has implications for the parameterization of variables

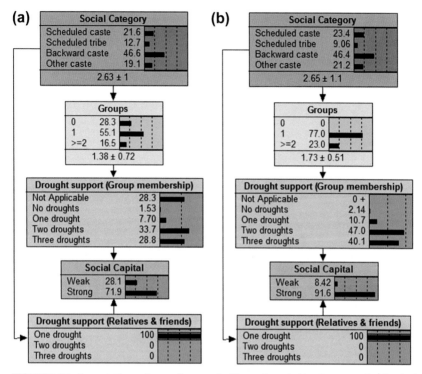

FIGURE 9.15 Impact of engaging small or marginal farmer households who report a low level of drought support ("One drought") related to their connections with relatives and friends in groups: (a) reported group membership and (b) assuming all households participate in on or more groups.

connected to household variables. This issue is of particular relevance when looking at combinations of variable states that include "Scheduled tribe" households ($n = 23$), which were from one hydrological unit, and all but one household were from the upstream village (S. Rangapuram). Models have been simplified as much as possible to avoid such issues.

In the social capital model, social category was considered to be more important than the position within the hydrological unit (i.e., downstream, midstream, or upstream). Despite this, 7 of the 36 combinations of parent states for administrative data have no information on the nature of the relationship. For example, there is insufficient information regarding the administrative connections of medium to large farming households from the Prakasam hydrological unit's control village. Moreover, of the 14 households from this combination of categories, all reported membership of one group. Hence, a uniform (or flat) distribution is shown for the other states of group membership (Figure 9.16). This lack of knowledge could potentially be addressed by targeting these households in follow-up surveys or through expert elicitation.

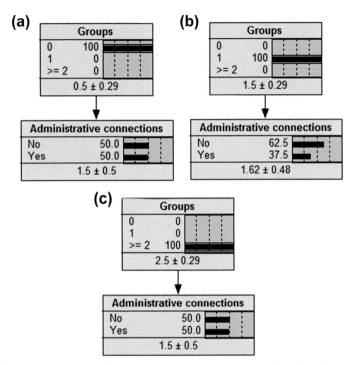

FIGURE 9.16 An example of data issue: administrative connections of medium to large farming households from the Prakasam hydrological unit who live in the control village: households who participate in (a) no groups, (b) one group and (c) two or more groups.

9.4.4 Relating Social Capital BN Results to Statistical Analyses of Social Capital Data

In Section 9.4.1, the likelihood of strong social capital is observed to be much higher than that reported in Chapter 7, despite the same rules as used to classify capital strength. The reason for this is the inclusion of the *Drought support (relatives and friends)*, *Drought support (administrative)*, and *Drought support (political)* in the definition of capital strength in addition to *Drought support(group)*. Most households (>90%) reported strong connections with relatives and friends in nearby villages, and thus the criteria for strong capital are met by such households. However, when these three variables are set to "no drought," and "one drought," the likelihood of weak capital is 50.6 and 2.4% (leaving all other variables unconditioned), respectively. This is comparable to the 50–80% incidence of weak social capital across the hydrological units and farm sizes reported in Table 7.25.

While the BNs should be roughly comparable to the tables in Chapter 7, as both analyses use the same data, we have explored a greater range of variables in the BNs that will explain some differences between the results.

Additionally, in most situations, only indicator stocks are directly linked to household variables (e.g., *Hydrological unit, farm size*) in each of the capital strength BNs. So, for example, the relationship between group membership, administrative connections, and the *Drought support (group membership)* variable was developed using all data, and not the data disaggregated by household category as presented in Chapter 7. Due to the way in which probabilities are propagated through the BN, as the chain, or number of variables between the driver (the household variables in the capital strength BNs) and the endpoint (capital strength) increases, the sensitivity of the endpoint to the drivers declines. For this reason, the component capital strength BNs have been kept as much as possible to the structure outlined in Figure 9.4a (i.e., a maximum chain length of five variables from the household variables to the capital strength endpoint).

9.5 SYNTHESIS

This chapter has provided an overview of the theoretical grounding behind BNs and introduced its application within the mesoscale project to relate the stocks of the livelihood capitals to the capacity of the households to survive consecutive droughts. The social capital model was used to demonstrate the utility of the BN models in analyzing social datasets and how scenario analyses can be implemented using the approach. The BN models form the basis for the integrated model described and used to run the biophysical and policy scenarios in Chapter 12.

REFERENCES

[1] Jakeman AJ, Norton JP, Letcher RA, Maier H. Integrated modelling: construction, selection and uncertainty. In: Jakeman AJ, Norton JP, Letcher RA, Maier H, editors. Sustainable management of water resources: an integrated approach. Cheltenham: Edward Elgar Publishing; 2006. pp. 263–86.

[2] Kelly (Letcher) RA, Jakeman AJ, Barreteau O, Borsuk ME, El Sawah S, Hamilton SH, JørgenHenriksen H, Kuikka S, Maier HR, Rizzoli AE, van Delden H, Voinov AA. Selecting among five common modelling approaches for integrated environmental assessment and management. Environ Model Software 2013;47:159–81.

[3] Varis O, Kuikka S. BeNe-EIA: a Bayesian approach to expert judgement elicitation with case studies on climate change impacts on surface waters. Climatic Change 1997;37:539–63.

[4] Ticehurst JL, Curtis A, Merritt WS. Using Bayesian networks to complement conventional analyses to explore landholder management of native vegetation. Environ Model Software 2011;26:52–65.

[5] Chen SH, Pollino CA. Good practice in Bayesian network modelling. Environ Model Software 2012;37:134–45.

[6] Koller D, Friedman N. Probabilistic Graphical Models: Principles and Techniques. Cambridge: Massachusetts Institute of Technology; 2009.

[7] Tsamardinos I, Brown LE, Aliferis CF. The max-min hill-climbing Bayesian network structure learning algorithm. Mach Learn 2006;65:31–78.

[8]   Kahn CE, Roberts LM, Shaffer KA, Haddawy P. Construction of a Bayesian network for mammographic diagnosis of breast cancer. Comput Biol Med 1997;27:19−29.

[9]   Aguilera PA, Fernández A, Fernández R, Rumí R, Salmerón A. Bayesian networks in environmental modelling. Environ Model Software 2011;26:1376−88.

[10] Bromley J, Jackson NA, Clymer OJ, Giacomello AM, Jensen FV. The use of Hugin® to develop Bayesian networks as an aid to integrated water resource planning. Environ Model Software 2005;20:231−42.

[11] Castelletti A, Soncini-Sessa R. Bayesian networks and participatory modelling in water resource management. Environ Model Software 2007;22:1075−88.

[12] Henriksen HJ, Rasmussen P, Brandt G, von Bulow D, Jensen FV. Public participation modelling using Bayesian networks in management of groundwater contamination. Environ Model Software 2007;22:1101−13.

[13] Calder I, Gosain A, Rao MSRM, Batchelor C, Garratt J, Bishop E. Watershed development in India. 2. New approaches for managing externalities and meeting sustainability requirements. Environ Dev Sustainability 2008;10:427−40.

[14] Newton AC, Marshall E, Schreckenberg K, Golicher D, teVelde DW, Edouard F, Arancibia E. Use of a Bayesian belief network to predict the impacts of commercializing non-timber forest products on livelihoods. Ecol Soc 2006;11:24 [online] URL, http://www. ecologyandsociety.org/vol11/iss2/art24/.

[15] Kemp-Benedict E, Bharwani S, de la Rosa E, Krittasudthacheewa C, Matin N. Assessing water-related poverty using the sustainable livelihoods framework. Stockholm Environ Inst Working Pap 2009:25.

Chapter 10

Justice and Equity in Watershed Development in Andhra Pradesh

Geoffrey J. Syme *, V. Ratna Reddy [§] and Ram Ranjan [¶]

Edith Cowan University, Perth, Australia, [§]*Livelihoods and Natural Resource Management Institute, Hyderabad, India,* [¶]*Department of Environment and Geography, Faculty of Science, Macquarie University, Syndney, Australia*

Chapter Outline

Integrated Assessment of Scale Impacts of Watershed Intervention
http://dx.doi.org/10.1016/B978-0-12-800067-0.00010-4. Copyright © 2015 Elsevier Inc. All rights reserved.

10.1 INTRODUCTION

This chapter examines the theory and practice of justice and equity in water resources and its relationship to watershed development (WSD) in Andhra Pradesh. It is organized into three sections, in addition to the introductory section.

The first section examines the evolution of the concepts of justice in the general context of water resources management, and compares and contrasts the use of these principles in recent Australian water reforms with those in India. In general, it analyzes the success of the Australian model in developing environmental allocation along with the introduction of property rights and markets in the Australian system, which does not necessarily represent a template for all Australian water resources management and may not be the path forward for India. A number of similarities as well as differences in justice formulations are presented in this section.

The second section explores alternative possibilities to traditional formulations of property rights in relation to the possible contribution of collective action in the Indian context—because of the central role of property rights as a perceived basis for economic development and their confused interpretation. It analyzes the existing property rights that failed to affect poverty alleviation, because equity and justice considerations are not included in their formulation. Since collective action has the capacity to change property rights, if its influence results in equity-based property rights these in turn are likely to motivate further community engagement and involvement with sustainable WSD. Thus, equity appears to be the critical factor in determining the effectiveness of collective action as well as property rights in addressing poverty.

While the first two sections of this chapter rely on reviews of published material in relation to property rights, justice, and water management, the final section examines the responses of the community to the WSD surveys in the two hydrological units (HUNs) and control villages. The questions relate to the influence of collective decision making on crop choice and perceived changes in water flows at different sites on the catchments—regarding any conflict in the WSD implementation and who the beneficiaries are. Both collective decision making and concerns about welfare issues seem to be weak in the current incarnation of WSD, but this may change markedly if the program is to be administered at a HUN level.

10.2 A COMPARATIVE EVALUATION OF THE CONCEPT OF JUSTICE IN WATER RESOURCES MANAGEMENT IN AUSTRALIA AND INDIA: THE ROLE OF PARTICIPATORY AND PROPERTY RIGHTS APPROACHES

In this section, we will examine the issues pertaining to equity, fairness, or justice associated with WSD and how these might change with scale. The possible evolution of these issues in India is examined in relation to the development of water policy and management in Australia. Further, in light of the current role of property rights and markets, the study evaluates whether or not the increase in population causes stress on water resources to increase through greater demand and exogenous variables such as climate change, and if there will be an inevitable evolution of carefully defined individual property rights and reliance on water markets in the Indian water management scenario.

The evolution of Australian water resources management has followed the path depicted in Figure 10.1. It has been assumed by some that the same path will inevitably be followed by developing (or south) countries [1]. The early development phase in which the priority was to augment supply has given way to an agenda for ensuring efficient management of water by considering water as an economic good that needs to be managed within the constraints of environmental sustainability. In many ways, this process reflects the assumptions of the environmental Kuznets hypothesis: evolution of efficiencies and economic prosperity leads to less social inequality and more inclusion of the environment itself [2]. These assumptions, however, have been contested by others outside the water reform process [3].

Nevertheless, property rights have been defined as a share of the resource whose absolute quantities vary according to water availability in a particular season. There has been a reliance on markets moving the water resource from less to more profitable uses, largely through the temporary purchases of others' entitlements. The provision for allocation for environmental purposes has also been met to a large extent by the Commonwealth Government purchasing water via the water market. There are, however, some difficulties in

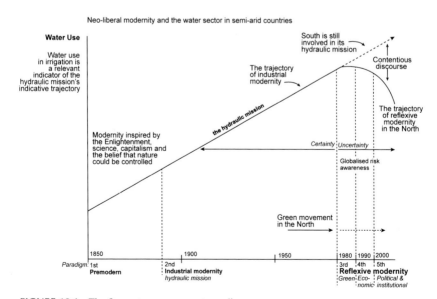

FIGURE 10.1   The five water management paradigms.

trading water from one state to another as different regulations may apply. There is also difficulty in defining property rights for differing uses such as recreation and transport rather than the traditional volumes, which are more apt for irrigation [4].

Progress, however, has been more checkered in relation to groundwater in Australia, which has had less systematic attention in regard to water reforms and has had less prominence in the water reforms debate. Conjunctive management of surface and groundwater has proven difficult to achieve. The contrasts between surface and groundwater management in Australia are shown in Table 10.1.

Because of the variable nature of groundwater aquifers and associated hydrogeology, land use markets have been difficult to establish in some cases [6], although there is the potential for developing some market-based instruments.

Nevertheless, whether irrigation is conducted using surface water, groundwater, or both, this process of reform has shifted the management of irrigation systems from government-run enterprises to those operated by the users. This has been common worldwide and has been the case generally in

TABLE 10.1 Characteristics distinguishing surface water and groundwater management in Australia [5]

Characteristic	Surface water	Groundwater
Primary nature of development infrastructure funding	Centralized	Decentralized
Management of flow	Linearly regulated	Unregulated
Public awareness	High	Low
Security of supply	Low	High
Water quality	High (managed)	Low
Physical extraction limit	Volume in storage	Bore capacity, draw down
Capacity to enforce legal limits	High	Variable
Monitoring and reporting	Regulatory and centralized	Variable
Primary financial costs of water use and entitlement	Levies	Infrastructure/operation
Markets	Yes—general	In development
Ease of monitoring and building resource data	Relatively high	Often low
Infrastructure funding	Publicly subsidized	Private

India as well. However, India has chosen an alternative route of relying on a participatory approach to irrigation management. This participatory approach is evident in the WSD approach [7], with the sustainable use of water resources (particularly groundwater) undertaken through the creation of water use committees in which participation is encouraged from both genders, communal monitoring of groundwater levels, public display of monitoring data, education about the water requirements of different crops, and communal negotiation in relation to individual land use by farmers.

The present irrigation systems in India have no property rights or formal markets. With groundwater, they are associated with the land and are sold informally by the landholder. In a recent contrast of the property rights-based management approach versus the participatory management approach in the Murray Darling Basin (MDB) and the Krishna Basin in India, Poddar et al. [8] suggested that the performance of the participatory approach adopted by the Indian policy is lower compared with the property rights market-based approach adopted in the MDB. Further, there were fewer equity issues identified for the MDB at the inter-farm or inter-regional level. This, according to the authors, was because the reforms in the MDB were driven from the farming communities (some with privatized irrigation schemes), while the formalized hierarchical structure of the water-use associations seemed to have weaker control of distribution issues, perhaps because they had been instigated by "top-down" government priorities. This has led some authors [9] to suggest that in the future control rights for user groups may improve efficiency such that evolution of markets is encouraged.

The maintenance of the participatory approach after the governments "seeding" funds had been exhausted was also perceived as problematic in India [10]. Reddy [11], however, had shown that WSD programs tend to work in villages with already strong informal institutions.

In the rest of this section, we examine whether "modernism" [12] is inevitable for water reforms, in the guise of individual property rights and associated markets. Boserup [13] and others in the property rights movement have seen that the evolution of property rights, envisioning an optimal response to resource costs in the form of gradually individualized rights and efficiency promoted by markets, are inevitable for development if the management of water resources is to become sustainable. The certainty associated with property rights is thought to encourage investment in infrastructure and innovations, thus creating an optimal situation for the future. This is considered to be the case when there is an increasing population and presumably where there is growing competition for the resource. This in many ways reflects the reforms in the Australian water management situation. Such developments are also supported by those who take a longer term view of prosperity [14].

There have been many studies examining whether this outcome of the introduction of property rights is as expected, and many authors, such as

Boserup [13], have found that in many situations this not true [15]. However, whether or not this evolution occurs and whether or not participatory approaches are preferred, their long-term outcomes depend on the acceptance of competing user groups. This acceptance is likely to depend on the perceptions of fairness and justice in the regimes that are established. While comparing the "modern" Australian reforms with the more traditional Indian approaches to water allocation and use, we examine the justice considerations and consider the notion that India too may follow an inevitable path to individual property rights and markets as suggested by the property rights movement.

To understand the justice issues involved in water allocation, we first examine the idea that water is a substance that needs to be managed with the awareness that multiple needs can be met by the same volume of water. It further needs to be acknowledged that these needs are related and that each may require a different management approach both procedurally as well as morally. There is also a need to accept that there has been an evolution of moral thought, not only in the Western world but also in the east, which governs people's assessment of the acceptability and equity of water management. We then describe these concepts and examine how they relate to the current institutions governing water management in Australia as well as in India. Finally, we discuss briefly whether from these case studies the "evolution" toward property rights and markets is likely to continue its development in Australia or will it be taken up in India as the process of water reforms evolves.

10.2.1 Water Benefits

Ideally, water management should be about the sphere of needs [16]. These needs vary from the utilitarian ones in the center of the sphere to the more cultural and spiritual ones on the outside (Figure 10.2).

As Syme et al. [16] indicated, the needs become more uncertain to manage as they become located in the outer rings of the sphere (Figure 10.2). This is evident for property rights [17]: while property rights associated with issues pertaining to irrigation can meet standard definitions for the essential components (such as specificity, security, exclusivity, enforceability, transferability, and divisibility [18]), it is more difficult when use benefits are defined in terms of recreation, aesthetics, culture, and spirituality. The latter benefits are also difficult to include in any form of market. Such benefits can more easily be defined in the public good area as can environmental flows.

Thus, individual property rights need to be balanced with respect to protection of the public interest, and sometimes the political economy ensures this. This is why even in Australia there is still investment in public infrastructure funded by the public purse ostensibly to conserve water either for distribution to the environment or to assist in maintaining resilience for local communities, even if such investments are inefficient [17] in economic terms. Public goods may be more appropriately managed by participatory processes,

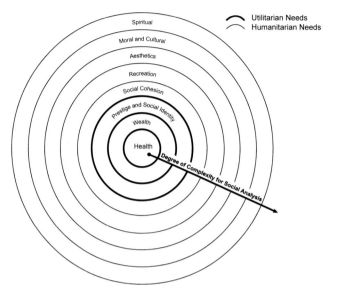

FIGURE 10.2 Sphere of needs (SON) met by water.

such as collaborative planning, which require the participation of community-based institutions.

These needs can be seen to be related; for example, healthy people are more likely to be able to take up the recreation opportunities provided by a volume of water [16]. Similarly, wealthy people are also more likely to use water to support their social prestige [19]. As Hoekstra et al. [20] have shown, the same body of water can provide multiple benefits as it passes through the catchment. This has led to the contention that it is better to allocate the benefits of the water resource rather than volumes of water [21].

This view is highly compatible with the social and economic objectives of programs such as WSD, but perhaps less comfortable with the individual property rights orientation. This approach tends to lead to volume definitions often largely based on the history of use, as seen in the Australian situation. The benefits of such orientation tend to aspire to issues such as community resilience and quality of life. However, regardless of the approach, there has been an evolution of moral or ethical thought over many years both from the philosophical and social science literatures [22,23] as to how to share water fairly.

10.2.2 Justice Principles and Water Allocation and Management

The principles of justice outlined in Table 10.2 have been derived from social, psychological, and philosophical studies largely based in the Western world.

TABLE 10.2 Principles of justice

1. Distributive justice (who gets what)	**1.1** Equity (you are rewarded in proportion to effort) **1.2** Equality (people get equal distributions) **1.3** Need (people who need it for basic reasons get first preference) **1.4** Self-interest (each individual prefers allocations which are best for themselves) **1.5** Efficiency (people who use the resource most efficiently are entitled to more)
2. Procedural justice (how decisions are made)	**2.1** Voice (having the opportunity to be listened to and have some influence) **2.2** Participation (being able to participate in decision-making processes) **2.3** Consistency (the rules are not abruptly changed) **2.4** Ethicality (lack of corruption, follows due process) **2.5** Impartiality (everyone has the same opportunities to participate) **2.6** Accuracy (the information provided for discussion is correct) **2.7** Error correctability (there is a chance to appeal in terms of facts) **2.8** Representativeness (there is a good range of stakeholders included)
3. Combining distributive and procedural justice	**3.1** Fairness (an overall judgment of both the process and the outcome)
4. Interactive judgment (do people feel they have been treated appropriately?)	**4.1** Trust (do people trust the government agency and other participants?) **4.2** Dignity (do people feel they have been treated with dignity?) **4.3** Respect (do people feel they have been appropriately respected?)
5. Justice philosophies	**5.1** Virtue theory (people who already have resources should retain it because they are inherently good) **5.2** Prior rights (people who have used the resource in the past or first-in-first-served) **5.3** Intergenerational justice (we need to think of the needs of future generations) **5.4** Environmental rights (the environment deserves its own allocation) **5.4** Property rights (individuals should be given rights to the amount of resources based on some contract) **5.5** Economic good (natural resources should be treated as economic goods with prices and markets) **5.5** Utilitarian theories (resources should be managed to maximize community welfare—there are many versions of this) **5.6** Moral imperative (people in one location have a duty to ensure they do not negatively affect people in other locations)

Although many have been empirically substantiated in community studies and in the water context [23,24], the procedural justice principles that have been derived to have a basis in the reasoned argument approach to fairness mooted by such authors as Rawls [25], who conceived fairness in water management as the result of logical or nonself- interested argumentation. Rawls and others observed that in practice, this achievement is an important component of underpinning prosperity and democracy [14,26].

It is important to note that these principles are not static criteria for assessing if justice has been achieved, but there are some of the common elements that are combined in different ways to construct the justice (or arguments based on fairness). These are the elements in which the justice "for who" arguments occur in normal democratic discourse by the different interests in decision making. A number of principles can be held with different emphases for different problems.

The word allocation in this chapter has been used in a generalized sense. In Table 10.2 the first four categories have not only been discussed philosophically but have also been the subject of formal social—psychological research. The latter approach can be justified by the ways in which people have actually behaved or decided in empirical studies, while others are more conceptual and have been analyzed in historical and current sociological or policy studies.

10.2.2.1 How Australia Uses the Justice Principles

During the recent water reforms there has been great emphasis on clarity regarding property rights (Table 10.2, Section 5.4), which have been underpinned by shares that can be placed on the market (Table 10.2, Section 5.5) either temporarily or permanently, subject to some rules. Historically, initial use was often determined by the allocations based on land use or a variety of forms of family privilege. The first allocations of property rights in Australia were substantially based on prior use (Table 10.2, Section 5.2) or the concept of bureaucratically defined beneficial use, which is perhaps the oldest in the system of water rights and possibly "a low cost way of providing an ownership claim" [27]. While environmental allocation is on the rise, it has not as yet attained the status of an environmental "right," and it is still under political debate.

Protection of water quality and quantity through land-based management relies more on voluntary effort sometimes prompted by small government incentives. Such catchment-based activities depend more on the principles outlined in Sections 2—4 of Table 10.2, and these are the principles that have resemblance to the justice as fairness principles outlined by Rawls [25] and developed by Sen [26] and others. Hence, the principles behind the use and protection of the resource are quite different—resource protection also does not separate land from water. Finally, studies of peoples' community-based catchment solutions founded on local knowledge and social preference tend to clearly include the categories in Table 10.2, Sections 1 and 2.

The quality processes designed by the principles in Section 2 in Table 10.2 tend to allow people to decide on specific solutions, which have a mix of the categories in Section 1 [23,28]. However, it is argued by some that reliance on a bottom-up approach has not necessarily led to enhancement of the overall public good; instead, this has resulted in the Federal Government expressing its own priorities and identifying sections of the community being funded to meet them [29]. Thus, community participation is prescribed in a utilitarian manner; for the perceived overall "good" of the society, local institutions are meant to organize themselves under the guidance of national priorities. At least this was the formulation of the last Australian Labor Government, which was the outcome of a comprehensive evaluation of the effectiveness of this approach in the light of different regional institutions in each state. The Australian resource protection scheme can thus be described as participative, to an extent [29].

At the community level there is also very strong support for the concept of intergenerational justice at a conceptual level, although there is a great deal of confusion when this concept is used in practical planning [30] as is the case on many occasions for the concept of the moral imperative. Everyone asserts that both principles are important but assume that others' definitions in practice can definitely be placed in the self-interest category (Table 10.2). This assignation of less worthy motives to others (or the group identity phenomenon) [31] is a common social—psychological observation, although it also demonstrates the dynamics of the use of morality or procedural justice in water allocation decisions. Nevertheless, these dynamics highlight the importance of the interactive justice component of decision making, which has been shown to be highly related to procedural justice in the social—psychological literature. The property rights and economic good category and participation in the market by the irrigating community are seen as a business tool rather than something related to the long-term social goals for water allocation or planning.

The Australian public service generally supports a utilitarian (Table 10.2, Section 5.5) stance that is underpinned by the role of objective science, which they presume gives justice to the outcome by not playing favorites (perhaps representing the "veil of ignorance" or the reasonable man of Rawls [25]). It is acknowledged, however, that there is a need for a discussion and there is support for the ability for people to participate in the decision-making process (the democracy as supported by Sen [26] and Acemoglu and Robinson [14]). It is suggested that public involvement in decision making is instrumental in nature, because the unspoken purpose of the involvement is to get the "right" answer [24]. Public programs are conducted for the purpose of reaching the science underpinning the utilitarian solution. In short, perhaps the only philosophy not used at least tacitly in water resource management is that of virtue theory, which basically asserts that people have more water because they are "better" people and therefore deserve it.

To summarize, it can be said that there are two houses to water resources planning and management: one that relates to use and allocation, which is designed to get the government out of making allocation decisions, and the other is a participative one for the long-term protection of the resource from a systems water cycle point of view. Property rights and markets protect the irrigation use by providing for long-term efficiency and a mechanism for governments to create "environmental water." The environmental market, however, will be prone to failure because it is subject to the influence of the political economy. The second approach protects the overall benefits obtained by the community for water resources. Theoretically, it is possible to develop a system of rights and markets if an "externalities" market was created (possibly in the form of pollution rights, but this would seem unlikely and very complex given the need for differential action depending on one's position on the catchment). As Skurray et al. [6] pointed out, this would be the case where groundwater is involved. Thus, it seems that this dual approach is likely to continue into the foreseeable future.

10.2.2.2 How India (mis)Uses its Philosophical Lineage to Perpetuate Inequity

Neither equity nor equality (Table 10.2, Sections 1.1 and 1.2) are given due consideration in the Indian water resource policies, although equity and justice (especially social) are among the grand objectives of planning. The British had established water rights—riparian laws in the case of surface water and laws linked to land in the case of groundwater (Easement Act, 1882)—and they ensured water use efficiency to some extent by pricing its usage [32]. Post independence, Indian planning also adopted these property rights religiously. However, over the years, economic efficiency and pricing of water were outmoded in favor of populist policies like free water, free power, etc. In the process, property rights were confined to land rights; i.e., if one owns land next to surface water bodies (rivers, streams, canal systems, tanks, etc.) or has subsurface water beneath their land, they have all the rights to the water resources [32]. This has become a *de jure* accepted norm over the years, which resulted in wide variations in access to water resources: while an enormous amount of water gets wasted at the head reaches of canal systems, crops perish at the tail end due to lack of water, in the absence of property rights or volumetric pricing; similarly, in the case of groundwater, those who can afford investments in groundwater pumping infrastructure are allowed to exploit as much as they want and even sell if possible.

Thus, Indian planning has conveniently ignored the fact that water is a common pool resource (CPR) and everyone has equal right, and this has been justified by reference to unwarranted transaction costs (i.e., efforts and resources required to frame and enforce appropriate policies and institutions that can ensure just access to water). India does not have an implementable water policy frame. It has only guidelines that are revised every two years.

In Australia, the Federal Government can only provide guidelines, although some guidelines are expressed as priorities that must be met for regional groups to become funded. This in some ways circumvents the power of the states. On the other hand, as per the Indian Constitution, water is a state subject; none of the Indian states have gone beyond adopting the federal guidelines in their respective water policies. However, even at the state level, water policies have remained as mere policy documents without any legal or legislative framework to implement them. Hence, to address the inequalities and water use inefficiencies, Indian policy makers have adopted participatory (Table 10.2, Section 2.2) approaches to water management, but in the absence of proper devolution of powers to the user associations, they are neither effective nor sustainable and have failed to deliver the intended results. Furthermore, even the participatory approaches are mostly limited to surface water resources.

Similarly, with groundwater, it is the private property approach that is prevalent, and inequality in access to groundwater is more pervasive. The unquestioned acceptance of such inequality in the access to water has its roots in the "karma" philosophy, which is close to the virtue theory (Table 10.2, Section 5.1); that is, a person or community is poorly endowed because of misfortune or destiny. This was clearly expressed in our interactions with communities regarding their access to groundwater and other natural resources [33]. While this philosophy has checked conflicts over resources, it has perpetuated inequalities and inefficiencies.

The state has a role in ensuring distributive justice, and efficient allocation of scarce resources is a constitutional mandate as these resources are common and everyone has equal rights over them. Creating awareness among communities that certain resources (like water) are not private properties is the responsibility of the state, given the people's ignorance in this regard. In some cases (e.g., canal water), conflicts over water sharing have been on the rise due to awareness after the advent of participatory management practices. This calls for evolution of appropriate institutional arrangements that ensure just distribution of water as per the mandate.

The recent developments in the Indian policy for making groundwater a common resource is in the direction of addressing unjust and inefficient water use. These policies need to be supported by a legal and legislative framework that ensures smooth enforcement. The main bottleneck in this regard is control over resources—while the state has control over surface resources, private people (often rich) have control over groundwater resources.

The experience pertaining to participatory water management over the last two decades clearly indicates that the state is not willing to hand over control to participatory organizations. While this has happened with forest resources to a large extent, it is yet to happen with water. In the case of groundwater, it is apprehended that making groundwater a common resource could prove politically detrimental, and the philosophical underpinnings are helping this slow process. Hence, awareness building and change management at the policy level, to begin with, is required.

10.2.2.3 Comparison of the Two Countries

Initially it seems as though Australia has experimented with a wider range of justice principles in water allocation and management than India. Nevertheless, Australia has achieved conceptual clarity only in terms of property rights and its markets mainly applied to surface water and in intrastate situations. The "modernistic" evolution by Australia to more clarity in and "better" justice in comparison with India is patchwork at best.

10.2.2.3.1 Groundwater

Both countries have found groundwater more difficult to govern than surface water. This is partly to do with the inherently more complicated nature of the differing aquifers and the tendency for groundwater to be developed as private property, initially in both countries. Moreover, monitoring groundwater usage is less clear than for surface water. Often groundwater usage is not metered in either country, leading to the lack of knowledge of what is happening to the resource. In India, to some extent it can be controlled by the limitations in access to electricity, while in Australia, this is the case in Western Australia, which in 2002, according to the National Land and Water Audit, had no groundwater monitoring at all. Although some metering has begun, there is still opposition to metering in this state despite significant overuse leading to problems for several key aquifers.

Thus, despite good intentions and guidelines from the national level and sometimes the state as well, there are few effective mechanisms for ensuring long-term justice either for the quantity or quality of groundwater—both need to develop institutionally to implement sustainable and just groundwater management through effective community action or property rights.

Markets, for hydrogeological reasons, can be difficult to implement. Some sections of the community are systematically advantaged/disadvantaged in both nations. For example, in Australia, allocation is often based on the early allocations made by estimates of the needs of areas of land and the perceived need for water governed by existing crop types. Such allocations were significant because of the government's insistence that the current allocations should be based on historical usage or prior rights. Therefore, in effect, the foundation for groundwater (and surface water) was largely influenced by the first-come-first-served system.

Groundwater irrigators want certainty of supply and, while water is not regarded as a private property per se, the amount of groundwater allocated is considered as known—it may be altered, but can never be taken away unless it is voluntarily sold. In India, groundwater is attached to land, and larger tracts of land have been owned by the more wealthy people who regarded the groundwater as private property, while karma has inhibited conflict to a great extent.

There are similar problems when cooperative water reforms or management are attempted in either country. Hence, both require institutional innovation and there is no evidence of justice issues in water planning for creating such institutions.

10.2.2.3.2 Surface Water

There is little doubt that the advent of tradeable property rights has created a degree of clarity and certainty regarding irrigation relying on surface water, but such clarity is lacking in India. This has led Poddar et al. [8] to conclude that Australia's property rights are more effective in delivering efficiency in this sector than the community-based social approaches in India. This may be the case overall, although there is considerable variability in the performance of community-based governance, perhaps depending on the leadership structure in the village [34] and the relationships between government institutions and the community [10]. This is also observed in other countries such as Spain where contrasting groundwater property rights emerged from irrigation communities in different locations [35]. While the vagaries of the local political economy and leadership structures can clearly create inefficiencies in voluntary community-based management in irrigation, it must also be noted that political boundaries—for example, between the Australian states—can result in market failure, as each region protects the viability of its own irrigators and thus prevents some movement of water. In many instances, the key relationships between property rights, participatory management, and markets lie in the "justice" inherent in the rules of a particular market; each does not act independently [36].

Outside irrigation, the role of property rights and markets becomes more fraught. Hence, creativity in developing alternative market-based instruments is required. As noted by Crase [17], recreational use seems to be ill suited to this concept of property rights as is the creation of cultural flows. Australia, like India, relies on the functioning of regional and community institutions for sustained protection of water resources. Hence, institutional economics is important and new institutions may be required to implement the government's priorities. Since in India the success of different models in terms of demonstrated overall improvement in the environment has been variable [34], generic lessons are difficult to derive [37] despite the progress made by Ostrom and others (e.g., [38,39]).

10.2.2.4 Conclusions from the Comparison

To summarize, we can say that Australia is often regarded as an example of effective water management. This is so in many cases because of the perception that both the environment and industry can be protected with a stable set of property rights that have developed gradually through procedurally just community involvement and increasingly evidence-based policies. Although not often overtly spoken about, there has been a gamut of justice principles applied in different situations to arrive at change through public discussion at different scales. Nevertheless, overt discussion about the social goals and the role of justice in meeting them has been rare outside the outcomes of water reallocation in the social impact sense.

On the other hand, such dynamics do not seem to occur as much in India, where karma, among other things, has acted as a ballast to prevent overt conflict. Despite the difficulties in implementing policy change, there are reasonably clear social goals for policies such as WSD, which aspires to benefit a whole community while protecting the water resource. However, clear criteria relating to justice to water allocation are not defined. While inequality occurs because often the well-off benefit most from improved water infrastructure [40], the social goal of improving equality is obvious. South Africa is another nation that has clearly identified the justice and ethical goals of its water management policies, but it is a rarity [41].

In short, the Australian water reforms process has identified some very useful approaches that appear to be gradually enhancing the sustainability of irrigation. However, they are not necessarily appropriate universally, and the clearest reforms relate comfortably only with one basin and one industry. Furthermore, resource protection remains an institutional problem. Thus, the modernist view of the evolution of justice is not applicable and is perhaps not even desirable, In his book on history of natural resource management law, Scott [42] reminds us that there have been cycles (Figure 10.2) on basic issues such as the separation of land and water management.

Finally, what can be learned from both India and Australia is that policies have not been well articulated in terms of justice aspirations and that there is a need to explicitly consider them both in the formulation and evaluation of policy implementation. Their construction will inevitably change over time, but if they are not made explicit the chances of institutional failure are increased.

Similarly, in the case of property rights who wins and who gains depends on the nature or state of the resource and the nature of change to the property right. "Most of the poor find themselves in poverty not due to the absence of property rights but due to their inability to change them through collective action" [43]. In the absence of justice consideration, collective action strategies tend to take on a zero sum game, which acts as a major barrier to change.

10.3 COLLECTIVE ACTION AND PROPERTY RIGHTS FOR POVERTY ALLEVIATION: A CONCEPTUAL FRAMEWORK BASED ON THE EXPERIENCE OF WSD IN SEMI-ARID INDIA

Property rights over natural resources are fundamental in shaping the livelihoods of the rural poor, who are often found to possess the weakest property rights, such as secure rights over land, water, trees, livestock, fish, and genetic resources. Although property rights are necessary and relevant in many circumstances, these alone may not result in poverty alleviation [44]; that is, property rights are necessary but not sufficient for poverty alleviation. Other resources (credit, human skills, infrastructure, markets, etc.) are needed to complement such rights to make sound investment decisions for overcoming poverty. In many cases, it is the poor who get excluded, such as in the WSD programs in India [40,45]. However, who loses and who gains depends on the type of resource and the nature of changes in property rights, i.e., moving from individual to group or vice versa [43]. By re-contracting rights in different ways, disadvantaged actors may create opportunities to amend their initial disadvantages into a more beneficial arrangement.

In most instances, the existing property rights embedded in political-economy systems are biased against the poor. Hence, collective action or social mobilization is an important channel for asserting the rights of the poor. Collective action could lead to poverty alleviation not only through asserting or changing rights over natural resources like land, water, etc., but also through asserting their right to information and sharing the developmental programs. Often the property rights framework considers the tangible assets or resources.

In the present form, the role of property rights is limited to alleviating poverty. Also, collective action to assert or change the property rights in resources or assets is more complex and costly, as it threatens the existing socioeconomic and political structures. The effectiveness of collective action in overcoming the socioeconomic and political dynamics depends on the relative strength of the collective group in changing the political fortunes; in the absence of such strengths, collective action may not necessarily guarantee success with respect to poverty alleviation. As long as socioeconomic inequities and "elite capture" are dominant phenomena in the system, institutional changes (including property rights) may not result in poverty alleviation, irrespective of the fact that these changes happen due to new ideas [46] or due to social justice concerns [47].

Although there is consensus regarding the role and importance of property rights and collective action in poverty alleviation, their importance, effectiveness, and relative strengths in varying resource, socioeconomic, and political situations is less understood. There also is no clarity regarding the linkages or synergy between property rights and collective action. These two are often treated as mutually exclusive rather than mutually inclusive or complementary. The question is can either of them or both of them ameliorate poverty and if so, under what socioeconomic and political situations? This section is an attempt to understand the intricacies in the relations between property rights, collective action, and poverty alleviation.

WSD in the Indian rainfed tropics forms the backdrop for understanding the complexities, as such regions epitomize the interplay of property rights, collective action, and poverty alleviation.

10.3.1 Concepts and Linkages

In this subsection the concepts of poverty, collective action, and property rights are defined in a manner to suit the watershed context. We define poverty in a comprehensive manner instead of limiting to income poverty. As per the Human Rights Office of the United Nations, poverty is defined as "a human condition characterized by the sustained or chronic deprivation of the resources, capabilities, choices, security and power necessary for the enjoyment of an adequate standard of living and other civil, cultural, economic, political, and social rights" (as quoted in [48]). Although management regimes differentiate between open access and other forms of property resources, when addressed together, they are termed as CPRs. CPRs are defined as natural or man-made resources with attributes of nonexclusion (large enough to exclude other users without cost or with low costs) and subtractability (consumption of the resource by one user will reduce its availability to others) [49,50]. In the context of WSD, land falling under all types of regimes (common, private, open access, etc.), forests, and degraded lands are covered. Surface water bodies are also treated as CPRs, while groundwater is treated as private property for all practical purposes, although it is common property "*de jure*."[1]

Collective action is defined as "an action taken by a group of individuals to achieve common interests" (as quoted in [48]), and often, participation is used

---

1. As per the Easement Act of 1882, groundwater rights are customarily attached to land ownership; hence, groundwater management is totally left to the private initiatives. Given the linkages between surface and subsurface water bodies, property rights are rather blurred in the case of groundwater. As a result, groundwater management becomes a stumbling block in addressing or resolving the dilemmas related to equity, property rights, and collective action.

synonymously with collective action. There could be varying modes of participation including nominal, passive, consultative, activity specific, active, interactive, and informed [51,52].[2] However, all these forms of participation do not result in collective action as defined above—mere contribution by members does not merit collective action. Participation can be equated with collective action as long as "individual costs of participation are more than that of individual benefits"; that is, there is an amount of "voluntary involvement and efforts for the sake of achieving a common good." Collective action institutions are understood as regularized patterns of behavior between individuals and groups in society, or complexes of norms, rules, and behaviors that serve a collective purpose [53, p. 556]. While such institutions could be either formal or informal (for a detailed discussion, see [54]), the institutions created for watershed management are purely formal.

An efficient system of property rights should have three features: (1) universality, (2) exclusivity, and (3) transferability (Posner, 1977 as quoted in [55]). It is argued that individuals, rather than the community, would be in a better position to allocate resources more efficiently and maximize societal returns. Property rights are developed to internalize externalities when the gains of internalization become larger than its cost (Demsetz, 1967, quoted in [56]). Although it sounds logical that clearly specified property rights lead to better and efficient allocation of resources, the individual property rights approach has some important drawbacks: First, it may not lead to an efficient allocation of resources because of the existing imperfections in capital and labor markets. Second, uneven distribution of rights would increase the ecological stress on the land if the majority of poor farmers were allotted rights in marginal and degraded lands [57]. It also would aggravate the existing inequalities due to the inequitable distribution of resources attached to land, such as groundwater (as is in Australia). Furthermore, distribution of such rights may be (dis)advantageous to certain communities/households due to their sociocultural background.

Groundwater and grazing lands in watershed management epitomize such anomalies. Heterogeneity in spatial distribution of groundwater not only creates the problem of assignment but also involves a further complication as land (under which groundwater lies) rights are privately owned. The intertwining of private and common resources results in further externalities, which can be termed as "legislative externalities" (Figure 10.3). They arise when there is no

---

2. The characteristics of these modes include: nominal, membership of groups with or without payment; passive, silent participation in meetings or getting information on decisions after the meetings; consultative, asked for opinions without necessarily being able to influence decisions; active-specific, volunteering to undertake specific tasks; active, proactively expressing view, taking other initiatives; interactive (empowering), with voice and influence on decisions; and informed (empowered), able to take into account information and opinions of external agents (experts) and make considered decisions.

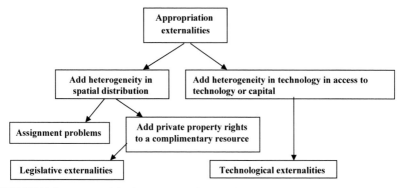

FIGURE 10.3 Property rights in the context of externalities. (*Source*: Adopted with modification from [49]).

clear-cut legislation demarcating and protecting different property regimes. While groundwater is a CPR in which rights are limited to use and deriving income, it is also sold and transferred along with land due to its link with the land. However, the legislation does not clearly specify how groundwater should be managed judiciously and distributed equitably. As a result, farmers make private investments thinking that they have absolute rights to the groundwater aquifer beneath their land. Similarly, the spatial distribution of check dams in a watershed may result in the exclusion of downstream communities, at least in the short run (assignment problems). On the other hand, common grazing lands are treated as CPRs in most watersheds. However, the benefits from these lands are disproportionate and often go to the large farmers due to the uneven distribution of livestock and consumption pattern [58]. As a result, the poor have no incentive to join the collective efforts. Such situations arise not only because of the nature of the resource but also because of the existing institutional arrangements. Therefore, the existing institutional arrangements for collective action and property rights are important in determining their impact on poverty.

10.3.2 Collective Action and Property Rights in Watershed Management

The distribution of benefits from the watershed to all sections of the community needs to be understood at the ground level. This becomes important because WSD as such does not guarantee the distribution of benefit across the community. Since watershed is a land-based technology, most of its benefits naturally accrue to the landed.

Although landless farmers are also expected to benefit from improvements to CPRs, the type and magnitude of the benefits accrued to the landless households depends on the existing property rights. Unless the benefits accrued to this section of the community are substantial in economic terms, it is unlikely that they will participate or evince interest in the program; in the absence of such rights to benefits, their likely apathy toward the program might jeopardize the sustainability of the program. In this subsection, we examine the context, the actors and the action arena (process of collective action), and their linkages in the backdrop of watershed management in rainfed regions.

10.3.3 The Context

The context is the WSD and the initial conditions in the rainfed regions. The initial conditions can be grouped as internal, external, and embedded. Internal conditions pertain to asset endowments of the local community, external conditions are those that influence the internal conditions from outside the community. and the embedded conditions are those that the community possesses traditionally and are difficult to alter in the short run. Embedded conditions are relevant in the context of the existing institutional arrangements, especially informal.

10.3.3.1 Internal

Internal conditions are mainly asset endowments, which include land, water, livestock, forests, etc. Apart from their quality, how these assets are distributed among households of the community is important for the poor. Dependence on the assets is relative—some households depend more and some less—and is linked to the quality of assets, property rights, and external sources like nonfarm income (including income from labor). A typical case would be in which most of the households possess land but do not have access to water, resulting in low productivity, high risk, vulnerability, and shocks. Similarly, livestock owners may not have access to grazing lands, forests, or even water. Economic dependence on the assets determines a community's stakes in managing it. The higher the dependence of the community on its natural assets such as land, water, and grazing lands, the higher the probability of collective action. Clearly defined property rights with specific justice criteria would not only broaden the base of stakeholders but also sustain the efforts in the long run.

Household circumstances and product markets reflect the level of economic development, which influences the opportunity costs of time of the stakeholders. Markets help to determine the allocation of time between different avenues, i.e., households can compare the returns based on the time invested in land improvements and labor markets. On the other hand, income from nonfarm employment would prompt the households to invest more in soil conservation activities [59]. In irrigated conditions, households choose to buy fodder and fuel wood from the market rather than growing fodder or harvest their farm trees due to the high opportunity costs of labor [60].

Smaller size is thought to facilitate better coordination and communication, resulting in lower transaction costs in organizing community action. The role of sociocultural heterogeneity in collective action is highly contested [61—63]. According to Olson [64, p. 45], the potential for collective action will be easier in a community with highly unequal degrees of interests. This may be true in the context of "privileged groups," where some of the members can bear the entire cost of providing public goods due to their self-interest. On the other hand, in situations where a dominant group has little or no interest in public good while a peripheral group has greater interest, the public good may not be provided. Furthermore, if the dominant group also happens to be in the majority, the potential for collective action is extremely dim. On the other hand, although a nondominant majority group (often poor) might have greater interest it lacks the ability to provide the collective good in spite of the existing potential. In most situations, it is observed that the poor have higher stakes and hence deeper interest when compared with the non-poor.[3] However, the potential for collective action is greater when the entire community has interests tied to the common resources. This is possible only when there is equity rather than inequity in the degree of economic interests of the community.

The importance of leadership and authority comes out clearly from the studies that have dealt with institutions in detail, i.e., studying the origins, process, and rules [65—70]. Institutional innovations are easier and sustained for longer periods in feudal societies where leadership is strong. The continuation of feudal norms in some communities explains the success of institutional arrangements even today [71]. The role of leadership is becoming more important as the politico-economic transformation in most of the societies is giving rise to individualistic and self-centered behavior. Also, the political factions at the community or village level are making it difficult for community action for the common good. This underscores the importance of leaders who can see beyond these short-run benefits and are strong enough to convince the groups to sideline their political rivalries for achieving the common good. A study of Rajasthan villagers showed that they wanted some outside support, either from a nongovernmental organization (NGO) or the state, to solve their self-centred problems relating to water, as there is no cohesion within the communities. Everybody was interested in cooperating but nobody was interested in taking the lead to initiate. Similarly, it is observed in the context of WSD that in some instances, the performance of watersheds is good in the presence of good leaders at the village level [72].

Transaction costs are present in formulating and changing property rights regimes as well as collective action strategies. The transaction costs for negotiation and organization are borne and shared within the community.

---

3. This is only to suggest that the poor depend more on natural assets and commons for their livelihoods. Their critical dependence, however, makes them vulnerable to changes in resource conditions.

Further, there are information costs pertaining to determining the quantity and quality of the resources, which arise when equity concerns are incorporated into property rights [73]. These costs tend to be lower in homogenous communities who share the same resource for similar productive purposes. Nevertheless, the process of negotiation could be long drawn due to information asymmetries. Often, transaction costs associated with equity and measurement disputes delay or block the moves toward effective property rights regimes [73]. This explains the existence of ineffective or inefficient (in terms of poverty alleviation) property rights arrangements. Despite high transaction costs and long-drawn process, it is necessary to move in the direction of pro-poor property rights regimes. The state should bear the transaction costs, as the social benefits outweigh individual benefits in this case.

10.3.3.2 External

External conditions include the policy environment and institutional environment. Policy environment is vital for institutional innovation as well as its sustenance. Policies can lead to institutional innovations or the disintegration of existing ones. For instance, colonial policies of viewing natural resources (water, forests, etc.) as sources of profit led to the decline of age-old community management systems in South India [32]. Similarly, even in independent India, centralized policies that did not involve the local people have not only led to the breakdown of local institutions but also resulted in the degradation of resources [74]. On the other hand, policy support for the involvement of local people in WSD and forest management has resulted in the innovation of successful institutions like the "watershed committees" and "forest protection committees" in India. These policies have provided incentives to the communities through usufruct rights to varying degrees on various benefit flows. Basically, policies either destroy or strengthen the existing property rights, and formal or informal, regimes; for example, the success of Taiwan's irrigation systems is mainly due to the state's policies of co-production and management [75].

While most of the policies are pro-poor in nature, their effectiveness in addressing poverty depends on the institutional environment fostering the policies. Institutional environment comprises of legal and governance structures. Unless policies are enforced through institutions, they may not have much impact on poverty. Hence, property rights need to have legal standing and institutions to enforce them. Pro-poor policies backed by legally honored property rights and enforcing institutional structures together make good governance.

Of late, decentralization of governance is often argued to be a more effective and efficient framework for delivering pro-poor programs [76,77]. It was observed that project costs are four times higher in centralized systems when compared with decentralized systems. Also, asset maintenance is much better in the decentralized systems [78]. However, it is cautioned that decentralization is not effective or efficient under conditions of greater inequality due to political and elite captures and nexus formation between interest groups [79]. While decentralized systems are found to be superior in terms of targeting intra-regional efficiency, their delivery systems are observed to perform better in low-poverty regions and worse in high-poverty regions [80].

10.3.3.3 Embedded

Embedded conditions pertain to the informal institutional arrangements existing in the village communities. These institutional organizations and their functioning might have evolved historically to address specific problems faced by the community. These are embedded in the community because they have evolved from within the system with little or no support from outside. Embedded institutions are found to be more robust and efficient when compared with the formal and imposed institutional arrangements [34]. Important conditions that make the informal institutions vibrant and sustainable include: (1) rules and regulations, such as operational rules, collective-choice rules, and constitutional choice rules as described by Ostrom [38]; (2) equity; and (3) quality of leadership.

Allocation of rights and distribution of costs and benefits (equity) are the most important among the rules and regulations. Free rider problems are more prevalent under institutional arrangements, where costs and benefits are distributed unequally across households. For instance, everybody apparently has equal right to the common grazing lands without any compensation. Equity problems arise when there is no equity in the distribution of cattle holdings across the households. Therefore, households owning a large number of cattle derive maximum benefits from grazing on commons without having to contribute anything extra. As a result, those who are on the other extreme (deriving least benefits) tend to assert their user rights inappropriately, such as encroaching; that is, when people view an established system as inequitable according to established social standards, they have incentives to undermine it [81]. This happens mainly because (1) inequity in the distribution of benefits as explained above and (2) absence of any stake in participation, as the households do not contribute to the maintenance of the CPRs. The principle of contribution (on an equity or proportional basis) seems to be important in some successful collective action situations [70,82]. In other words, when use rights are not "gratis," there are better chances for institutional sustainability.

Unless sanctioning is used against rule breakers, there is no other way to maintain rule compliance, which is termed as quasi-voluntary compliance. This is termed as "quasi-voluntary because the non-compliant are subject to coercion … if they are caught" (Levi, 1988 as quoted in [38]). Who, insider or outsider, should carry out monitoring and enforcement effectively is an important aspect that needs to be addressed. Ostrom [38] observed that monitoring and sanctioning are undertaken by the participants rather than by any external authorities in long-enduring institutions. However, in the context of the changing socio-political conditions, insiders (appropriators) are reluctant to carry out monitoring and sanctioning activities. Internal monitoring can be fearless and effective under a strong and effective leadership; in the absence of such conditions, one should not under-rate the importance of external authorities. Many appropriators seem to prefer external forces to doing it themselves [74]. A mix of internal and external forces may also prove a better alternative, as is the case with the "Panipanchayat."

In Panipanchayat, appropriators undertake monitoring, while sanctioning is enforced by the water council consisting of appropriators with the help of the trust (NGO), an external organization [70]. Quality here is not limited to leaders' skills and assets; instead it includes their commitment (to the cause) and character (upholding the societal values) in discharging duties. These qualities are emulated by others in the community to uphold the tradition of following rules and norms in the long run. However, in the absence of such leaders, self-interest can override the societal or common interests and result in degeneration of institutions.

10.3.4 The Action Arena

From the point of poverty alleviation, the community consists of two main actors: the resource rich and resource poor. Poverty alleviation requires pro-poor changes in policy and property rights: in the process of reallocation, the resource rich ought to forego some of their benefits for the overall improvement in the socioeconomic conditions. Therefore, consensus between the rich and poor is required to arrive at an acceptable trade-off. While bringing different actors together is central to collective action and poverty alleviation, we discuss in this subsection the conditions that would facilitate collective action.

In watershed-related literature, participatory management is defined as a process, which operates at various levels right from planning through execution and finally ensuring sustainability for the future. This, however, excludes the necessary and sufficient conditions as well as different forms and layers of participation. Pimbert and Pretty [83] identified a few forms of participation, ranging from passive to self-generated. These forms are, however, not exhaustive and can change in priorities and intensity according to the situation [84,85]. A good example of this process can be seen in Sukhomajiri watershed study [69]. Cohen and Uphoff [86] identified basic four layers beginning with decision making and going through implementation, distribution of the accrued benefits, and finally evaluating for correction of the shortfalls in the process of implementation. A common feature observed from the above classification of forms is that these authors have specifically concentrated upon the functional aspect while the utility of these aspects was limited to testing the extent of participation and the depth of involvement [70].

Broadly, in the real world, one could distinguish three types of collective choice situations: (1) potential situations for collective action or institutional innovation, (2) conditions for initiating collective action or institutional innovation and change, and (3) conditions for sustaining collective action or institutional sustainability. All the communities faced with resource management problems might not have the potential for institution building, in the absence of which the potential resources tend to degrade. Similarly, all potential situations need not necessarily lead to institutional innovation, and all the institutional innovations are not necessarily sustainable in the long run. This distinction is important as different factors tend to operate at different levels and the degree of their importance also differs across situations, although they overlap at times.

As is clear from the literature (for a review, see [54]), some of the assumptions, especially behavioral, made in various theories are not very realistic and do not reflect the rural communities with which we are dealing. This calls for the articulation of assumptions necessary to understand rural societies. These assumptions include:

1. Individuals are "boundedly rational," i.e., individuals act rationally within the limits of the information available to them. Hence, under conditions of low awareness and information lags, assumption of a priori expectations would be erroneous. Moreover, calculating costs and benefits pertaining to natural resources that continuously change needs a higher order or computational ability that is beyond the capability of ordinary people (without training) let alone rural people.

2. Individuals are often "intentionally irrational (economic)." An individual's actions, sometimes, may not reflect even bounded rationality. This is mainly due to socio-political factors, which are increasingly becoming central to rural societies. Here group rationality overtakes individual rationality, and the implicit rationale is political gain for the group as a whole, although a particular individual may or may not get any tangible benefits in the immediate future. What is rational for the group may be irrational for an individual, and the individual may simply follow the group, even if she/he is aware of the irrationality of the (economic) choice. This may be attributed to expected socio-political benefits in the future.

3. All people in a community are "not opportunistic" as defined by Williamson: "self-interest seeking with guile" [87, p. 47]. They trust one another although we do not rule out the possibility of opportunistic behavior, whatever the size of such a group. It is observed that trust, truthfulness, and acting with justice are essential lubricants of societies [88,89].

4. Individuals are aware of the behavioral pattern of others, as they have known each other for a long time and would have faced a variety of situations together (involved in "repeated games"). They also "communicate freely" even in a

faction-ridden community; hence they are neither prisoners nor victims of isolation.

5. In general, a majority of individuals are "trend followers rather than trend setters."; that is, individuals are willing and ready to cooperate but not ready to take initiative or lead in creating institutions.

These assumptions largely reflect the basic characteristics of rural communities, and one has to find ways to bring them together. According to Rajendra Singh, (the Magsaysay Award winner for his seminal work on organizing rural communities to develop water-harvesting structures in Rajasthan), there are five types of groups among rural communities: (1) base (influential due to their background, credibility, character, etc., and boundedly rational); (2) negative (oppose everything and anything—intendedly irrational);(3) opportunists (always take the side of majority); (4) target group (vulnerable and no voice); and (5) motivated (eager to do something—trend followers). Singh synthesizes that the task in the collective action strategy is to bring the base and motivated together. Once these two are together, involving the target group is easy. Together, these three groups become the majority in the villages and hence attract the fourth group, i.e., the opportunists. The negative group then gets marginalized, leaving them with the option of either joining or quitting.[4] The three collective choice situations identified above arise due to the variations in the composition of these groups in the community, i.e., communities with greater proportion of negative and opportunist groups, etc. The conditions discussed (external, internal, and embedded) influence or determine the three types of collective choice situations.

In the context of WSD, as is the case with other commons, the focus has been mainly on the participation or collective action. The discussion has focused on why collective action failed to take off in the majority of cases while it is effective in others. Most of the fact finding has been performed on the basis of the implementation process and identifying the project implementing agencies that have done/not done a proper job. However, the investigation stopped at the implementation level, as fewer watersheds were sustained after the implementation phase, irrespective of who implemented them. The reason could be that property rights issues are either avoided or taken for granted; i.e., the approach has been to honor the existing rights. In the process, equity issues are sidelined. Hence, the linkages between property rights and collective action on one hand and equity and collective action on the other are less understood. This is true at the policy as well as implementation levels.

---

4. Based on the discussion with Rajendra Singh on February 6, 2005, at Tarun Bharat Sangh, Alwar, Rajasthan.

10.3.5 Policy Directions

This section set out to understand the relative importance, effectiveness, and strengths of property rights and collective action in alleviating poverty in the backdrop of WSD in rainfed regions. In the course of the analysis we tried to understand the linkages or synergy between property rights and collective action; how necessary and sufficient the conditions of property rights and collective action in poverty alleviation are, whether either or both of them ameliorate poverty, and if so, under what socioeconomic and political situations. It is argued that the importance and strengths of property rights and collective action depends on the nature and type of resource and the existing property rights.

Existing property rights are often found to be biased against the poor. This is mainly due to the absence of equity concerns in the existing property rights. As a result, these property rights are not effective in addressing the issues of poverty, and this is one of the reasons for widespread failure of collective action strategies in natural resource management like WSD.

While collective action could initiate changes in property rights regimes, incorporating the equity issues into property rights involves transaction costs. On the other hand, equity-based property rights facilitate collective action strategies. Thus, equity (or justice) appears to be the critical factor in determining the effectiveness of collective action and property rights in addressing poverty.

The issue of how to secure the rights and entitlements of poor people to access water and common property resources in WSD needs to be resolved. Experience from the Panipanchayat approach developed in Maharashtra and from examples of successful community forestry and CPR management can be helpful in this process. Equity and equality in the distribution of economic gains among the community members is as important as the equity in WSD coverage. While the latter is concerned with the equity in access, the former pertains to equity in outcomes.

Justice issues pertain to the neutrality of technology in terms of location (different geographic locations of the watershed) and well-being (economic distribution) of the participants. Inequity in the former case is purely technical while the latter is institutional. Since no technology has an in-built bias toward a particular class or caste, the bias is always due to the existing institutional (property rights) structures (agrarian structure, credit markets, social structure, etc.). Therefore, inequalities could be minimized through more egalitarian institutional arrangements and legislations that demonstrate procedural justice. In other words, technical inequalities can be corrected by compensating the participants from the disadvantaged locations, while distribution bias can be reduced by correcting distortions in land, labor, water, credit markets, and property rights regimes.

It is necessary to protect the interests of the disadvantaged sections of the community such as landless families, the landed poor, and women. The most pressing issue is access to CPRs, especially water, to all sections of the community. As indicated earlier, access to water can be ensured only by delinking the water rights from land rights fostered with clearly defined property rights on water. This requires an appropriate legal framework and effective institutional arrangements. Although this dimension is often brushed aside at the policy level b as a difficult task, there is a need to move toward this direction [11].

Linkages between collective action and property rights must be explored and understood, as collective action thrives on equity-based property rights. Unless future research focuses on understanding these ground level realities and integrates them with policy, policies will continue to be *ad hoc* and ineffective in addressing the poverty issues. Therefore, future development of watershed policy needs to both reflect as well as influence the wider policy environment, especially by including agricultural, poverty relief, and other linked policies. It needs to be a two-way process. How watershed policy is developed needs to be based on an understanding of these linked policies. At the same time, this understanding should be used to influence reforms in the other policies as well; a situation where the implementation of one key policy is undermined by the effects of other policies is not acceptable.

10.4 COMMUNITY VIEWS ON COLLECTIVE ACTION AND THE EQUITY OF THE WSD PROCESS

Having reviewed the use of justice principles and the role of property rights in Australia and India, and the relationship between communal property rights and collective action, we can interpret these insights through the behavior and opinions of those involved with WSD in the two HUNs of Andhra Pradesh involved in the survey (see Chapter 2). Data from both the first and second surveys are used for the following tables.

Ideally, WSD is designed to promote community action through a water use committee with representative membership to ensure that the outcomes are equitably distributed. It is hoped that the community will monitor rainfall and groundwater levels and discuss the water requirements of differing crop types. Communal decisions regarding what should be grown and where are desired. Locally based rules, such as "if a high water using crop is chosen then the area allowed to be planted is reduced to ensure sustainable water use" can evolve. We therefore asked those who had been in WSD at different points on the HUN about their crop decision making and their views of equity-related issues associated with the implementation of WSD.

10.4.1 Collective Decision Making

Respondents were asked how they made their decisions for crop choice. According to general WSD principles, there should be a collective influence to ensure sustainable water use. The results are shown in Table 10.3.

TABLE 10.3 Decision-making: crop choice (% of HHs)

Type of watershed	HUN1 (Anantapur/ Kurnool)		HUN2 (Prakasam)	
	Own decision	Collective choice	Own decision	Collective choice
Upstream	77	23	97	3
Midstream	76	24	98	2
Downstream	59	41	96	4
Watershed villages	69	31	97	3
Control village	100	0	100	0

TABLE 10.4 Planted prohibited crops because of profitability (% of HHs)

Type of watershed	HUN1 (Anantapur/ Kurnool)	HUN2 (Prakasam)
Upstream	13	11
Midstream	29	0
Downstream	46	2
Watershed villages	34	5

It can be seen that in the higher rainfall HUN (Prakasam) there is minimal collective influence on decision making—less than a quarter of the farmers in the upstream and midstream operated by collective choice and in downstream a larger proportion of landholders operated collectively, although still less than half.

The proportion of people who planted nonpreferred or prohibited crops is shown in Table 10.4. It can be seen that this frequency was modest for the Prakasam HUN, although it is more than 10% in the upstream location. Similarly, close to half those downstream in Anantapur/Kurnool were prepared to grow prohibited crops for profit even though that location claimed to take collective choice into account. This may reflect the immediate need from chronic low yields; only one farmer claimed that the community had forced him to abandon a particular crop.

Finally, Table 10.5 shows how prevalent wider group welfare was in participating in the process of collective decision making. It can be seen that

TABLE 10.5 Concern about the welfare of the larger group during a collective decision over crop choice and groundwater use (% of HHs)

Type of watershed	HUN1 (Anantapur/Kurnool)	HUN2 (Prakasam)
Upstream	15	50
Midstream	5	0
Downstream	13	0
Watershed villages	11	25

these concerns are not prevalent and perhaps only occur to a significant extent in upstream Prakasam.

To summarize, we see that while there were collective decision-making processes, the substantial majority of landholders made their decisions based on individual needs for profit. A significant minority were prepared to ignore community rules, particularly in downstream Anantapur /Kurnool. On the other hand, wider or Kantian (or moral inclusion) type welfare judgments were not a significant feature of communal decision making. Since the responses related to WSD, as it was perceived in their own area, on the issue of sharing at a meso or HUN level given the current WSD processes, seem to reflect problems, it is likely that new institutional structures will have to be created.

10.4.2 Procedural and Distributive Justice and WSD

A major distributive concern associated with WSD is access to water. Householders were therefore asked their impression of any reduction in water flows as the result of upstream WSD. The results are shown in Table 10.6.

TABLE 10.6 Reduction in water flows into the village water bodies as a result of WSD in the upstream villages (% of HHs)

Location of the watershed	HUN1 (Anantapur/Kurnool)	HUN2 (Prakasam)
Upstream	4	0
Midstream	44	57
Downstream	33	50
Total	31	29

It was found that a sizeable minority of the midstream and downstream households felt that they had observed a decrease in flow, which they attributed to the advent of WSD. This could be the basis for some equity disputes if the implementation of WSD becomes more widespread.

In terms of the procedures for decision making, there seemed to be very little conflict, with only 3% of the householders reporting it. These conflicts were mostly dealt with by the village elders, although there was some involvement of political leaders. Householders were also asked whether they thought the WSD process endeavored to represent the majority of the community or whether the elite were advantaged. Most people felt that the process was representative (85%) while the remainder expressed that the elite profited most. Finally, the majority (72%) thought WSD was helpful to the poor while the rest felt that this was not the case.

To summarize, we can say that equity issues do not seem to be a high profile issue within WSD. There also seems to be limited attention to consider overall welfare with regard to WSD, and there appears to be little conflict in its implementation. Collective decision making also seems to have little influence in crop selection, which is mainly driven by the need for profit, especially in the downstream areas of the Anantapur/Kurnool HUN. It would seem then that the communal requirements of WSD have not been adopted universally with much enthusiasm and this is likely to mute the effectiveness of WSD outcomes.

Nevertheless, the interviews about the meso-implementation have indicated that the equity issues currently indicated by changed flows and the lack of meso-level institutions may sharpen the sense of competition among those in the catchment. It is notable in this regard that the findings of Chapter 8 indicate that WSD programs may have led to an increase in inequality as higher caste farmers reported higher drought resilience in the post-WSD era as compared with the pre-WSD era. Concern has also been expressed about some villages on the catchment dominating others. Hence, planning for meso-catchments will require close attention to the design and support of community-based institutions [33]. This is likely to be a vital component of social capital, which has a direct link to resilience (see Chapter 12) as a primary outcome of WSD.

10.5 CONCLUSION

Each section of this chapter concluded that the long-term management of WSD must consider how collective decision making and action can be maintained at an appropriate hydrological scale in both Australia and India. While property rights and markets can assist, there is no natural evolution to sustainability through these vehicles. Although both can be helpful, they need to be underpinned by concerted community action, which needs to be based on distributive and procedural justice. It is clear from the survey that currently landholders are motivated by individual profit needs; they either rely largely on their own judgment or follow lead farmers when choosing the crop type. In the long term, this trend will result in the ongoing deterioration of the resource in terms of quantity as well as quality.

While these issues of communal approaches to groundwater management are not currently of great priority to the community in Andhra Pradesh, the move to mesoscale WSD will require careful attention to how justice principles can be used to promote sustained community action and appropriate property rights. In this regard, the eight "rationalities" or criteria for the successful delivery of WSD identified by Crase et al. [90] provide a very useful evaluative tool. These rationalities include social, political, organizational, and government rationality, all of which are highly pertinent to the achievement of justice and cooperation at the local level and will be crucial if meso-institutions are required to be designed and created.

REFERENCES

[1] Allan JA. Water in the environment/socio-economic development discourse: sustainability, changing management paradigms and policy responses in a global system. Government Opposition 2005;40(2):181—99.

[2] Magnani E. The environmental Kuznets Curve, environmental protection policy and income distribution. Ecol Econ 2000;32:431—43.

[3] Leigh R. Economic growth as environmental policy. Reconsidering the environmental Kuznets Curve. J Public Policy 2004;24(3):327—48.

[4] Crase L, Gawne B. Coase-coloured glasses and rights bundling: why the initial specification of water rights in volumetric terms matters. Econ Pap 2010;30(2):135—46.

[5] Turral H, Fullagar I. Institutional Directions in Groundwater Management in Australia. In: Giordano M, Villholth KG, editors. Agricultural Groundwater Revolution. Oxfordshire, CABI International; 2007. pp. 320—61.

[6] Skurray JH, Roberts EJ, Pannell DJ. Hydrological challenges to groundwater trading: lessons from south-west Western Australia. J Hydrol 2012;412-413:256—68.

[7] GoI. Guidelines for Watershed Development. New Delhi: Ministry of Rural Development; 1994. October.

[8] Poddar R, Qureshi ME, Syme G. Comparing irrigation management reforms in Australia and India—special reference to participatory irrigation management. Irrigation Drainage 2011;60:139—50.

[9] Meinzen-Dick R, Mwange E. Cutting the web of interests: pitfalls of formalising property rights. Land Use Policy 2009;26(1):36—43.

[10] Koontz TM, Sen S. Community responses to government defunding of watershed projects: a comparative study in India and the USA. Environ Manag 2013;51:571—85.

[11] Reddy VR. Water Security and Management: Lessons from South Africa. Econ Political Wkly 2002;37(28):2827—81.

[12] Giddens A. Consequences of modernity. Cambridge, MA: Polity Press; 1990.

[13] Boserup E. The Conditions of Agricultural Growth. London: Allen and Unwin; 1965.

[14] Acemoglu D, Robinson J. Why Nations Fail: The Origins of Power, Prosperity and Poverty. London: Profile Books; 2012.

[15] Heitberg R. Property rights and natural resource management in developing countries. J Econ Surv 2002;16(2):189−214.

[16] Syme GJ, Porter NB, Goeft U, Kington EA. Integrating social wellbeing into assessments of water policy: meeting the challenge for decision makers. Water Policy 2008;10:323−43.

[17] Crase L. The Murray Darling Basin Plan: An Adaptive Response to Ongoing Challenge. Econ Pap 2012;31(3):318−26.

[18] Garry T. Water markets and water rights in the United States: Lessons from Australia. Macquarie J Int Comp Environ Law 2007;4:23−60.

[19] Strang V. The Meaning of Water. Berg: Oxford; 2004.

[20] Hoekstra AY, Savinije HHG, Chapagain AK. An integrated approach towards assessing the value of water: a case study in the Zambesi basin. Integr Assess 2001;2:199−208.

[21] Syme GJ, Nancarrow BE. Justice and the allocation of benefits from water. Soc Altern 2008;27(3):21−5.

[22] Wenz PS. Environmental Justice. Albany, NT: State University of New York Press; 1988.

[23] Syme GJ, Nancarrow BE, McCreddin JA. Defining the components of fairness in the allocation of water to environmental and human uses. J Environ Manag 1999;57:51−70.

[24] Lukasiewicz A, Davidson P, Syme G, Bowmer K. Assessing government intentions for Australian water reform using a social justice framework. Soc Nat Res 2013;26:1314−29.

[25] Rawls J. A Theory of Justice. Cambridge MA: Harvard University Press; 1971.

[26] Sen A. The Idea of Justice. London: Penguin; 2009.

[27] Anderson TL, Scarborough B, Watson LR. Tapping water markets. Hoboken: Taylor and Francis; 2012. pp.55−74.

[28] Syme GJ, Nancarrow BE. The determinants of perceptions of fairness in the allocation of water to multiple uses. Water Resources Res 1997;33(9):2143−52.

[29] Robins L, Kanowski P. Caring for country: eight ways in which 'caring for country' has undermined Australia's regional model for natural resource management. Aust J Environ Manag 2011;18(2):88−108.

[30] Syme GJ. Struggling with uncertainty in Australian water allocation. Stochastic Environ Res Risk Assess 2014;28:113−21.

[31] Tyler TR, Blader S. The group engagement model: procedural justice, social identity and cooperative behavior. Pers Soc Psychol Rev 2003;7:349−66.

[32] Reddy VR. Irrigation in Colonial India: A Study of the Madras Presidency during 1860−1990. Econ Political Wkly 1990;Vol. XXV.(Nos. 18 and 19):1047−54. May 5−12, 1990.

[33] Reddy VR, Syme G, Ranjan R, et al. Scale Issues in Meso-Watershed Development: Farmers' Perceptions: Designing and Implementing Common Guidelines. LNRMI Working Pap 2011. No. 2, Hyderabad, India.

[34] Reddy VR, Reddy PP. Water institutions: Is Formalisation the Answer? (A Study of Water User Associations in Andhra Pradesh). Indian J Agric Econ 2002;57(3):519−34.

[35] Guillet D. Reconsidering institutional change: property rights in northern Spain. Am Anthropologist 2000;102(4):713−25.

[36] Strauch AM, Almedon AM. Traditional water resources management and water quality in rural Tanzania. Hum Ecol 2011;39:93−106.

[37] Hukkinen J. Institutions in Environmental Management: Constructing Mental Models in Sustainability. London: Routledge; 1999.

[38] Ostrom E. Governing the Commons: The Evaluation of Institutions for collective Action. Cambridge, U.K: Cambridge University Press; 1990.

[39] Loehman ET, Kilgour DM. Designing Institutions for Environmental and Resource Management; 1998. Cheltenham, Elgar.

[40] Reddy VR, Reddy MG, Galab S, Soussan J, Baganski OS. Participatory Watershed Development in India: Can it Sustain Rural Livelihoods? Dev Change 2004;35(2):297—326.

[41] Francis R. Water justice in South Africa: natural resources policy at the intersection of human rights, economics and political power. Georgetown Int Law Rev 2005;18:149—96.

[42] Scott A. The evolution of resource property rights. In: Scott A, Coustalin G, editors. Rights over flowing water. OXFORD SCHOLARSHIP ONLINE; 2008. http://dx.doi.org/10.1093/acprof:oso/9780198286035.003.0003.

[43] Reddy VR. Watershed development for sustainable agriculture: need for an institutional approach. Econ Political Wkly 2000;35(38):3435—44.

[44] Mwangi EN. Conceptions of poverty and the contributions of property rights and collective action to poverty reduction (overview of concepts), 1st session, e-conference organised by CAPRi on Poverty, Property Rights and Collective Action (CAPRi-talk); 2004.

[45] Meinzen-Dick R. Conceptions of poverty and the contributions of property rights and collective action to poverty reduction (overview of concepts), 1st session, e-conference organised by CAPRi on Poverty, Property Rights and Collective Action (CAPRi-talk); 2004.

[46] Bromley DW. Actors, Institutional Change and Poverty Reduction, 5th session, e-conference organised by CAPRi on Poverty, Property Rights and Collective Action (CAPRi-talk); 2004.

[47] Birner R. Actors, Institutional Change and Poverty Reduction, 5th session, e-conference organised by CAPRi on Poverty, Property Rights and Collective Action (CAPRi-talk); 2004.

[48] Di Gregorio M. Conceptions of poverty and the contributions of property rights and collective action to poverty reduction (overview of concepts), 1st session, e-conference organised by CAPRi on Poverty, Property Rights and Collective Action (CAPRi-talk); 2004.

[49] Ostrom E, Gardner R, Walker J. Rules, games, and common-pool resources. Ann Arbor: University of Michigan Press; 1994.

[50] Singh K. Managing Common Pool Resources: Principles and Case Studies. New Delhi: Oxford University Press; 1994.

[51] Agrawal B. Participatory Exclusions, Community Forestry, and Gender: An Analysis of South Asia and A Conceptual Framework. World Dev 2001;29(10):1623—48.

[52] GoI. From Haryali to Neeranchal: Report of the Technical Committee on Watershed Programmes in India, Department of Land Resources. Ministry of Rural Development, Government of India; 2006. January.

[53] de Janvry Alain. State, Market and Civil Organisations: New Theories, New Practices and their Implications for Rural Development. World Dev 1993;Vol. 21.(No. 4). April.

[54] Reddy V, Ratna. Managing the Commons in Transitory Economies: Towards a Theory of Collective Action. Paper presented at the International Conference of the European Society of Ecological Economics; 1998. Geneva, (4—7 March).

[55] Bromley DW. Economic Interests and Institutions: The Conceptual Foundations of Public Policy. New York: Basil Blackwell; 1989.

[56] Baland JM, Platteau JP. Halting Degradation of Natural Resources: Is there a Role for Rural Communities? Oxford, U.K: Clarendon Press; 1996.

[57] Gans O. Economic Analysis of Ecological Disequilibrium in Industrialized and Developing Countries' in Gins Oscar. In: Environmental and Institutional Development: Aspects of Economic and Agricultural Policies in Developing Countries. Saarbruken: Vela breitenbach Publishers; 1989.

[58] Reddy VR. Reviving the Traditional Systems for Sustainable Rural Livelihoods: A Study of Tank Renovation Programme in Rayalaseema, Project Report, Centre for Economic and Social Studies, Hyderabad; 2001.

[59] Shiferaw B, Ratna Reddy V, Wani SP, Rao GDN. Watershed Management and Farmer Conservation Investments in the Semi-Arid Tropics of India: Analysis of Determinants of Resource Use Decisions and Land Productivity Benefits. Socio-economics and Policy Working Paper Series No. 16. Patancheru 502 324, Andhra Pradesh, India: International Crops Research Institute for the Semi-Arid Tropics (ICRISAT); 2004.

[60] Reddy V, Ratna, et al. User Valuation of Renewable Natural Resources: A Study of Arid Zone, Project Report. Jaipur, India: Institute of Development Studies; 1997.

[61] Heckathorn DD. Collective Action and Group Heterogeneity: Voluntary Provision versus Selective Incentives. Am Sociological Rev 1993;58(3):329−50.

[62] Varughese G, Ostrom E. The contested role of heterogeneity in collective action: some evidence from community forestry in Nepal. World Development 2001;29(5):747−65.

[63] Poteete A, Ostrom E. Heterogeneity, group size, and collective action: the role of institutions in forest management. Dev Change 2004;35(3):435−61.

[64] Olson Jr. Mancur. The Logic of Collective Action: Public Goods and the Theory of Groups. Cambridge: Harvard University Press; 1965.

[65] Jodha NS. Population Growth and Decline in Common Property Resources in Rajasthan, India. Popul Dev Rev 1985;vol. 11.(No. 2):247−64.

[66] Guha R. Scientific Forestry and Social Change in Uttarakhand. Econ Political Wkly 1985;vol. 20. Special Number, November.

[67] Guha R. The Unquiet Woods: Ecological Change and Peasants Resistance in the Himalaya. New Delhi: Oxford University Press; 1989.

[68] Wade R. The Management of Common Property Resources: Collective Action as an Alternative to Privatisation or State Regulation. Camb J Econ 1987;Vol. 11:95−106.

[69] Chopra K, Gopal K, Murty MN. Participatory development, people and common property resources. New Delhi: Sage Publications; 1990.

[70] Deshpande RS, Reddy VR. Differential impact of watershed based technology: some analytical issues. Indian J Agric Econ 1991;46(3):261−9.

[71] Wade R. Village Republics: Economic Conditions for Collective Action in South India. Cambridge University Press; 1988.

[72] Reddy V, Ratna U, Hemantha Kumar D, Mohan Rao. Watershed Management for Sustainable Agriculture: Need for an Institutional Approach. Hyderabad: Mimeo, CESS; 2005.

[73] Libecap Gary D. A Transactions-Costs Approach to the Analysis of Property Rights. In: Brousswau Eric, Glachant Jean-Michel, editors. The Economics of Contracts: Theory and Applications. Cambridge, UK: Cambridge University Press; 2002.

[74] Reddy VR. A Study of Willingness and Ability to pay for Water, Project Report. Jaipur, India: Institute of Development Studies; 1996.

[75] Lam WF. Institutional Design of Public Agencies and Co-production: A Study of Irrigation Associations in Taiwan. World Dev 1996;vol. 24.(No. 6). June.

[76] Manor J. Madhya Pradesh Experiments with Direct Democracy. Econ Political Wkly 2001. 3 March;36(9):715−16.

[77] Seabright P. Accountability and decentralisation in government: an incomplete contracts model. Eur Econ Rev 1996;40 (January), 61−89.

[78] Bardhan P. Decentralized Development. Indian Econ Rev 1996;31(2):139−56.

[79] Bardhan P. Decentralization of governance and development. J Econ Perspect 2002;14(4):185−205.

[80] Bardhan P, Mookherjii D. Capture and governance at local and national levels. Am Econ Rev (Pap Proc) 2000;90(2):135−9.

[81] Hanna S. Designing Institutions for the Environment. Environ Dev Econ 1996;1(1):122−5.

[82] Agrawal A. Rules, Rule Making, and Rule Breaking: Examining the Fit between the Rule Systems and Resource Use. In: Ostrom Elinor, Gardener R, Walker J, editors. Rules, Games and Common-Pool Resources. Ann Arbor, MI: University of Michigan Press; 1994. pp. 267–82.

[83] Pimbert M, Pretty J. Diversity and Sustainability in Community based Conservation, Paper Presented at the UNESCO – IIPA Regional Workshop on Community Based Conservation; 1997. New Delhi, February.

[84] Deshpande RS, Nikumh. Treatment of Uncultivated Land under Watershed Development Approach: Institutional and Economic Aspects. Arthavijnana 1993;vol. 35.(No. 1).

[85] Sharma J. Joint Forest Management – Some Fundamentals Reviewed. The Indian Forester 1997;vol. 123.(No. 6). June.

[86] Cohen JM, Uphoff NT. Participation's Place in Rural Development: Seeking Clarity through Specificity. World Dev 1980;8(3):213–35.

[87] Willimson OE. The Economic Institutions of Capitalism. New York: The Free Press; 1985.

[88] Dasgupta P. The Economics of the Environment. Environ Dev Econ 1996;vol. 1. Part 4, October.

[89] Sethi R, Somanathan E. The evolution of social norms in common property resource use. Am Econ Rev 1996;86(4):766–88.

[90] Crase L, Ghandhi V, Clement F. Enhancing Institutional Performance in Watershed Management in Andhra Pradesh, India. Final Rep 2013. FR2012–10.

Part IV

Integrating Science into Policy and Practice

Chapter 11

High Stakes—Engagement with a Purpose

T. Chiranjeevi*, Geoffrey J. Syme[§] and V. Ratna Reddy*

* *Livelihoods and Natural Resource Management Institute, Hyderabad, India,* [§] *Edith Cowan University, Perth, Australia*

Chapter Outline

11.1 INTRODUCTION

Getting the stakeholders' buy-in for implementing project recommendations requires a much deeper partnership at the level of methodology selection and model formulation than is traditionally needed for research projects. Moreover, such purposeful interactions would require the involvement of all members of the research team, and not just the project head or the most senior team members—every member of the team needs to know and understand the stakeholder needs and expectations from the project. Thus, delivering outputs in line with the stakeholders' requirements and time lines should become integral to project objectives and purpose; such a deeper commitment requires more than a perfunctory reference to stakeholder engagement in the project documents.

Integrated Assessment of Scale Impacts of Watershed Intervention
http://dx.doi.org/10.1016/B978-0-12-800067-0.00011-6. Copyright © 2015 Elsevier Inc. All rights reserved.

There are many articles [1−5] emphasizing the importance of stakeholder engagements in a variety of contexts. Articles on the subject have dealt with a range of issues starting from the definition of stakeholders to defining the typologies of stakeholders. Studies also use different types of tools to engage and communicate with different stakeholders. For example, Welpa et al. [5] examined the relevance of three theoretical frameworks: rational actor paradigm, Bayesian learning, and organizational learning. According to the authors, science-based dialogs have the potential to help identify socially relevant and scientifically challenging research questions, act as tools for a "reality check," incorporate ethical considerations, and provide access to data and knowledge that could otherwise be difficult to access. Defined as a "structured communicative process of linking scientists with actors that are relevant for the problem at hand," science-based dialogs may be very similar to other forms of dialog such as policy, multi-stakeholder, and corporate, although the objectives behind these dialogs could differ. The authors contend that early and regular involvement can create a sense of ownership of the research process so that the research results are more likely to be used by the stakeholders.

Although there are quite a few definitions [2] of "who" or what a stakeholder is, in general, a stakeholder could be considered as any person, group, or agency that has an interest in the issue under consideration or is impacted by it. In this project, although it was not explicitly defined, the stakeholders include those agencies that could benefit from the research findings as well as the groups or individuals who could potentially be affected by any policy or implementation changes resulting from the adoption of the project recommendations. These include policy makers, implementing agencies, and the local communities. Given such a broad definition of who a stakeholder could be, it is quite a task to identify the key stakeholders among the variety of stakeholders. As Carney et al. [2] pointed out, stakeholder engagement is not a one-sided process—both the researchers and the stakeholders make choices. Both parties make choices in terms of when, how much, and how frequently interaction is required. It is not uncommon to find a single stakeholder targeted by different projects of the same funders, probably because she/he is more obliging, the projects have some common threads, or some of the teams are involved across different projects. Such situations sometimes run into the risk of tiring out or confusing the stakeholders with information overload.

When a new project approaches, the stakeholder response could be quite unenthusiastic initially, and the efforts to engage the attention of the stakeholders may convert into a contest of influence among different projects, almost resulting in a lose−lose situation. This project too experienced something similar as there were quite a few projects going on at the same time in inter-related areas [watershed development (WSD) and climate change] funded by the same funders (Australian Centre for International Agricultural

Research). Therefore, it was not easy for the project (probably, other projects as well) to connect with the stakeholders (here, the Department of Rural Development, DRD) at a level where they could show a sense of ownership for the project outputs.

The main objective of engagement with the DRD was to develop linkages at the staff level to first understand, disseminate, and train in the usage of the integrated approach developed by the project. The endeavor was to create a sustainable watershed model and decode it for easy adoption in planning, design, and implementation, in addition to monitoring the activities of the WSD. While the focus was on developing a methodology to help better evaluate the socio-economic impacts of mesoscale watersheds, the challenging task set by the project was to develop a model that could seamlessly integrate the hydro-geological, biophysical, and socioeconomic aspects of a mesoscale watershed. A second important aspect of this project was to consider the impacts across different streams (upstream, midstream, and downstream of a hydrological unit, HUN) from an integrated perspective, i.e., to understand the surface and subsurface flows and their influence on the stream-level impacts.

The project also explored design-related issues such as the impact of different treatments at different stream levels and shaping the interventions to minimize the externalities and maximize the scale-related benefits. The process, method, and tools developed as part of this approach proved to be capable of providing a scientific basis to delineate watersheds, enabling design for sustaining impacts in the long term, and ensuring equity and efficiency through proper design and implementation. The study also had developed an alternative approach to evaluating impacts in situations where baseline data were not available: by considering resilience at the household level as a measure for composite impact of watersheds, the method developed by the study avoids the issues involved in "before and after" and "with and without" methods of impact evaluation. The integrated approach to design and imple-mentation and the new methods, templates, and tools have been converted into a package of inputs for a workshop that could help build the knowledge and skills of the people involved in watershed-related work who use this integrated approach.

The design, implementation, and impact assessments are mostly performed by the implementing agencies: either a government organization (GO) or a nongovernment organization (NGO). Thus, it is very important to create awareness about this integrated approach and an understanding at that level. However, this awareness also needs to be built at the community level, because it needs to be involved in maintaining and sustaining the watersheds to sustain the favorable impacts as well as minimize conflicts. Furthermore, equity and efficiency, which are highly dependent on the design and implementation of a watershed, would be difficult to achieve without the informed participation of the local communities. Thus, the project wanted to place equal emphasis on educating, creating awareness, and getting a buy-in for the integrated approach from stakeholders at the level of local communities such as water user associations and farmers cooperatives, the GO and NGOs at the implementation level, and all the departments involved in WSD and monitoring. The project's experience in engaging with the different stakeholders for this purpose, right from the field to the implementing and monitoring agencies, provides lessons in making research outputs socially relevant and effective in supporting better utilization of resources. The lessons learned are used as a basis for putting together a systematic approach for engaging different stakeholders and creating ownership for project results among the policy and implementing agencies.

This chapter is divided into three main sections in addition to the introductory section. The first section describes the process of stakeholder engagement as it actually happened in the project. The second section describes the outcomes of this process and the lessons learned. The third section focuses on developing an effective stakeholder engagement model along with any practical issues that may need to be dealt with in order to follow it.

11.2 ACTUAL PROCESS ADOPTED IN THE PROJECT

Some of the reasons for the general skepticism or stakeholder apathy toward research outcomes include not understanding the complex models used by researchers, difficulty in discerning the multidirectional relationships that are depicted, and/or lack of belief in the underlying database used. Experience in this study proved that stakeholders cannot be effectively engaged unless the interactions go beyond the high-level presentations and sharing of results through policy briefs. Complex models need to be broken down into easy-to-use rules of thumb, tools, and templates that simplify the adoption of research outputs from a practical point of view. The interactions that the project team had with the potential users of the research outputs at various stages of the project not only helped understand this need but also paved the way for extending their usage to a wider group of users. This section tries to provide an account of this journey, which is definitely not linear, and provides insights into what worked and why.

One of the most important aspects of this journey is the evolution of the team from a set of individuals to a team that realized the interdependencies of their respective areas. Members of a multidisciplinary project usually differ in the extent, intensity, level, and type of stakeholder exposure they have. They also have different notions about the relevance and importance of engaging with different stakeholders. Thus, projects would do well to bring their teams onto one platform by creating a shared understanding of the purpose, frequency, and type of stakeholder engagements planned, and the responsibility of each team member to keep those commitments.

Commitment to the stakeholders should be as much individual as it is collective. Such a distributed approach would help each researcher to constantly question the practical relevance of his/her work and develop relations with end users, as well as adhere to time lines. Constant dialog with the relevant stakeholders also ensures an understanding of the constraints and issues faced by the end users at various levels. Although such an interaction happened quite spontaneously for the team, it did prove to be a major turning point in changing team perceptions about who the stakeholders are, what they need, and how the project could make itself useful to them. This is the other important aspect of the journey, which tells us the importance of planning for such experiences and timing them appropriately.

The reason the team incrementally discovered its real purpose rather than through a planned process needs to be explored more in depth to learn from such experiences. The project's main objective is integration of different sciences to develop a model for assessing the scale impacts in the context of a meso-watershed. The team members coming from different disciplines felt it necessary to spend substantial time, internally working out the linkages within and across disciplines, in an attempt to build an impact assessment model for a mesoscale watershed project. For example, issues like scale had different connotations to different streams of sciences, which had to be clarified even before the project site could be identified.

Thus, for a major part of the project the team hardly had any interaction with the stakeholders. The team was so focused on the technical aspects of the problem that it paid little attention to identifying the stakeholders apart from the DRD as possible users of the project outputs. The DRD, to its credit, had taken an active interest in setting up expectations and shaping the key objectives of the project. However, beyond the DRD the project identified few stakeholders who could possibly use its outputs. Such high-level focus is commonly adopted by research projects, because reaching out to the lower levels would require the project to move away from the theoretical models and toward practicable outputs. However, even the policy-level stakeholders often find the recommendations from research projects to be of limited use for policy purposes.

The model, which was a conjoined conglomeration of different sciences and models, turned out to be quite complex to comprehend as well as explain to anyone outside the team. Hence, the project team hesitated to share the model with outsiders without first understanding the full implications of this integrated approach. Therefore, the team never sought out input or feedback on the relevance and usefulness of such a huge and complex model from a practitioner's point of view. Without such an articulated need for understanding the stakeholder's perspective, the stakeholder communication plan (Appendix 1) became a mere formality to be completed rather than a tool for harnessing the stakeholders' viewpoint. The plan was mainly prepared to identify all the key stakeholders to assess the project's impact at the policy and implementation levels. The plan also lists the impacts or areas of results relevant to each stakeholder group as well as the teams responsible for engaging them. What was lacking in the plan was a stepwise process and specific deliverables for each stakeholder group listed. Therefore, the interactions had no requirement to lead to any specific commitments or outcomes either from the project team or from the stakeholders.

Another issue that kept the team occupied internally was trying to understand the technical language from the different teams. Because of this, it became difficult to focus on finding the linkages among the different models making up the integrated model. The team did not realize how important it was to break through this language barrier by using a systematic approach; instead, it was an individual's problem to either understand or ignore the difficult terms. Thus, the technical language obstructed the effective integration of the different models and prevented the team from appreciating the full implications of the integrated approach. Without fully understanding these implications, the teams tended to focus more on their part of the model hoping that it would somehow fit into or lead to an integrated model. Such a phenomenon seems to be true for many multidisciplinary teams, which with encouragement it can be alleviated to some extent as interpersonal relationships develop.

Sensing that without a proper tool for integration the models could end up with outputs that may not fit into each other, a new team member expert in Bayesian network (BN) modeling was added. Thus, the team started building a networked model using the Bayesian method. This added to the learning cycle of the team with the members trying to understand how the BN approach worked and then providing data necessary for running the model. On the positive side, the tool pushed the team to think of integration first rather than as an end result. The original model was more like an amoeba without any specific parts. Nevertheless the livelihoods framework, which was originally chosen to integrate the outcomes from WSD, provided a natural starting template. Thus, the team, which is also the first stakeholder in the project, evolved from an individual level to an integrated level of approach. However, they were still largely focused on the technical aspects until that providential visit to the case study sites. What follows provides an account of that visit and its results.

The livelihoods component of the socioeconomic module became an automatic choice for leading the integration process, mainly because the data were readily available. The socioeconomic team was among the first to initiate data collection from the study sites and also the first to have close interactions at the field-level with stakeholders. The socioeconomic team had to collect information directly from the individual households and local communities, such as the farmers and the water user associations. The team closely interacted with them through different data collection methods like focus group discussions, which provided the team with the opportunity to understand the needs and expectations of the local communities. Other teams, which could depend on the data collected by the socioeconomic team or on secondary sources of information collected either by the local NGOs or the relevant government departments, had little need to interact directly with the local communities. Although they had to collect some of the data directly from the field, it did not require much interaction with the local people.

As the outputs from the Bayesian model started flowing in and each of the teams started to share their findings, the teams started checking with each other to find better explanations for the results they were getting from their modules. These interactions at the result level helped the team as a whole perceive the larger picture a little more clearly than they could when they were working as individuals. Encouraged by this change in perspective, the groups decided to explore the field together. The visit was expected to help verify and validate their understanding of the linkages between surface and subsurface flows, land use patterns, and livelihood impacts.

The field visit was a real eye-opener for all team members in terms of understanding how the linkages actually played out in real situations and impacted the effectiveness of a watershed project. At the same time, the researchers also realized how difficult it would be for the local communities to see these linkages without being educated about some of the technical aspects of hydrogeology. The group could also visualize the areas for capacity building for different stakeholders starting from the field level to the policy level. The responses of the local communities to the group's efforts to share simple but valuable technical information made researchers feel the work they were doing was useful. At the same time, the interaction with the locals also helped scientists appreciate the role of local knowledge and information in shaping the scientific thinking and approach to problem resolution. The field visit also made it obvious for the different scientists to see how their fields could be brought together to provide a better explanation for the present situation as well as help create a better future for the local communities.

The group came back from the field with a better understanding of the integrated approach, which they were unable to do after many long hours of discussions and debates in meeting rooms. More important, the team actually came back with an understanding of the kind of outputs required for making the project more useful on the ground. Issues of equity, which need to be defined within the cultural context of the local communities, could be better articulated when the team could actually see the socioeconomic conditions of different communities living in different locations of the upstream, midstream, and downstream areas of an HUN. Overall, the field visit provided the team with a very good perspective of all the issues the project was trying to address. Although the above benefits were realized without much planning going into the interactions, the team could understand the value of interacting with the local communities with better preparation and also the need for repeating the process in other study locations as well.

While at the ground level educating the people on the technical aspects of WSD was needed, at the policy level the issues were more to do with the scale of operations as the WSD policy shifted from small-scale watershed projects of ~500 ha to medium or mesoscale projects of 5000 ha and above. Here the project's main aim was to explore the issues involved in the process of scaling up the watershed areas and provide an integrated framework to better measure the impacts of mesoscale WSD.

While the increase in scale was expected to reduce the issues arising out of externalities, it could potentially lead to problems of administration and implementation. Issues of equity were also expected to become more complicated with the involvement of larger populations and different locations within the HUN; the project was more focused at the policy level because of its regular interactions with the stakeholders at that level. From the beginning, the team engaged the DRD (the nodal agency for the implementation of watersheds in the state) in a dialog, which helped shape the objectives of the project. Similarly, at the national level, although with less frequency, efforts were made to register the project objectives with the relevant groups in the National Rainfed Area Authority (NRAA). Figure 11.1 summarizes the levels of interaction and identifies the stakeholders.

After the field visit, the group realized the full implications of the stakeholder plan and started discussing the strategy to reach out to various stakeholders. The first step was initiated with the team briefing the NRAA director on the integrated approach to mesoscale WSD. With the encouragement they received from the director, the team decided to engage the state-level nodal agency for WSD, the DRD, to expose their staff to the

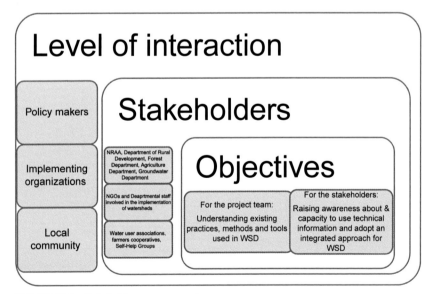

FIGURE 11.1 Stakeholder engagement plan.

integrated approach. The team took extra care to understand their current practices and pitch the integrated approach at the design level. The workshop adopted a participatory approach and made the learning fun and highly interactive. The workshop was well received and the staff recommended the program to their colleagues and subordinates (Appendices 2 and 3). The department further encouraged the project team by endorsing the integrated approach and asked them to help adopt the approach in their upcoming projects. Subsequently, wider endorsement has been received from external reviewers as well as NGOs to make this approach adaptable from a practical point of view.

Thus, the journey, which started with the objective of building an integrated model, was successfully completed with the effective engagement and productive interactions of developing and propagating an integrated approach to WSD. The question was whether there were any short cuts or any planned processes that could have saved time and effort, or helped reach more stakeholders. The team's feedback on the stakeholder engagement experience in the project does reflect this kind of thinking (Appendix 4), and the development of this activity is ongoing.

Next, a survey was conducted among the entire membership of the project group to understand how they benefited from the stakeholder engagement process and what they felt should be done to improve in this area.

The results of the survey are as follows. The team felt that identifying champions in the policy, action, and research arenas is critical for the success of any transdisciplinary research. Hence, more efforts need to be made to include as many organizations as possible and engage stakeholders from diverse rainfall regions. Engaging with ongoing Integrated Watershed Management Program (IWMP) sites could have been more beneficial as well. Further, research questions for any new work should be derived from, or at least verified by, the stakeholders since having stakeholders on board makes it an enriching experience and helps the project. However, to get the most out of future projects, stakeholder engagement should be a two-way process with the stakeholders enlisted early in the process and adequately convinced of the project benefits and viability so that they can get a sense of ownership and carry forward the relevant implementation of the project outcomes.

11.3 OUTCOMES OF THE PROCESS ADOPTED AND LESSONS LEARNED

Toward the final phase of the project, the team managed to achieve some important milestones from stakeholder involvement and engagement. One of them was consent from the DRD to put a team of field-level officers involved in WSD through a workshop based on the technical aspects of the integrated model developed by the project. The second achievement was the interest expressed by the director of the NRAA to extend the integrated approach to the watershed projects implemented in various states through capacity building of the implementing agencies. The third and main achievement was a request from the NGOs involved in WSD to develop a tool to delineate watershed areas using the integrated approach.

These opportunities would not have been possible had the team not realized the need for simplifying and expressing the technical recommendations to meet the requirements of various stakeholders, and this realization could not have happened if the teams had not visited the study site as one group. More important, the group's continued effort to think together and analyze the design aspects together helped them actually conceive their outputs in an integrated fashion. This approach also helped them to produce more outputs within shorter time lines. Thus, a project that experienced multiple challenges—such as the unwieldy integrated model, frequent exit of team members, unreliable sources of secondary data leading to data inconsistencies, teams struggling to learn unfamiliar technical language, and lack of proper stakeholder engagement strategies— finally proved itself worthy of the time, money, and resources invested. The real achievement for the group was its capability to bring together different disciplines as one scientific approach and address any problem from all relevant angles.

The important lessons learned during this project include: (1) focusing on the integrated approach within the team during the early phases would have helped in following a more systematic approach to engagement; (2) transdisciplinarity is a big challenge, and achieving it requires a team that is open to other disciplines—team composition is very critical in this regard; (3) WSD is a centrally sponsored program that requires more effort to engage the national-level agencies, which would have provided more scope for policy changes; (4) involvement of appropriate NGO partners would have helped to produce a more effective engagement with the farming communities; (5) the action research format should have been adopted to make it an effective tool of communication at the policy as well as implementation level.

11.4 MODEL FOR EFFECTIVE STAKEHOLDER ENGAGEMENT—NEED FOR AND ISSUES INVOLVED IN CLOSER ENGAGEMENT WITH STAKEHOLDERS

A revelation for the team was that integration does not come through connecting modules/models but through connecting thoughts and thought processes. Connecting at the thought level requires the team to work together rather than independently sharing the results. Working together requires more frequent interactions of short duration rather than long duration sessions at sporadic intervals. Getting the team members to share thoughts in person is definitely not a small task as most members are partly engaged in the research and not present 100% of the time. Thus, finding ways, using technology, and more significantly, realizing the importance of thinking together is essential when different disciplines are brought together to analyze a situation or a problem. In addition, trust, confidence, and mutual respect can help the team learn and grow together. Team members differ not only in subject matter expertise but also in terms of research experience and exposure to the problem at hand. When members look at multidisciplinary projects not just as problem-solving activities but also as opportunities for learning and mentoring, more value will be created than originally hoped. The value created in terms of nurturing compatible and competent teams, which are mutually trusting and synergetic, lasts beyond a single project and helps reduce time lines for future projects, which are increasingly going to be multidisciplinary.

Although the project started with the intention of building an integrated model for WSD, in the end, it not only came up with an integrated model for WSD, but also an integrated approach to develop mesoscale watersheds. The experience in this project demonstrated that the two tasks required two different types of thinking and team work: while the integrated model needed data from different experts, the integrated approach needed ideas and explanations from different disciplines. Similarly, while the integrated model depended on the method to validate the linkages and dependencies, this approach depended on interactions with the stakeholders and validation and verification from the field realities.

The effort to build an integrated model could culminate in a complex model, which may also be quite abstract because of the need to integrate different disciplines. Such a model would be very useful for getting an overall view of the impacts and developing a theoretical understanding of the linkages. However, its applicability in reality could be limited by the amount and variety of data required for its estimation and the degree to which it is intelligible to the stakeholders.

The integrated approach, on the other hand, would require frequent interactions within the larger project team as well as with the stakeholders. The interactions could make the understanding of the problem at hand and the possible deliverables quite dynamic. Thus, it would require plenty of patience, perseverance, and careful planning of interactions to ensure that the stakeholders' faith in the project and its relevance and usefulness are not adversely impacted. Stakeholder apathy and skepticism could be avoided by planning a sequence of outputs delivered in line with the requirements of the identified end users. The deliverables need not be exotic but should definitely be useful from the practitioners' point of view. They could be in the form of easy-to-use processes, templates, methods, and tools reflecting the integrated approach to the problem.

While there may have been advantages in the earlier engagement of the stakeholders in this project, it may be because of the evolution of the project that the researchers will tend to start with the questions and methods they know well from their own discipline and that the transition to the integrated process progresses as the trust and sharing of knowledge grows. When the team is not operating in an "integrated" way, the interactions with the stakeholders will be less productive, so care must be taken to avoid stakeholder "burnout" as coherent messages cannot be well communicated at this time. Choosing integrative methods from the beginning will hasten team cohesion and the team will be confident about early engagement.

The requirements for systematically engaging with stakeholders based on our experience with this study are as follows:

1. Each team member should identify the stakeholders who could benefit from the project findings, use the output, or provide useful input to the project.
2. Engagement should be divided into major areas for interaction:
   a. Problem specification
   b. Identification of research questions
   c. Specification of outputs along with outcomes
   d. Scheduling outputs
   e. Feedback on and validation of outputs
   f. Implementation plan
   g. Scaling up or expansion plan
   h. Resources allocation
   i. Areas for collaboration (data, templates, methods, tools, etc.)
3. Team members should be involved in the identification of relevant stakeholders and take ownership for engagement with a planned output schedule.
4. Every team meeting should be preceded by member-level stakeholder interactions; actions should be identified and the team should discuss and prioritize the action list.
5. Monitoring and follow-up on commitments to the stakeholders should be taken seriously.
6. Greater stress should be placed on the usability of the research findings compared with technical sophistication.
7. Output format should be decided taking into consideration the stakeholder requirement; the output should be delivered to their satisfaction.

The above process may not be complete, but it is based on our experience and what we believe could have helped the project to engage the stakeholders in a mutually beneficial way.

REFERENCES

[1] Elizabeth Allen, Kruger Chad, Fok-Yan Leung, Jennie C. Stephens. Diverse Perceptions of Stakeholder Engagement within an Environmental Modeling Research Team. J Environ Stud Sci 2013;3:343−56. http://dx.doi.org/10.1007/s13412-013-0136-x.

[2] Sebastian Carney, Whitmarsh Lorraine, Nicholson-Cole Sophie A, Shackley Simon. A Dynamic Typology of Stakeholder Engagement within Climate Change Research. Working Paper 128 Tyndall Centre for Climate Change Research 2009. January.

[3] Domenico Dentonia, Brent Ross R. Towards a Theory of Managing Wicked Problems through Multi-Stakeholder Engagements: Evidence from the Agribusiness Sector. International Food and Agribusiness Management Review 2013; Vol. 16(Special Issue A).

[4] Robert Strand. The Stakeholder Dashboard. USA: University of Minnesota Carlson School of Management; 2008. p. 23−36. www.Strand%20(2008)%20-%20The%20stakeholder.pdf.

[5] Martin Welpa, de la Vega-Leinerta Anne, Stoll-Kleemannb Susanne, Jaegera Carlo C. Science-based stakeholder dialogues: theories and tools. Global Environmental Change 2006;16:170−81.

APPENDIX 1: DETAILS OF STAKEHOLDER COMMUNICATION PLAN

Group	Farming community	Development practitioners	Policy makers (state and national)
Objectives: Determine the need for communication: What do you want to achieve with this group?	Expand the context of WSD to cover the HUN-level impacts (to expand the horizon of farmers' thinking) Learn from and increase the awareness levels regarding scale-related issues in WSD Better management of meso-level WSDs through informed participation of farmers Expand support base for the project activities and get a buy-in or build consensus for project recommendations	Enhance their understanding on the hydrogeological aspects of mesoscale WSDs Learn from and increase awareness on scale issues in the context of meso-WSDs Facilitate building of appropriate institutions for informed participation of the communities Expand the NGOs' thinking horizon to include HUN-level impacts	To increase sensitivity to scale issues surrounding the HUN level Adding hydrogeological context to the current policy thinking on mesoscale IWMP Generate policy debate on socioeconomic (including equity) valuation of ecosystem services of the watersheds

Relationships: Understanding the group: What are their perceptions about you? What are their concerns? What are their communication needs?	Right now, it is a one-way relationship of collecting information. Plan is to involve local NGOs to build rapport for sharing information	Existing relationship is not very conducive to the stakeholder engagement. Plan is to identify local NGOs to involve in the process for the remaining period of the project	DRD is a formal partner. Engage DRD more closely in the process. Involve other state-level departments, such as groundwater, agriculture, etc., to have an integrated approach for meso-watershed management. Engage with national-level stakeholders like the Ministry of Rural Development, National Rainfed Authority, Planning Commission, etc.
Messages: Design communication messages: What are your three key messages for this group?	Need for considering compensation at mesoscale/HUN level and in the context of long-term costs and benefits. Key messages for improving farmers' decision making on crop choices. Key messages on building resilience	Key message regarding use of hydrogeological information in watershed-related activities. Key messages dealing with equity issues. Key messages relating to change management (resilience vs adaptation). Key message on considering scale issues relating to community-based groundwater management	As the scale increases, equity issues assume greater importance for better participation and management. Need to build institutions in a cascading manner from village to the HUN level. Scientific approach to community groundwater management practices

(Continued)

—cont'd

Group	Farming community	Development practitioners	Policy makers (state and national)
Activities: Choose communication activities: How does this group like to receive information?	Farmers' club and Self Help Group (SHG) meetings Pamphlets, posters, etc. Focus group discussions Feedback	Interactive workshops Case study material demonstrating linkages in the meso-watershed context Explore the possibility of demonstrating groundwater management at the HUN level through modeling using the real-time technical data and also on resilience models Feedback	Interactive sessions Case studies Audiovisual presentations Feedback
Evaluation: How will you evaluate as you go? How will you evaluate at the end?	Pre- and post-awareness levels Observed changes in farm practices Observed changes in groundwater management Changes in resilience-related issues Feedback	Change in implementation modalities Changes in in-house (NGO) mandate Feedback	Policy changes New guidelines issued New GOs issued Changes in enforcement of existing GOs Feedback
Roles, responsibilities: The action plan	LNRMI with Local NGOs	LNRMI with support from project partners and local NGOs	LNRMI + ECU + ACIAR IWMI + CRIDA + NGRI

LNRMI, Livelihoods and Natural Resource Management Institute; IWMI, International Water Management Institute; CRIDA, Central Research Institute for Dryland Agriculture; NGRI, National Geophysical Research Institute; ACIAR, Australian Centre for International Agriculture Research; ECU, Edith Cowan University.

APPENDIX 2: STAKEHOLDER ENGAGEMENT DETAILS

Stakeholders	Level	Activities	Outcomes
Policy makers	State and national DRD and NRAA	Project updates	Inputs for project objectives, briefing on policy context, expected scenarios, and feedback on project outputs
Implementers	DRD staff (Project Directors, Project Officers, and Technical Officers)	Workshop	Appreciation for the approach, usage of scientific methods, tools for Detailed Project Reports, etc.
Local communities	Farmers in sample villages	Focus group discussions	Exchange of information on local conditions and the scientific explanation for the same
Scientific community	Water forum, conferences, publications, and project team meetings	Papers, articles, presentations	

APPENDIX 3: DETAILS OF WORKSHOP ORGANIZED BY THE PROJECT TEAM IN COLLABORATION WITH DRD

Workshop Title: Integrated Approach to Design an IWMP

Workshop Design

What is included in the training:

- Limiting the present training program to design-related inputs
- No lecturing in the first half of the first day
- Exploring what the participants already know and what they are currently doing in various stages of watershed (WS) design

First half of the first day:

- *Quiz*: To test the knowledge on terminology, WSD-related information basically regarding hydrogeology, biophysical, and rainfall information, etc.
- *Large group discussion*: Elements or building blocks of WS design, factors impacting WS design, data sources, analytical tools, etc.

- *Reading material*: Quiz answers, WS design related, sustainability, equity, scale, etc.

Second half of the first day:

- *Case study*: At least two different types of HUNs
- Presentation of the case studies from the WS design point of view by the participants

First half of the second day:

- Lecture on methods to estimate the number, type, and placement of structures
- Participants to use the methods and estimate the structures

Second half of the second day:

- Discussion on detailed project report format
- What information is required
- How the reports can be improved

Participant profile:
Staff involved in WS-related activities from the DRD (Andhra Pradesh, India)
Level of participants:

- Additional Project Directors (PDs): 5
- Project Officers (POs): 5
- Technical Officers (TOs): 5

Total number of participants: 15
The group was a mix of experienced and new staff.
Districts covered: Mahbubnagar, Anantapur, Kurnool, Kuddapa, Chittoor, Khammam, and Medak.

PARTICIPANTS' FEEDBACK ON THE WORKSHOP

Quantitative Feedback

Question	Response score	Response
How relevant has the training been to your work?	4.6/5	Very relevant to relevant
To what extent would you be able to apply what you have learned during the training in your work?	4.14/5	Apply very frequently to frequently
How much did you benefit from the training?	4.29/5	Benefit very much to benefit a bit

Qualitative Feedback

What parts of the workshop have been most relevant for your work and why?

Case study, group discussions, and presentations
Socioeconomics is very relevant
Socioeconomics, biophysical details
Designing with new approach
Two case studies
Incorporating socioeconomic and biophysical data and geology data into Detailed Project Report (DPR) preparation; enough attention was not paid earlier; this is important learning Activity 5: designing an IWMP and quizzes 1, 2, 3, and case studies 1 and 2
Feedback on present implementation of IWMP and quiz by Central Research Institute for Dryland Agriculture facilitators
Presentations on hydrogeology and design of certain factors in IWMP
Case studies are relevant for my work for preparation of DPRs
Socioeconomic studies relevant to improve the economic status and put planning into DPR
Preparation of DPR and implementation of program
We are working in WS and this is helpful for preparing DPR in a better way
Case study presentation and group discussion
Socioeconomics to study the social status of farmers and increase their financial status; geohydrology to selection of sites for execution of work

Would you recommend this training to any others? If yes to whom?

We recommend to TOs
Yes; recommended to newly joined TOs who are working in DPR preparation
Other ground staff in an easy way
POs of IWMP with more case studies
POs and JEs that actually do the DPRs. At their level, duration may be 3 days.
Yes; recommended to TOs working under IWMP
All POs and TOs because they have to prepare the DPR and implement the program
Yes, additional PDs, POs, and junior engineers (JEs)
Yes, other staff of District Water Management Agency (DWMA) and IWMP
Our team members like POs, TOs, and TO Institutional and Capacity Building (I&CB)
To all IWMP staff
To field staff (TOs)

Would you like to attend one more training to continue your learning on these subjects?

Yes (12)
Yes, if any additions are there in future
Natural Resource Management and Population Services International (PSI) agriculture

What could be covered in the next training that is not already covered in this training?

Social mobilization
Add some more relevant maps for better understanding
Biophysical information
Runoff, groundwater aspects
Estimation of runoff and an insight into indigenous structures

(Continued)

—cont'd

Basics on geology and hydrogeology
Design and estimation of water-harvesting structures under IWMP works
Detailed planning on water budgeting and taking account of geological features and groundwater storage availability
Geohydrological subject
Evaluation and monitoring
PSI agriculture aspect
Works to small and marginal farmers—deep into subject is needed
Geohydrological features and how to select the site locations with maps

APPENDIX 4: PROJECT TEAM'S PERCEPTIONS ABOUT STAKEHOLDER ENGAGEMENT

Question 1: What in your view will define the success of this project?

Responses:

(ECONOMIST): Integration of hydrogeology and biophysical aspects into WSD planning and designing; WSD design and implementation (type and density of structures) as per hydrogeology and biophysical aspects.

(BIOPHYSICIST): Systematic integration of hydrology and economic impacts with livelihoods; better understanding of these aspects and their interlinkages in watershed.

(HYDROGEOLOGIST): Giving a practical solution and effective strategy to utilize the available amount of water resources in the hard rock mesoscale watersheds.

(HYDROGEOLOGIST): Giving effective awareness, which can be followed by the farmers voluntarily; implementing site-specific recommendations such as the methods to be followed for artificial recharge measures.

(HYDROLOGIST): Sharing the knowledge of the project and the new findings that are relevant to policy, especially IWMP, with a wide range of stakeholders.

(ECONOMIST): Continued application of project findings for future analysis, as a standalone project of this scale will have limited impact.

(BAYESIAN NETWORK EXPERT): Positive engagement of agency staff with the key messages and methodologies from the projects with workshops and meetings; establishment of research teams that respect each other and work well together in future integrated projects; robust tools to support IWMP (e.g., guidelines for what interventions are required and where; guidelines for developing biophysical and socioeconomic monitoring programs for future IWMP; hydrogeological characterization; suite of tools for analyzing social survey data).

Question 2: Who do you think are the project's stakeholders?
Responses:
(ECONOMIST): National-level policy makers (Department of Land Resources (DoLR), National Rainfed Area Authority (NRAA)); state-level nodal agency (DRD); district-level implementing agencies (DRD; other line departments; NGOs, etc.); watershed and village communities.

(BIOPHYSICIST): National and state-level project implementing authorities (district level and NGOs); field practitioners of watershed management.

(HYDROGEOLOGIST): National and state-level policy makers who are involved in rural development, irrigation (minor irrigation).

(HYDROLOGIST): DRD and line agencies, local NGOs, NRAA.

(ECONOMIST): Local government, local administrators, NGOS, village panchayats, etc.

(BAYESIAN NETWORK EXPERT): Government agencies and NGOs involved in designing and implementing watershed interventions.

Question 3: What is the purpose of stakeholder engagement in this project?
Responses:
(ECONOMIST): To provide policy-relevant and scientific knowledge to bring in policy changes; it is the socioeconomic component of training.

(BIOPHYSICIST): From communities—to receive feedback on the impacts of watershed management through hypothesized interventions and implications on resource availability, use and livelihood impacts, as well as biophysical and hydrological impacts of watershed management; From practitioners—ways and means to implement the interventions to put the watershed in a win—win scenario on resources use.

(HYDROGEOLOGIST): Stakeholder acts as implementation machinery for effective implementation and evaluation of any new scientific approach for the agricultural sector. Local NGOs facilitate extending the project results to the other areas.

(HYDROLOGIST): Direct the attention of the stakeholders to the project, and communicate the findings.

(ECONOMIST): Learn from their experience, avail the existing information, facilitate better communication with the farmers, and convey the project findings in local terms to farmers and NGOs.

(BAYESIAN NETWORK EXPERT): Capacity building—improving the understanding of WSD design and assessment; information provision to research team; scoping issues.

Question 4: What benefits have you got as a researcher through the project's stakeholder engagement?
Responses:
(ECONOMIST): The state-level IWMP nodal agency has limited capacities to train the district-level implementing agencies and adopt the integrated approach. They need to develop proper training material, methods, and tools.

(BIOPHYSICIST): There is less understanding of the integrated impacts of the watershed management program. The practitioners could not visualize the impacts of the WSD program if implemented on uniform scale across rainfed regions. There is a need to develop training material for proper design of WSD methods; researchers need to develop robust methods for inclusion of many aspects of WSD to understand the impacts.

(HYDROGEOLOGIST): A practical link has been established with the farmers through the project stakeholder engagement involving exchange of local and scientific knowledge.

(HYDROGEOLOGIST): Provided services for local user agencies, which facilitated in obtaining important field parameters.

(HYDROLOGIST): Deeper understanding of the issues as well as the practical realities on the ground.

(ECONOMIST): An appreciation for the challenges faced in the policy implementation phases and a better understanding of the ground realities; understanding the management issues as well as the resources available to deliver WSD; understanding the decision-making processes of landholders and their implications for meso-delivery of WSD.

(BAYESIAN NETWORK EXPERT): An appreciation for the serious consideration given to the social factors in WSD and natural resource management in general; the sheer scale of investment in, and complexities of, WSD implementation in Andhra Pradesh.

Question 5: What could have been done in this project that was not done in terms of stakeholder engagement?

Responses:

(ECONOMIST): Identifying and partnering with the right kind of action research NGO.

(BIOPHYSICIST): Although the study sites could be limited to one or two, for stakeholder engagement we could have planned for more organizations/NGOs.

(HYDROGEOLOGIST): Implementation of project components without area-specific recommendation may lead to poor results as well as reduced cost-benefit ratio.

(HYDROLOGIST): Perhaps following through on the IWMP planning process.

(ECONOMIST): Stakeholders at the central level should have been engaged early on.

(BAYESIAN NETWORK EXPERT): I was not involved in the initial phases of the project or in the later stakeholder engagement so I may not be the best qualified to answer this.

Question 6: What are your recommendations for future projects?

Responses:

(ECONOMIST): Identifying the champions at the policy, action, and research arenas is critical for the success of any transdisciplinary research.

(BIOPHYSICIST): More efforts required including as many organizations as possible to get into stakeholder engagement from diverse rainfall regions; engaging with ongoing IWMP sites could be more beneficial.

(HYDROGEOLOGIST): Prior implementation of area-specific information regarding soil depth, excess rainfall runoff components available for artificial recharge, subsurface geology, and aquifer geometry need to be assessed and accordingly the work plan in terms of type of structures, cropping patterns etc., needs to be decided; post-project monitoring is an important parameter to assess the effectiveness of the project; the life of the structures of importance must also be included in the maintenance of the structures proposed for artificial recharge.

(HYDROLOGIST): The research questions for any new work should be derived from, or at least verified by, the stakeholders.

(ECONOMIST): Having stakeholders on board makes it an enriching experience and helps the project. However, to get the most out of future projects, this should be a two-way process, with stakeholders enlisted early on and adequately convinced about the project benefits and viability so that they get a sense of ownership and carry forward the relevant implementation of the project outcomes.

(BAYESIAN NETWORK EXPERT): This was a really interesting project to be involved in. The level of interest shown by the limited number of stakeholders, LNRMI, NGRI, and CRIDA in particular, was very high.

Chapter 12

Exploring Implications of Climate, Land Use, and Policy Intervention Scenarios on Water Resources, Livelihoods, and Resilience

Wendy Merritt *, K.V. Rao §, Brendan Patch *, V. Ratna Reddy ¶, Geoffrey J. Syme ‖ and P.D. Sreedevi #

*Fenner School of Environment and Society, The Australian National University, Canberra, ACT, Australia, §Central Research Institute for Dryland Agriculture, Hyderabad, India, ¶Livelihoods and Natural Resource Management Institute, Hyderabad, India, ‖Edith Cowan University, Perth, Australia, #CSIR-National Geophysical Research Institute, Hyderabad, India

Chapter Outline

12.1 INTRODUCTION

The potential of integrated assessment methodologies to understand and support resource management problems is increasingly promoted in the scientific and associated management literature (e.g., [1−3]). This is because of the need to take a "whole of system" approach to complex environmental problems, where complex and interacting processes exist within a system. Integrated Assessment Models (IAM) provide a framework that represents the current understanding of

Integrated Assessment of Scale Impacts of Watershed Intervention
http://dx.doi.org/10.1016/B978-0-12-800067-0.00012-8

the system, identifies key relationships along with critical knowledge gaps, and explores likely trade-offs between environmental, economic, or social outcomes under alternative management options [4]. In this chapter, outputs from the hydrogeological, hydrological, biophysical, and socioeconomic analyses presented in earlier chapters are used to explore the possible impacts of biophysical and socioeconomic scenarios on capital stocks and drought resilience.

12.2 ANALYSIS TOOLS

This section provides an overview of the analyses described in previous chapters and a recap of key findings.

12.2.1 Disciplinary Models and Analyses

Hydrogeological characterization and any implication of hydrogeology on the use and management of groundwater were described in Chapter 3 for the two study watersheds. This research entailed the use of geological surveys (specifically electrical resistivity tomography and electrical resistivity logging), recharge estimation using the lithologically constrained rainfall method, and assessment of changes in groundwater storage over time. It identified critical constraints to groundwater storage and availability for use in the study watersheds. The study revealed that extraction of groundwater has exceeded recharge into the system despite interventions, and groundwater storage mostly declined over the period of investigation, particularly from 2009 to 2012, although an increase in storage was identified between June and December 2012. Further, in both watersheds, most shallow (dug) wells dried out in summer and, because of the shallow nature of the aquifer, most bore wells dried out in summer in the midstream regions of the Peethuruvagu watershed and the upstream regions of the Vajralavanka−Maruvavanka watershed. In Chapter 3, Sreedevi et al. developed artificial recharge zones across the watersheds that identified appropriate interventions based on hydrogeology (see Figure 3.8). According to the authors, injection wells are most suitable for achieving artificial recharge in areas where groundwater is tapped from the deep fractured zones. In contrast, interventions such as check dams, percolation tanks, and farm ponds are useful in areas where the thickness of weathered zone exceeds 5 m.

A distributed surface water and groundwater model based on water balance principles was described in Chapter 4. This model was developed to represent key hydrological processes while limiting data requirements compared with the more complex surface or groundwater hydrology models such as SWAT and MOD-FLOW. Input data requirements included climatic data, topography, soil properties, land use, crop characteristics, hydrogeology, storage capacities of watershed development (WSD) interventions, and calibration data (surface runoff and groundwater levels). The model was applied to the Peethuruvagu watershed to demonstrate its potential in modeling hydrological processes and scenarios of alternate management options.

An alternate and less data-intensive approach to modeling the impact of WSD on water resources was described in Chapter 5. Recognizing that data availability is a significant problem for mesoscale studies in rural India, the IHACRES rainfall−runoff model was modified to represent the water-harvesting interventions. Initially developed for a small (2.5 km^2) catchment in West Bengal, the model was adapted to the conditions in Andhra Pradesh by testing and modifying it using data from the gauged Lakshmipuram catchment (2750 km^2 in area). The model was then applied to the ungauged Gooty catchment in Anantapur/Kurnool districts, which encompasses the Vajrala-vanka—Maruvavanka watersheds, and the Vendutla catchment, which covers the extent of the Peethuruvagu hydrological unit (HUN). The input data necessary for the model require the catchment area and proportion, the storage capacity, and infiltration rate for each land class, along with the evaporation/infiltration threshold parameters for the catchment moisture deficit module, and rainfall and potential evaporation time series. Because it is a lumped model, the data requirements are significantly lower than the methods applied in Chapter 4. However, because the data are often lacking, uncertainty in model predictions is high; this must be taken into consideration while using the model outputs to support the design of WSD interventions and other policies.

In Chapter 6, the currently existing WSD interventions within the Peethur-uvagu (Prakasam District) and the Vajralavanka—Maruvavanka (Anantapur/Kurnool districts) watersheds were mapped, and cropping patterns in the water-shed were related to deficit, normal, and above normal rainfall years. In Anantapur/Kurnool, the proportion of area cropped during the kharif season matches the rainfall distribution with reduced areas cropped during the deficit rainfall years, i.e., 2006 and 2011. On the other hand, the relationships were not as clear in Prakasam, as the low crop areas in some "normal" rainfall years potentially re-flected non-ideal distribution of rainfall during the monsoon. In the study, delin-eated sub-watersheds were used along with calculations of hypsometric integral, drainage density, and stream network to identify suitable sub-watersheds for watershed interventions. The stability (or equilibrium) of many sub-watersheds and the greater length of drains per unit area indicated the potential in the water-harvesting systems in the Prakasam HUN. However, the interactions identified between the upstream, midstream, and downstream locations suggested a need for careful planning of the interventions as excessive upstream in-terventions can affect the downstream sub-watersheds. Using a plot-scale water balance model, Rao ct al. assessed the influence of watershed interventions on resource conservation by running scenarios without watershed development, and with WSD interventions of 50 and 100 m^3/ha. It was found that at this scale, runoff could be reduced by 30% on average with interventions due to increased evapo-transpiration and recharge in both clayey and gravelly soils.

Chapter 7 analyzed the socioeconomic impact of watershed interventions using two approaches: (1) measurement of socioeconomic indicators based on the sustainable rural livelihood (five capitals framework) and (2) assessment of household resilience in the context of WSD. The study revealed that little evidence was found to support significant and positive impacts of watershed interventions using the five capitals approach. On the other hand, in the resilience approach, probit regression analyses identified significant relationships between watershed interventions and higher household resilience (Table 7.27). Furthermore, stream location was important—downstream villages had higher resilience when compared with midstream and upstream villages. The HUN in which resilience was measured was also found to be important as households in the villages from the Prakasam HUN were more resilient than households from the Anantapur/ Kurnool HUN. While it appears that the expected trends of improved resilience in downstream locations and treated villages were generally met, the resilience methodology was also able to identify villages that did not conform to these trends and for which there were valid hydrogeological or land use practice explanations (Table 7.29).

A parametric and semiparametric analysis was undertaken in Chapter 8 to evaluate the determinants of farmers' perceived drought resilience. The study tested the perceived resilience both with and without WSD through a resilience survey from all participating households from villages that had received WSD. Consistent with the analyses in Chapter 7, households in the villages from the Prakasam HUN were found to have benefited more from watershed interventions compared with the households from the Anantapur/Kurnool HUN. However, downstream users of groundwater in villages that have received interventions expressed a lower perceived resilience to drought. The analyses identified several critical factors in determining household resilience, including education levels and expenditure, skill levels, location within the watershed, and expenditure on wells and tractors.

The Bayesian network (BN) approach was used in Chapter 9 to develop submodels of the five capitals as well as perceived resilience. The BNs were selected as the integration tool in this project partly because they can integrate different types of data from different disciplines. This was considered an essential feature for an integrated model that links knowledge and model outputs from hydrological and biophysical analyses to livelihood indicators and resilience (drought survival). The structure and parameterization of the BN models is based on the resilience survey dataset that was used in the analyses in both Chapters 7 and 8. While the results from the BNs are roughly comparable to other analyses of the resilience survey dataset, there are some philosophical differences in the approaches:

The aim of the BN was to explore aspects of causality in the sustainable livelihoods framework and link the capitals to household resilience. The model is structured to define relationships between (Figure 9.4a).

- household variables (e.g., HUN, farm size) and indicator stocks
- indicator stocks and indicator drought support (i.e., how many consecutive droughts the stocks of an indicator will support a household's ability to survive drought)

- indicator drought support and capital strength
- capital strength and drought resilience

Given the large number of indicators that could be related to resilience, the capital strength variables act as intermediate nodes between the indicator drought support variables and the perceived capacity to survive consecutive droughts. This is a point of contrast between the analyses in both Chapter 7, which considers the influence of WSD interventions on resilience separately from the impacts on the five capitals, and Chapter 8, which relates indicator stocks directly to perceived resilience.

12.2.2 Integrated Model

To explore the possible impacts of WSD and other scenarios on capital stocks and resilience, hydrogeological and biophysical outputs from the work described in Chapters 3−6 are linked to variables in the socioeconomic BNs outlined in Chapter 9. A loosely coupled approach is adopted; that is, the hydrological and water balance models developed within the project have not been coded along with the BNs into one integrated model.

The relationship between the outputs from the biophysical work and the resilience BNs is illustrated in Figure 12.1. We see that there are links between the biophysical models and the natural and financial capital BNs. Within the integrated BN model, variables in natural capital are linked to variables in the financial and physical models primarily through irrigated and rainfed land area and access to common pool resources (CPRs; see Table 9.1 for all variables common to two or more submodels). There are also links from the human capital submodel to the financial capital submodel.

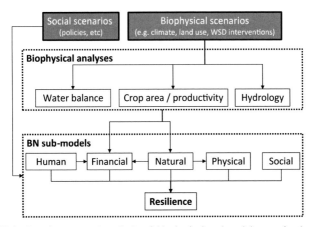

FIGURE 12.1 Development and analysis of biophysical and social scenarios by integrating outputs from biophysical analyses with the BN models of livelihood capitals and drought resilience.

Two broad types of scenarios are envisioned: social scenarios run through the BN models only and biophysical scenarios run through the biophysical models. The outputs from these biophysical models—primarily relating to groundwater availability and crop area and productivity—are then input in the BN submodels. Examples of each scenario type are provided in the next two sections of this chapter.

12.3 BIOPHYSICAL SCENARIOS

This section demonstrates the biophysical scenarios comparing the actual and ideal density and placement of WSD interventions in the study watersheds.

Chapter 7 identified the issues in the design and implementation of WSD in the study HUNs that resulted from the failure to account for hydrogeological and biophysical aspects. WSD in the treated villages focused mainly on check dams with some evidence of on-farm treatments. The higher concentration of check dams without understanding the geological conditions that influence the possibilities of groundwater recharge in these upstream regions was not very effective. For example, in S. Rangapuram, the groundwater situation has not improved despite the construction of check dams, while the rainfall during the "normal" climate years only supports the shallow wells. Because of the very shallow basement rock, this village has limited groundwater potential; the more sustainable and beneficial option would have been to invest in on-farm interventions in conjunction with judicious land use planning. Similarly, the importance of land use planning was demonstrated in the midstream and downstream regions of the Prakasam HUN where the large-scale promotion of horticulture was not sustainable. Although check dam interventions can improve recharge to the groundwater in these areas, which are characterized as having moderately shallow basements, the recent below normal rainfall years led to most of the horticultural crop drying out resulting in losses for the farmers.

Watershed interventions will result in land use changes, although these need to be considered together with other policy interventions and developmental programs such as conversion of lands from wasteland or forestry to cultivable lands. Changes in crops for cultivation could happen because of the perceived benefits of the WSD program by individual farmers. With assured water availability from additional recharge, or from the addition of new bore wells (at least in some farmer plots), the shift in the cropping pattern is mostly from rainfed to irrigated crops. Although the WSD-implementing agencies' intended use of water captured by the water-harvesting structures is to drought proof the pre-existing rainfed crops, farmers tended to use this water to cultivate irrigated crops, which actually exposes them to more uncertainty and risk rather than drought proofing them. For example, shifting from sorghum/ groundnut-based systems to cotton increases farmers' risk as cotton requires more water and fluctuations of market prices are also high. Additionally, the introduction of Bt Cotton, a variety of cotton genetically modified to self-produce an insecticide, reduced the drought hardiness of cotton crops compared to earlier hybrids or varieties and reduced the number of harvests to a maximum of two to three. Similarly, growing vegetables on a continuous basis has increased the dependence on additional water compared with short-duration pulses, which have traditionally been grown in rainfed systems.

In this section, we model some of the likely impacts of the "ideal" design and implementation of watershed development in the study villages (ideal) and compare the results with those obtained in the actual type and location of structures ("implemented"). We explore how results vary across the treated (WSD) villages and for households under the broad land types (rainfed and irrigated). The relevant variables and their relationships are shown in Figure 12.2 for the natural (green box) and financial (orange box) capital models (for the full models, see Figures 9.6 and 9.8, respectively). Variables not directly relevant to this analysis are not displayed in Figure 12.2.

12.3.1 Scenario Description

Given that the available rainfall is low to medium in these HUNs, it is not possible to drought proof the entire system regardless of whether the implemented or ideal WSD interventions were installed. Further, even during the high rainfall years, the water available through additional recharge would be

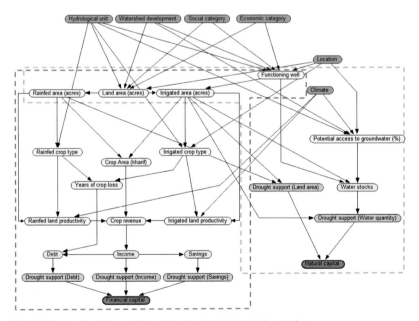

FIGURE 12.2   Network structure relevant to the biophysical scenarios.

used in the following rabi season as the area under various crops is increased and the net water recharged by the commencement of the following kharif season would be zero.

Under the implemented scenario (i.e., current conditions), it is possible to grow crops during the kharif season across the study area. Crops can be grown during both the kharif and rabi seasons in parts of the area, given the suitable climatic conditions within a year, while horticultural crops can only be grown in limited areas. Farmers with access to irrigation infrastructure irrigate these crops, whatever the available quantum of water; they typically irrigate, in the order of preference, rice followed by vegetables, horticultural crops, and groundnut-based systems. Irrigated areas are reduced during the rabi season, and during a drought year there may only be enough water for horticultural crops. Farmers typically plan for crops during the rabi season, based on monsoon rainfall and possible recharge. Rice crops can be planted when the rainfall is good throughout the rabi season. Under normal rainfall conditions, farmers would reduce their area under paddy and may opt to irrigate horticultural crops (if existing in the area). Rainfed farmers may opt for short-duration pulses, which can be grown on residual moisture. When there is a delay in the onset of the monsoon (in the kharif season), contingent crops are planned and farmers may plant crops such as chick pea during the rabi and forego growing crops during the kharif season (observed in Prakasam District). In Anantapur/Kurnool, on the other hand, where hardy groundnut-based systems predominate, farmers can sow groundnut crops up to the first week of August, while for non-groundnut-based systems, the cropped area would reduce with the delayed onset of the monsoon and as a result, paddy (rice) may not be grown.

In the study area, the WSD program should have placed more emphasis on *in situ* conservation methods and farm bunding on agricultural cropped areas as well as ensuring water flows to existing tanks to increase recharge in the existing and available wells (both open and bore wells). This recharge could have happened in all wells, at least during high rainfall events. In addition to technical interventions, WSD should have been aimed at sustaining the rainfed crop production, encouraging the farmers to choose crops that have similar water requirement and that are selected based on market demands. Instead, the farmers have preferred to grow crops with higher water requirement thus exposing them to more risk (as described earlier). Under the ideal scenario, it is possible to grow crops during the kharif season across the study area.

Crops can be grown during both the kharif and rabi seasons in downstream areas, partly in midstream areas, but they cannot be grown in the upstream areas. Kharif, rabi, and summer or horticultural crops can be grown in downstream and midstream areas and sometimes in the upstream areas. In downstream areas, growing crops across the three seasons along with horticultural crops could be possible on about 70% of the area, although this is not guaranteed on a year-to-year basis, and the possibility for growing horticultural crops across years could reduce to $\sim 30\%$ of the area, below which it is not sustainable.

Some aspects of the implications of the implemented versus ideal scenarios are modeled in this section using the BNs described in Chapter 9. We focus on the impacts of the scenario on water resources (represented as *Potential access to groundwater (%)* in Figure 12.2) and crop-related indicators, namely, changes in crop area and crop type and the propagation of these changes through the model to *Years of crop loss* (as a surrogate for risk), land productivity, and crop revenue. By necessity, the representation of cropping decisions within the financial capital BN is simplified, compared with the decision-making processes described previously. Thus, we focus primarily on the cropped area on rainfed and irrigated land, assuming the distribution of the broad crop types (e.g., paddy, groundnut-based systems, and vegetables) is common between the two scenarios. We assume a normal onset of the monsoon and thus, planting during the kharif season, although the area cropped will depend on the amount of rainfall (deficit, normal, and above normal).

In the natural capital submodel described in Chapter 9, the indicators for natural capital are the area and quality of land that households own, the quality of water resources that the household can access, and the direct value of CPRs accessed by each household. In the survey used to populate the BN, the participants were asked to rate the adequacy of their household water stocks. This is not a biophysical variable and the responses reflect how households use (or wish to use) water and whether the available water resources support these activities. The *Potential access to groundwater (%)* variable relates household responses to this question regarding the biophysical resource and is defined as the maximum percentage of wells within a village area from which water could be extracted under different climate years. This variable has been populated for both the implemented as well as ideal scenarios, and for each study village using expert elicitation from hydrologists and water resource scientists in the project team (Table 12.1). The *climate* variable is based on the percentage departure from the average annual rainfall, and has the states of deficit, normal, and excess.

The maximum percentage land area that is cropped during the kharif season is shown in Table 12.2 for both scenarios. The calculation of the percentages for the ideal scenario using the survey data was based on the area of rainfed and irrigated land. For landholders with more than 10 acres of rainfed land, there were limited survey data to determine the percentage cropped area, although the available data suggest a similar pattern to landholders with 5−10 acres or rainfed land.

12.3.2　Results and Discussion

The elicited data in Table 12.1 were converted into conditional probability tables (CPT) for the *potential access to groundwater (%)* variable in the

TABLE 12.1 Maximum percentage access to groundwater resources in the study villages

Climate	Deficit			Normal			Above normal		
WSD design	Implemented	Ideal		Implemented	Ideal		Implemented	Ideal	
Anantapur/Kurnool									
S. Rangapuram	0 (DW)	10 (DW)		40 (DW)	60 (DW)		70 (DW)	80 (DW)	
Utakallu	15 (DW)	20 (DW)		30 (DW)	40 (DW)		40 (DW)	50 (DW)	
	25	30		45	50		75	75	
Basinepalle	40	50		90	90		100	100	
Prakasam									
Thaticherla	10	20		30	40		50	65	
Penchikalapadu	<5	<15		15	25		30	40	
Vendutla	15	30		40	50		80	90	

Note: All data are specific to bore wells except for dug wells, which are labeled as DW.

TABLE 12.2 Maximum land area (%) cropped by households under the ideal scenario

Climate (rainfall year)	Deficit	Normal	Above normal
Farming households with only rainfed land			
0−5 acres land	100	100	100
5−10 acres land	100	100	100
≥10 acres land	100	100	100
Farming households with <5 acres irrigated land and			
0 acres rainfed land	100	100	100
0−5 acres rainfed land	100	100	100
5−10 acres rainfed land	70	100	100
≥10 acres rainfed land	60	90	100
Farming households with ≥5 acres irrigated land and			
0 acres rainfed land	100	100	100
0−5 acres rainfed land	100	100	100
5−10 acres rainfed land	70	100	100
≥10 acres rainfed land	60	90	100

natural capital BN (Figures 12.3 and 12.4). Improvements in groundwater resources under the ideal scenario in Basine Palle and Utakallu fall within the same state as for the implemented scenario and, consequently, the natural capital BN will not be sensitive to the scenarios (Figures 12.3). For S. Rangapuram (Figure 12.3) and the villages in the Prakasam District (Figure 12.4), the *Potential access to groundwater (%)* variable for the ideal scenario increases relative to the implemented scenario, across all climate scenarios. This increased groundwater resource availability has mixed impacts on the adequacy of water stocks for households from the different villages and with different areas of rainfed and irrigated lands.

The variable *Water stocks* generally has a moderately higher likelihood of being classified as "adequate" or "more than adequate" under normal rainfall years compared with dry or drought years under both scenarios and an increased likelihood of these states under the implemented scenario. For example, for households in Penchikalapadu with rainfed land only, the summed likelihood of these two states is 41 and 47% under deficit and normal settings, respectively, for the ideal scenario compared with 38 and 43% for the implemented scenario (Figure 12.5). An exception to this pattern occurs for scenarios as well as villages where there is a greater likelihood of falling within the "60–100%" state of the *potential access to groundwater (%)* variable. For this state, the relationship between the expert elicited *potential access to groundwater (%)* variable and the survey-based *water stocks* variable suggests reduced adequacy compared with the adjacent states ("20–60%"). In the dataset used to learn the CPT shown in Table 12.3, this state corresponds to *Water stocks* for Basinepalle households under the normal climate years. In trying to develop relationships in the network between and

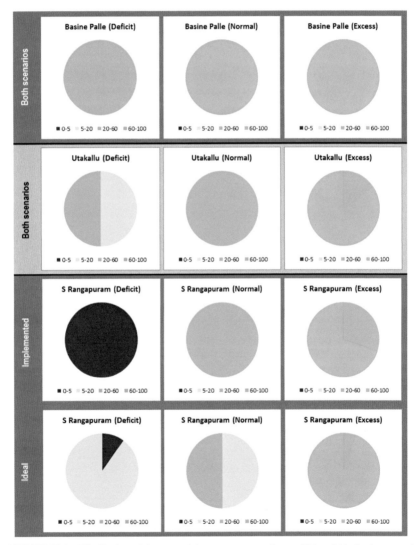

FIGURE 12.3 Distribution of the *potential access to groundwater (%)* variable in Anantapur/Kurnool district. Note: The distribution of this variable is the same under both the implemented and ideal scenarios.

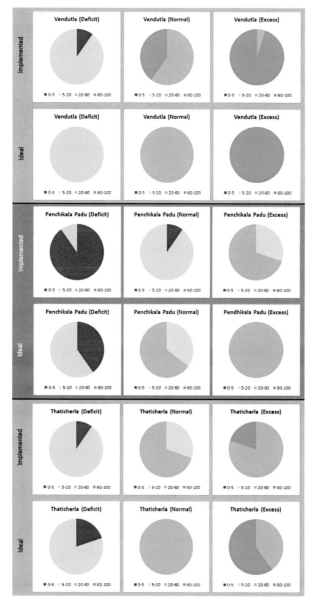

FIGURE 12.4 Distribution of the *potential access to groundwater (%)* variable in Prakasam District.

the village scale groundwater resource, household ownership of functional wells, and irrigated area variables, there may be other factors at play that are not represented in the model structure and that result in some non-intuitive entries in Table 12.3.

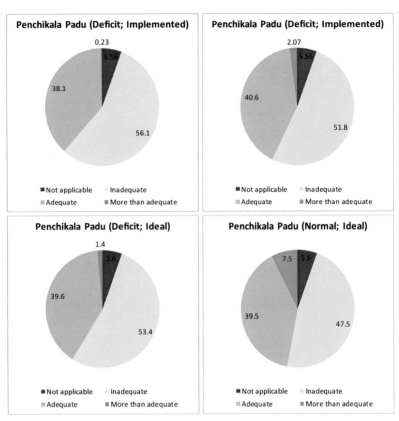

FIGURE 12.5   Modeled adequacy of household *Water stocks* for rainfed only farmers in Penchikalapadu.

The model does suggest some differences with respect to the type and amount of land that the surveyed households own. In Figure 12.6, the difference between the implemented and ideal scenarios for each state in the *Water stocks* variable is shown for the villages in the Prakasam District. We see that for rainfed farmers, the implemented scenario does have a minor benefit, represented as a 5% or less increase in the "more than adequate" state. This has negligible impact on the contribution of *Water stocks* to household capacity to survive consecutive droughts. For households with functioning wells and less than 5 acres irrigation, there is no difference between the scenarios under a deficit rainfall climate for any village. On the other hand, under a normal rainfall climate, the likelihood of more than adequate is considerably greater under the implemented scenarios in both Penchikalapadu and Thaticherla. This results in an increased likelihood that *Water stocks* could support household survival for two or more consecutive droughts (Table 12.4).

TABLE 12.3 Conditional probability table for the *Water stocks* variable

Func-tioning wells	Irrigated area (acres)	Potential access to ground-water (%)	Not applicable	More than adequate	Ade-quate	Inade-quate
No	0	0−5	0.0	0.0	40.0	60.0
No	0	5−20	0.0	2.4	43.3	54.3
No	0	20−60	0.0	10.9	41.0	48.1
No	0	60−100	0.0	18.9	43.8	37.4
No	0−5	0−5	0.0	0.0	52.6	47.4
No	0−5	5−20	0.0	9.5	64.6	25.9
No	0−5	20−60	0.0	*0.0*	82.8	17.2
No	0−5	60−100	0.0	20.3	71.9	7.8
No	≥5	0−5	0.0	0.0	100.0	0.0
No	≥5	5−20	0.0	0.0	79.3	20.7
No	≥5	20−60	0.0	0.0	96.3	3.7
No	≥5	60−100	0.0	18.2	81.8	0.0
Yes	0	0−5	100.0	0.0	0.0	0.0
Yes	0	5−20	100.0	0.0	0.0	0.0
Yes	0	20−60	100.0	0.0	0.0	0.0
Yes	0	60−100	100.0	0.0	0.0	0.0
Yes	0−5	0−5	0.0	0.0	100.0	0.0
Yes	0−5	5−20	0.0	0.0	100.0	0.0
Yes	0−5	20−60	0.0	58.6	41.4	0.0
Yes	*0−5*	*60−100*	*0.0*	*46.2*	*0.0*	*53.8*
Yes	≥5	0−5	0.0	9.8	80.3	9.8
Yes	≥5	5−20	0.0	*6.4*	93.6	0.0
Yes	≥5	20−60	0.0	14.2	85.8	0.0
Yes	*≥5*	*60−100*	*0.0*	*26.3*	*29.3*	*44.5*

Note: Potentially spurious relationships are indicated in italics.

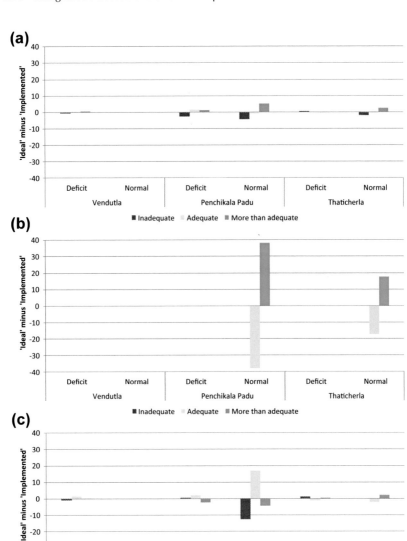

FIGURE 12.6 Difference in *Water stocks* states between the implemented and ideal scenarios in the Prakasam District: (a) rainfed only farmers, (b) irrigator farmers with less than 5 acres of irrigated land and functioning wells and (c) irrigator farmers with more than 5 acres of irrigated land and functioning wells.

TABLE 12.4 Contribution of *water stocks* to drought support in Penchi-kalapadu and Thaticherla under the implemented and ideal WSD scenarios for households with functioning wells and less than 5 acres irrigation (% likelihood)

Drought support (water quantity)	Penchikalapadu		Thaticherla	
	Implemented	Ideal	Implemented	Ideal
Not applicable	2.9	1.8	1.7	1.2
No droughts	0.0	0.0	0.0	0.0
One drought	28.8	18.0	17.0	11.9
Two droughts	49.0	60.6	61.7	67.2
Three droughts	19.2	19.5	19.5	19.7

The cropped area for the implemented scenario listed in Table 12.2 is used to explore the possible impacts of changed cropping area on the aspects of financial capital for households that own only rainfed land or only irrigated land. For irrigators, the rainfed area in the BN was conditioned (set) to "0" and each category of irrigated area was investigated. The results are shown for Basine-palle and Vendutla in Figures 12.7 and 12.8, respectively. Similarly, the irrigated area is conditioned to 0 for rainfed-only farmers and the impact of the imple-mented scenario was examined for each size class. The results are shown for Basinepalle and Vendutla in Figures 12.9 and 12.10, respectively. For both irrigator and rainfed farmers, the most sensitive variables to changes in crop area are those directly linked to the *Crop area* variables. Variables further down the network such as *Debt*, *Financial capital*, and *Drought support* variables are not greatly influenced by changes in crop area. The key variables that change (shown in the figures below) are *Crop revenue*, *Income*, and *Years of crop loss*.

For farmers with up to 5 acres of irrigated land and no rainfed land, there is a 60% likelihood that "2.5−5" acres is cropped (Figure 12.7, top, left). A similar proportion of irrigators with more than 5 acres of irrigated land cropped "5−7.5" acres. Under the ideal scenario the cropped area is conditioned 100% to the 2.5−5 acres and >7.5 acres categories, respectively. This has a reasonably large impact on crop revenue, with a shift toward higher revenue states compared with the implemented scenario. This is demonstrated in Figure 12.7 where the left side shows the probability distribution for the implemented scenario and the right side shows the difference between the probability of the ideal states and the imple-mented states. A negative value indicates a reduced probability of that state occurring under the ideal scenario. The greatest increase in revenue is for irrigated households in deficit climate years where the predicted likelihood of revenue ">64,500" is about 0.6 compared with the value that is nearly 0.4 under the

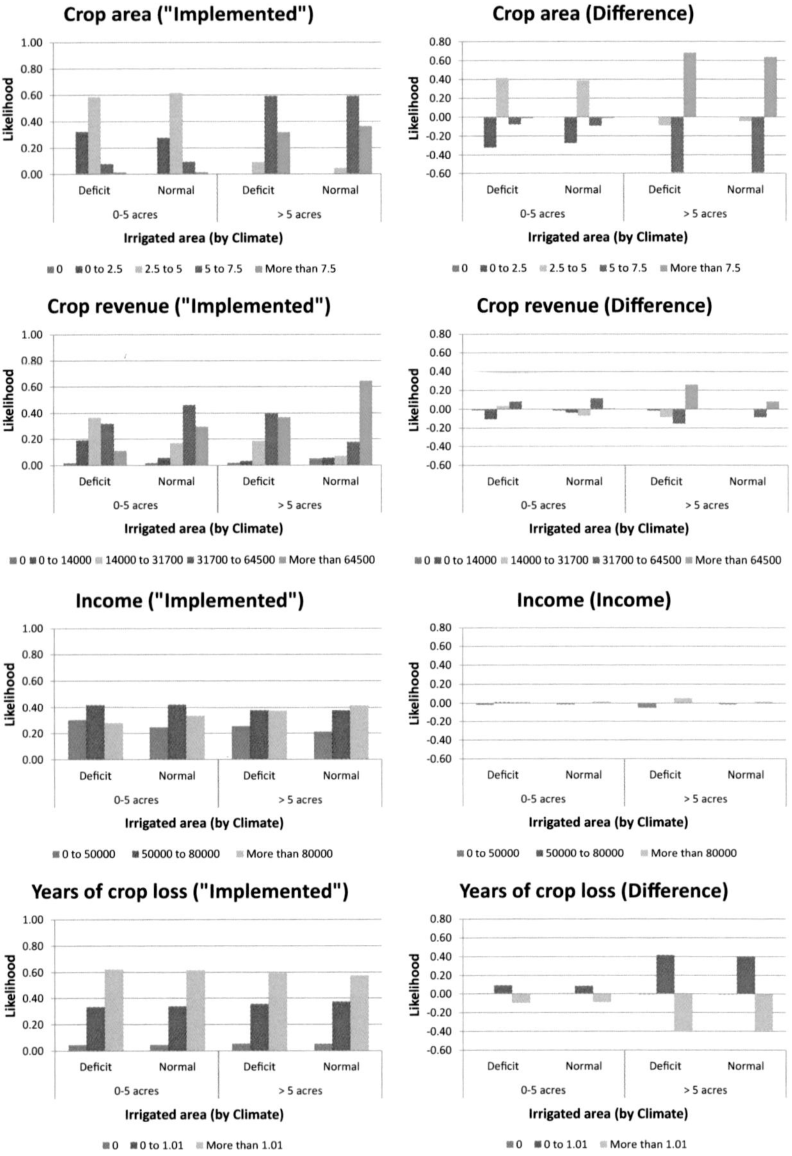

FIGURE 12.7 Probability distribution of key crop-related variables, for irrigated-only households in Basinepalle, in the financial capital submodel. Note: The implemented scenario is shown on the left side and the difference between the probability of the ideal states and the implemented states is shown on the right side.

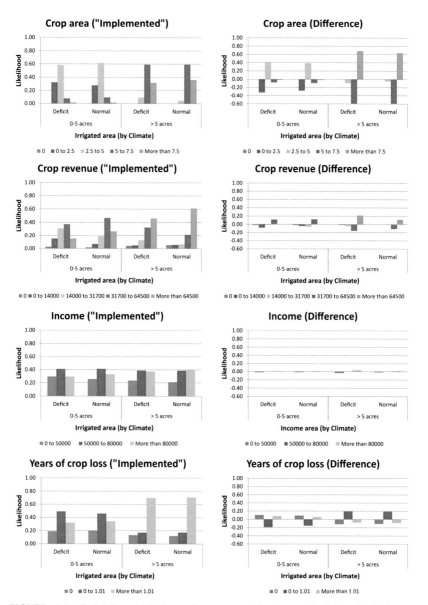

FIGURE 12.8 Probability distribution of key crop-related variables, for irrigated-only households in Vendutla, in the financial capital submodel. Note: The implemented scenario is shown on the left side and the difference between the probability of the ideal states and the implemented states is shown on the right side.

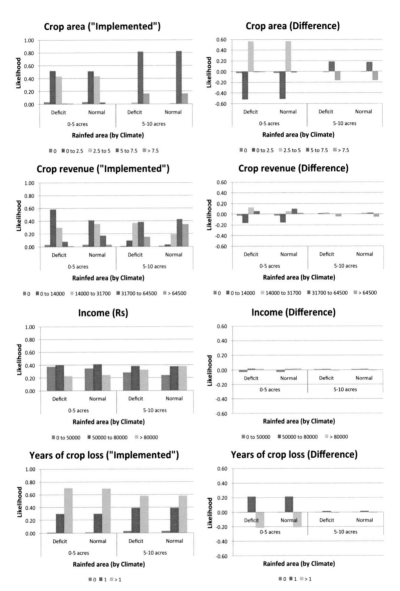

FIGURE 12.9 Probability distribution of key crop-related variables, for rainfed-only households in Basinepalle, in the financial capital submodel. Note: The implemented scenario is shown on the left side and the difference between the probability of the ideal states and the implemented states is shown on the right side.

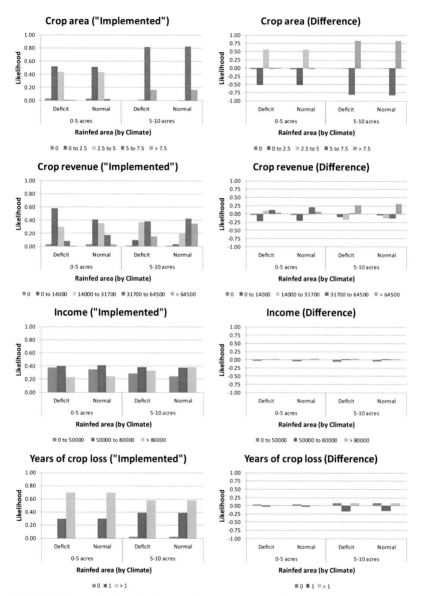

FIGURE 12.10 Probability distribution of key crop-related variables, for rainfed-only households in Vendutla, in the financial capital submodel. Note: The implemented scenario is shown on the left side and the difference between the probability of the ideal states and the implemented states is shown on the right side.

implemented scenario. Impacts of changed crop area on the *Income* variable are more muted than for crop revenue with a maximum decrease in the lowest income state ("0–50000") of 0.06. Further, the likelihood of crop failures reduces under the implemented scenario reflecting the survey data that suggested farmers who cropped larger areas of land were less likely to have a history of failed cropping in recent years.

In Figure 12.2, Crop area (Kharif) is the "child" of both *Rainfed area (acres)* and *Irrigated area (acres)* and has no other parent variable. As such, by conditioning *Rainfed area (acres)* to 0 and *Irrigated area (acres)* to the 2.5–5 acres or >7.5 acres categories, depending on land area, the distribution of the crop area variable is the same for Basinepalle as well as Vendutla (and all villages represented in the BN). The differences between the villages are because the *Irrigated crop type* variable is linked to the *Hydrological unit* and *Stream location* variables. In Basinepalle, irrigated crops predominantly fall within the paddy category (~65%) compared with almost all households in Vendutla classified under the "other" crops category. *Crop revenue* is linked to *Rainfed land productivity*, *Iirrigated land productivity*, and *Irrigated crop type*, and is most sensitive to changes in the productivity variables in the model. However, the *crop revenue* variable is slightly higher in Vendutla under the implemented scenario (Figure 12.8) compared with Basinepalle (due to the broad crop type), although the shift toward increased crop revenue with the cropping area assumed under the ideal scenario is of a similar order between the two villages, as are the changes in the *Income* variable. The *Years of crop loss* variable differs between the two villages, reflecting its strong sensitivity to the crop type variable where growing paddy crops on irrigated land is associated with a lesser recent history of crop failure compared with growing other irrigated crops. With the broad crop type in Vendutla, larger crop areas do not ensure against crop failure (Figure 12.8) contrary to the results suggested for Basinepalle (Figure 12.7).

For farmers with up to 5 acres of rainfed land and no irrigated land, there is an ~50% likelihood that "0–2.5" acres was cropped (Figure 12.9, top, left), and nearly 80% of the rainfed-only farmers with 5–10 acres of irrigated land cropped "5–7.5" acres. Under the ideal scenario, the cropped area is conditioned 100% to the 2.5–5 acres and >7.5 acres categories, respectively. As for the irrigated farmers, this increased crop area results in a shift toward higher revenue states compared with the implemented scenario in both Basinepalle (Figure 12.9) and Vendutla (Figure 12.10). In Basinepalle, smaller farmers (<5 acres) gain a greater increase in crop revenue under the assumed cropping area, although the impacts of changed crop area on the *income* variable reflect to a lesser extent the lower incomes that these households tend to have compared with irrigator households. In contrast, farmers in Vendutla with >5 acres of land had the largest increase in crop revenue. For Basinepalle, the likelihood of crop failures is reduced under the implemented scenario, reflecting the survey data that suggested farmers who cropped larger areas of land were less likely to have a history of failed cropping in recent years.

In Vendutla, on the other hand, there is an increased likelihood of both 0 crop failures and >1, which may reflect the broad cropping types.

12.4 SOCIAL SCENARIO

This section models the possible impacts of removing access to CPR forests from landholders.

12.4.1 Scenario Description

The distribution of land to weaker sections of the society is one of the major policy initiatives in Andhra Pradesh that has been implemented during the last decade. Under this initiative, "excess" and "wastelands" including CPR areas have been identified and allocated to landless households in villages. As a result, large stretches of marginal lands were brought under cultivation. Although some of these lands are not suitable for viable crop production activities, they are being cleared of bushes, scrub, etc., which is expected to impact surface water flows and groundwater recharge. While runoff could be expected to increase and groundwater recharge could decrease with removal of bushes and shrubs, deep plowing could mitigate these impacts by encouraging infiltration into the soil. Furthermore, the redistribution of these lands to landless households may also negatively impact other households in the villages that access CPR lands to support their livelihoods. The latter is examined in this section using the natural and physical capital strength BNs.

In the models presented in Chapter 9, *Access to CPR forests* is the "child" of the *Hydrological unit* and *Stream location* variables. It is a binary variable with the states of "yes" or "no," where "yes" corresponds to a household having access to CPR forests (424 of the 522 survey respondents) and "no" means that the household does not have such access (98 of the survey participants). In Figure 12.11, the relevant variables and their relationships are shown for the natural (green box) and physical (orange box) capital models (for the full models see Figures 9.9 and 9.10, respectively). Variables not directly relevant to this analysis are not displayed in Figure 12.11.

To explore the possible impacts of converting CPR land to private land, we model the effect of reducing landholder access to CPR forests on key indicators of natural and physical capital. Households with access to forests are first compared with households without access in terms of natural and physical capital indicators and the overall natural and physical capital strength indicators. We then explore if the results vary across different villages, HUNs, or economic categories. The analysis is only performed when a particular stratification of the data includes households with and without forest access.

12.4.2 Results

Households with forest access are found to be more likely to have stronger physical capital strength compared with those households without forest access,

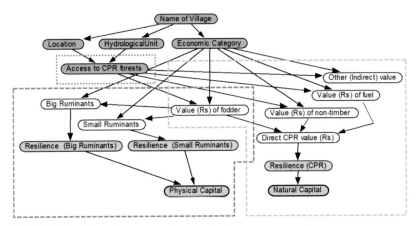

FIGURE 12.11 Network structure relevant to the forest access scenario.

with a 65% likelihood of being in the strong category, compared with 47% (Figure 12.12a). Further, landholding households without forest access are less likely to depend on physical capital for their livelihoods compared with those landholding households with forest access. This is indicated by a likelihood of 15 and 7%, respectively, in the "not applicable" category of the *Access to CPR forests* variable; households with no forest access are also less likely (47%) to have strong physical capital compared with those with access to CPR forests (65%). Removing access to CPR forests could result in weakening of the landholding household's physical capital. In Figure 12.12b, landholding households are found less likely to be in the "strong" category of *Physical strength* (∼18% decrease) and that the probability is displaced to the "weak" (9% increase) category and "not applicable"

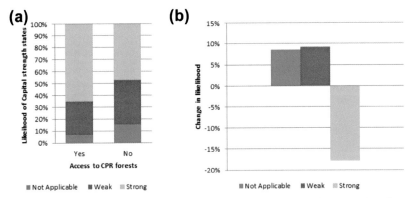

FIGURE 12.12 (a) *Physical capital* strength for landholding households with and without forest access. (b) Change in the states of the *Physical capital* variable when *access to CPR forests* variable is changed from yes to no for landholding households.

(9% increase) category. The increased likelihood of the not applicable category may indicate that removing access to forests will stop some households from accessing physical capital to sustain their livelihoods.

The changes in Figure 12.12b are disaggregated by the HUN in Figure 12.13, and the results suggest that the difference between medium-large households with and without forest access is more significant in Prakasam compared with Anantapur/Kurnool. However, very few landholder households in Prakasam ($\sim$2%) reported no access to CPR forests, meaning that the uncertainty in the model estimates is much higher for this HUN. Because of this uncertainty, the remaining results presented in this section are presented for the whole sample.

The strength of the *Physical capital* variable is determined by four drought support indicator variables: *Big ruminants*, *Small ruminants*, *Wells*, and *Agricultural tools*. Impacts from changes in access to CPR forests occur because of the effects on the ruminant stocks and drought support (ruminant) variables. The response of small-marginal and medium-large households to removal of forest access is similar in terms of *Drought support (big ruminant)*, with a strong increase in the not applicable category and decreases across all other categories (Figure 12.14). Similar but reduced impacts for small ruminants are shown in Figure 12.14.

In contrast to the *Physical capital* variable, the likelihood of stronger natural capital for households with access to CPR forests is not much higher than for those households without forest access, with a 98 and 91% likelihood, respectively. All households in the survey reported stocks of at least one of the capital indicators that would support survival for two or more consecutive drought years. As such, although removing access to CPR forests may reduce the likelihood of strong capital strength (by 7% in Figure 12.14), the households will still have some capital available to support their livelihood activities. In terms of the farm size, the small-marginal households are more likely to be

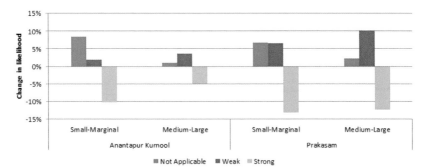

FIGURE 12.13 Change in the states of the *Physical capital* variable when the *Access to CPR forests* variable is changed from "yes" to "no" for small-marginal and medium-large landholder households (by HUN).

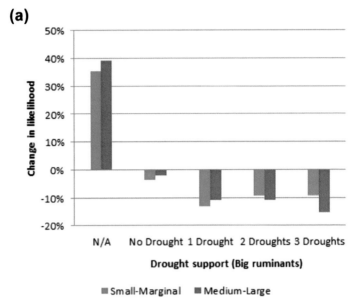

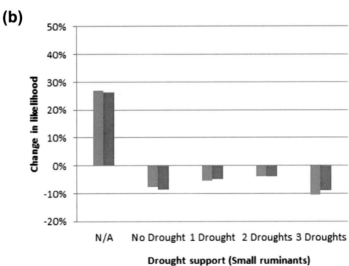

FIGURE 12.14 Change in the states of the ruminant drought support variables when the *Access to CPR forests* variable is changed from "yes" to "no" for small-marginal and medium-large landholder households: (a) big ruminants and (b) small ruminants.

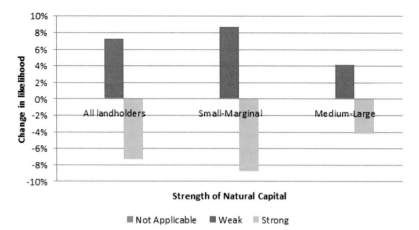

FIGURE 12.15 Change in the states of the *natural capital* strength variable when *access to CPR forests* variable is changed from yes to no for all landholders and small-marginal and medium-large households.

impacted than the medium-large households with an 8.7 and 4.2% reduction in the likelihood of "strong" natural capital, respectively (Figure 12.15).

The strength of the *Natural capital* variable is determined by four drought support indicator variables: land area and quality, water quantity, and CPRs. In the BN, impacts from changes in access to CPR forests occur from the effects on the *Drought support (CPR)*. The response of the small-marginal and medium-large households to removal of forest access is similar in terms of the *Drought support (CPR)* variable with a strong increase in the not applicable category and decrease across all other categories (Figure 12.16). With removed access to CPR forests, the landholder households stand to lose a substantial direct value of fuel, fodder, and (to a lesser extent), non-timber resources collected from CPR forests. Of the landholder households that access CPR forests, more than 50% reported an annual direct CPR value of more than Rs.4000 (Figure 12.17). Additionally, CPR forests provide indirect value (e.g., erosion mitigation) to a sizeable percentage of small-marginal (52%) and medium-large households (23%). Hence, any conversion of CPR forests to non-forest lands is likely to impact the existing landholder households.

12.5 SYNTHESIS

In this chapter we used the BNs developed in Chapter 9 to explore the possible impacts of biophysical and socioeconomic scenarios on indicators of the livelihood capitals. The biophysical scenario considered some of the impacts that the ideal density and placement of WSD interventions in the study watersheds could have had on natural capital (through impacts on increased

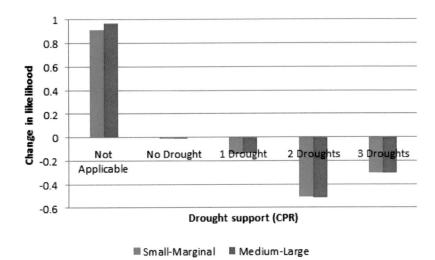

FIGURE 12.16 Change in the states of the *CPR drought support* variable when *access to CPR forests* variable is changed from yes to no for small-marginal and medium-large landholder households.

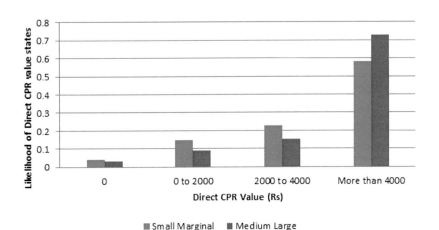

FIGURE 12.17 Value gained from direct use of CPR forests for small-marginal and medium-large landholder households.

adequacy of water stocks) and financial capital (through increased cropping area and impacts of this on crop revenue and household incomes).

Biophysical aspects and the quantity and type of watershed interventions are represented quite simply in the BNs, and there is limited capacity to use the BNs to explore complex alternate interventions and patterns of land use as well as their effectiveness given the hydrogeological constraints. Rather, these biophysical aspects are captured in the model by the hydrological and location (e.g., upstream) variables, broad crop types and area, and the expert-derived *potential access to groundwater (%)* variable. However, the model was able to demonstrate the possible benefits to crop revenue and income under the ideal scenario for both rainfed and irrigated farmers and, in general, increased adequacy of water stocks.

An advantage of the BNs for integrated modeling is their versatility. For example, they offer the capacity to investigate different social policies (such as income support) in combination with WSD and other biophysical scenarios. The social scenario examined in this chapter explored the possible impacts of removing access to CPR forests from landholders through a policy of redistributing lands to poor people. While a simple representation of a complex issue, the scenario identifies the potential for substantial impacts of such policies on landholders and the need to weigh these outcomes up against the benefits to the landless households that receive the redistributed land.

As they stand now, the BN models are more suitable for the research team to explore interactions and relationships in the resilience survey dataset and undertake scenario analysis such as those presented in this chapter. However, the development of a simpler BN tool that synthesizes the knowledge from the project team and incorporates the critical factors (or stocks) affecting household resilience could allow users to explore some of the interactions between household assets, the biophysical context in which villages are situated, and the household capacity to survive consecutive droughts. One possible role of such a tool could be to promote group discussion of watershed issues and implications of Integrated Watershed Management Program (IWMP) design and implementation either as part of training workshops or during early stages of IWMP planning.

REFERENCES

[1] Simpson MC. An integrated approach to assess the impacts of tourism on community development and sustainable livelihoods. Community Dev J 2009;44:186−208.

[2] Varela-Ortega C, Blanco-Gutiérrez I, Swartz CH, Downing TE. Balancing groundwater conservation and rural livelihoods under water and climate uncertainties: an integrated hydro-economic modeling framework. Global Environ Change 2011;21:604−19.

[3] Portoghese I, D'Agostino D, Giordano R, Scardigno A, Apollonio C, Vurro M. An integrated modelling tool to evaluate the acceptability of irrigation constraint measures for groundwater protection. Environ Model Software 2013;46:90−103.

[4] Jakeman AJ, Letcher RA. Integrated assessment and modelling: features, principles and examples for catchment management. Environ Model Software 2003;18:491−501.

Chapter 13

Summary and Conclusion

V. Ratna Reddy* and Geoffrey J. Syme[§]
*Livelihoods and Natural Resource Management Institute, Hyderabad, India, [§] Edith Cowan University, Perth, Australia

13.1 BACKGROUND

Watershed development (WSD) has a long history as a soil and water conservation technology. While most of the developed and developing countries have been implementing WSD to protect their river basins and watersheds, Indian policy has given WSD the much bigger role of providing stability to its vast regions of rainfed agriculture. Although WSD has had a long history in India, its adaptation at the policy level as a developmental intervention began in the 1980s. Introduced as a soil and water conservation intervention aimed at stabilizing agricultural productivity in the rainfed regions, the WSD program has transformed into a rural development intervention over the last three decades. WSD is among the flagship programs of the Government of India with substantial annual budgetary allocations.

Thus, WSD has become critical for developing rainfed agriculture, which accounts for 60% of the cropped area in India. Over the years, about one-quarter of the rainfed areas have been covered under the WSD program. However, the impact of WSD on the productivity or stabilization of agriculture has been marginal. Earlier evaluation studies have pointed out that people's participation and collective action is a prerequisite for effective implementation and impact of the program. Thus, the implementation process is rather intensive and demands substantial human resources, apart from financial resources. Guided by these studies, implementation guidelines for participatory watershed development were introduced in 1995. Since then, about nine variations of these guidelines have been developed to improve the implementation of the program. While

Integrated Assessment of Scale Impacts of Watershed Intervention
http://dx.doi.org/10.1016/B978-0-12-800067-0.00013-X Copyright © 2015 Elsevier Inc. All rights reserved.

frequent procedural changes in the implementation guidelines resulted in confusion among implementing authorities at the cutting-edge level, the guidelines did bring about some changes in the social capital indicators such as participation in the program. However, no substantial improvement has been observed in the implementation and impacts of the program over time.

While inadequate implementation is often identified as the root cause of the poor performance of the program, a number of other reasons have been flagged: being a government program, WSD has all the management constraints associated with such programs, including lack of sufficient time, delays in fund releases, and so on. Although these are common to all the developmental programs, the intensive nature of WSD cannot absorb such drawbacks. For instance, the guidelines provide a 12 month time frame for organizing the communities and ensuring their participation, but only 3 months are allowed for the implementing agencies to perform the process on the ground. Engaging communities like nongovernmental organizations (NGOs) that have built rapport with the local communities prior to the program, or who could spare more time (by allocating more human resources), has proved to be more effective in revealing the WSD impacts compared with the involvement of the government departments. On the contrary, the responsible departments for implementing watersheds are often constrained by limited available human resources coupled with the demands of their other mainstream responsibilities, which impede progress. Given this, it is often concluded that NGO-implemented watersheds perform better when compared with government-implemented watersheds; 80% of the watersheds are implemented by government departments, thus, the overall performance of the program has remained low.

The concentration of effort in identifying the factors responsible for greater impacts and fixing the basic characteristics so new guidelines could be developed has resulted in sidelining the original purpose of watershed interventions. While WSD is a technology meant for soil and water conservation that would strengthen the natural resource base for the farming systems and improve its resilience, the focus has been on improving crop yields and agricultural incomes. More important, the interlinkages between different and dependent natural systems such as biophysical and hydrogeological systems have been totally neglected.

Biophysical aspects, such as rainfall, soils, and land use, determine the nature and intensity of impacts. Similarly, the hydrogeological features of a watershed determine groundwater storage potential and its sustainability in the short and medium terms. However, these aspects are hardly considered while designing or assessing the impacts of the WSD programs. In the absence of information on these aspects, WSD interventions (type as well as intensity) have been uniform across locations.

Similarly, common indicators of impact assessments, such as irrigation, crop yields, and income, have been used irrespective of the variations in biophysical and hydrogeological attributes of the watershed or location. This results in (1) interventions that may not be effective as they are not in line with the hydrogeology and biophysical requirements; (2) impact assessments that are not comprehensive because they do not account for the externalities associated with hydrogeology (groundwater) leading to under- or over-estimation of the cost-benefit ratios; (3) variations in the impacts of WSD at different locations (upstream and downstream) are not captured in the context of meso-WSD (about 5000 ha); and (4) the impact of WSD on the resilience of the households is not getting assessed, although it is the main impact expected in any situation.

The chapters in this book attempt to address these missing aspects in WSD impact assessments. They are based on the scientifically designed approach of selecting sample watersheds located in a hydrological unit (HUN). Technical data have been generated on biophysical and hydrogeological aspects through monitoring the wells, collecting long-term rainfall data, geo-referencing the water bodies, and watershed interventions. Modeling was used to capture the rainfall–recharge and groundwater–surface water linkages, and the socio-economic data were collected using scientific and representative sampling methods complemented by qualitative research. The sustainable rural livelihoods (SRL) framework (five capitals) was adopted to assess the watershed impacts, along with a separate resilience survey to assess the resilience of the farming households.

Household resilience was used as an indicator for WSD impact. Resilience has been explained with the help the five capitals of the households and modeled to identify the factors influencing resilience. All these aspects (hydrogeology, biophysical attributes, five capitals, and resilience) have been integrated to assess the linkages using Bayesian networks (BNs). A consistent stakeholder engagement process was adopted to communicate the findings at the policy, implementation, and community levels. Stakeholder engagement was used to influence the policy (state and national-level policy makers), implementation (implementing agencies), and validation of the findings (farmer level).

This chapter pulls together and synthesizes the analyses from all the chapters and provides an overview of the impacts of WSD from a hydrogeology and biophysical aspect. The aim is to provide a central theme of argument that is drawn from the analysis in various chapters. Apart from summarizing the main arguments, this chapter also provides policy guidance based on the analysis and the policy environment in general and in India in particular.

13.2 HYDROGEOLOGY AND BIOPHYSICAL ASPECTS

The rainfed regions of the Deccan Plateau are hard rock aquifers characterized by shallow, deep, fractured, and non-fractured zones. The characteristics of the aquifers vary widely across and within (especially mesoscale) the watersheds.

Such wide variations result in diversity in the potential and availability of groundwater resources in the region. Groundwater systems in these regions mostly depend on rainfall, which occurs during limited periods of the year. Of late, the yearly rainfall variations have gone up due to climate changes, adding to the temporal dimension to groundwater variation. These regions depend extensively on groundwater for drinking as well as irrigation purposes, so variations in the availability of groundwater become an important determinant of agriculture and related livelihoods. Thus, the supply side of groundwater is associated with high variability due to the nature of the aquifer system and the changing rainfall pattern.

The supply-side variations would not have been a serious concern had demand remained constant. As long as the demand for groundwater remains within the limits of recharge from rainfall, the supply constraints are hardly noticed. However, the demand for groundwater during the last two decades has outstripped the supply, i.e., beyond the rainfall recharge. This has caused severe constraints on the availability of water, even for drinking, in these regions. Often this has resulted in over exploitation of the resource and deterioration of groundwater quality. The first victims of this resource degradation were the communities located on shallow aquifers. On the other hand, communities located on deep aquifers have resorted to capital-intensive deep bore wells. As a result, access to groundwater has been privy to capital-rich large and medium farmers in these regions. Thus, depletion of aquifers has aggravated inter-regional as well as intraregional inequities.

Watershed interventions are expected to enhance groundwater recharge artificially. Given the huge demand for groundwater in these regions, communities as well as the watershed implementing agencies have given priority to on-stream interventions (mainly check dams) without understanding the aquifer geometry, water level trends, groundwater recharge, and changes in groundwater storage. The watershed interventions have been based on surface drainage pattern and do not improve the recharge in an optimal way. Therefore, a more rigorous assessment of the hydrogeology is required to optimize watershed interventions. The geophysical investigations coupled with rainfall–recharge estimates performed at the study sites have helped to assess the groundwater availability at various space and timescales (Chapter 2).

Given the geometry of the aquifer system, i.e., soil cover, weathering thickness, etc., differential watershed interventions are required across the locations. For instance, areas tapping the first fracture can be treated with water-spreading methods (e.g., check dams) and areas tapping deep fractures should have injection wells. Thus, a complete knowledge of the system with details on the varying weathered thickness and presence of fractures as well as the groundwater storage capacity helps in judiciously planning the watershed interventions (Chapter 3).

Further, the integrated surface water–groundwater modeling simulation tool has provided assessments of the availability of surface water and groundwater resources on a monthly basis for a range of watershed interventions, land use, and climate-related scenarios. This model clearly indicates that there is scope for a pragmatic broad-scale approach for developing more robust and equitable WSD interventions instead of the presently followed uniform interventions. This tool is simple in formulation; tries to be as generic as possible; and requires limited amounts of data for climate, topography, soils, land use, hydrogeology, and watershed interventions, which can usually be met from secondary sources. The model can help in shedding light on designing and implementing improved watershed development strategies that can be used by the relevant government and nongovernment agencies to support planning and decision making (Chapter 4).

Another model of rainfall–recharge linkages that was developed for the West Bengal study site was tested on an ungauged study site as well as a gauged catchment adjacent to the ungauged study site in Andhra Pradesh. The modified model in the ungauged study sites resulted in a decreased modeled runoff (and a lower rainfall–runoff coefficient). The watershed interventions resulting in higher storage created in the upstream of the watershed reduced the runoff; hence, more rainfall is needed to cause the same amount of runoff. The calibrated values of the gauged catchment mainly influence the exfiltration, infiltration, and percolation of the area. The small difference (increase) in rainfall–runoff coefficients of the gauged catchment and the nongauged study site could mostly be related to a change in rainfall, while the more intensive rainfall events increased the modeled runoff. This model can be used to estimate the effects of watershed developments in this region. A very complex model structure and model processes including spatial variability could have been chosen but for the data constraints. Therefore, a model structure and the processes defined in a manner as simple as possible were chosen for performing this research. Improved data availability could help produce more precise assessments in future research and planning (Chapter 5).

As is the case with surface and subsurface hydrology, other biophysical aspects such as climate, soils, and land use not only vary within and between watersheds but also influence watershed interventions. With uniform technological interventions under low and medium rainfall zones, the interventions may create new problems, especially with the mesoscale WSD programs. Interventions on every land parcel—namely "net planning" for water conservation intervention mainly through farm bunding and water absorption trenches for land use patterns such as scrub lands—not only render the investments unproductive in the immediate term but also raise new hydrological issues such as reduced flows into the existing water bodies. This can create conflicts within communities.

To overcome these problems, it is necessary to estimate the water availability under different scenarios such as with and without watershed interventions. Water conservation efforts with a certain quantum of water harvesting in a modeling framework would provide valuable insights into water availability. Based on the available water after conservation efforts at the farm level, additional storage could be planned on streams as *ex situ* conservation interventions after accounting for the existing storage capacities using tanks. Modern tools such as Geographical Information Systems coupled with the high computing power available and public datasets enhance the capabilities of project implementing agencies in visualizing the watershed features and key parameters representing erosion status and runoff potential. This helps with making informed decisions when prioritizing the sub-watersheds within the mesoscale HUNs (Chapter 6).

13.3 SOCIOECONOMIC IMPLICATIONS

In the absence of appropriate or optimum design and implementation of watershed interventions, the expected positive socioeconomic impacts may not be evident. Moreover, they would vary across locations depending on the aquifer geometry, type and nature of aquifer, rainfall—runoff and recharge, surface—groundwater recharge, extent of surface water storage, and land use pattern. The externalities of hydrogeology could be captured in the upstream/downstream context at the mesoscale, when watersheds are placed within a HUN. Also, the lag between the implementation and impact assessment could influence the impacts, as measured from the experience of the households—the greater the lag the higher the risk of households missing the linkages between the interventions and their impacts. Unlike the earlier impact assessment studies, the sample watersheds are purposively selected from a HUN by the hydrogeology and biophysical scientists.

Technically, watershed interventions are expected to strengthen the natural resource base and improve the resilience of the farming system. Hence, the resilience of the household is included as an indicator of watershed impact. In addition to the standard approach of measuring the impacts on various socioeconomic indicators, the SRL framework of five capitals has been adopted to provide a holistic assessment.

As expected, the standard approach of impact assessment failed to provide any clear evidence because of time lag. With five capitals, the impacts are observed to be subdued but statistically significant. On the other hand, reported resilience provided clear evidence of impact when compared with the five capitals approach. Resilience, measured in terms of household capacity to withstand a number of droughts, is positively associated with the rainfall (HUN), location (downstream), and watershed (treated area); that is, the unit (HUN2) with better rainfall is more resilient than the one with lower rainfall (HUN1), downstream locations are more resilient than upstream and midstream locations, and villages treated with watershed interventions are more resilient than untreated (control) villages. These findings support the formulated hypotheses.

However, there are deviations to this logical pattern. The extremely poor performance of the upstream village in the low rainfall zone (HUN1), despite being a model watershed (acclaimed as a best-implemented watershed), and the unexpected poor performance of a watershed in the relatively better rainfall zone (midstream village in HUN2), despite the shifts to high-value horticultural crops, seem to defy standard explanations.

The explanation for these deviations lies in the hydrogeology of the locations. The hydrogeology of the upstream village in the low rainfall zone (HUN1) is very shallow (basin) and does not support any on-stream interventions for groundwater recharge. As a result, despite well-constructed and maintained check dams, this village could not benefit from groundwater recharge. Hence, it continues to depend on shallow wells and the situation worsens during years with less than normal rainfall.

On the other hand, in the midstream village in the better rainfall zone (HUN2) the land use pattern is not in line with the groundwater potential. This village is characterized by a moderately shallow basin with limited groundwater potential. Due to the nature of the aquifer, groundwater swells and depletes faster during good as well as bad rainfall years. Because of the absence of this hydrological information, horticultural crops were promoted, and when the demand for water surpassed supply the potential wells started failing and the crops started drying up. This was because groundwater was exploited beyond its potential (sustainable yields or rainfall–recharge coefficient). Therefore, as long as the demand and supply is balanced, cultivation of water-intensive crops such as horticultural crops is sustainable, as observed in parts of the low rainfall zone (HUN1).

These two cases clearly demonstrate the role and importance of hydrogeology and land use practices when explaining and understanding watershed impacts. In the absence of this information, the impacts are often attributed to the quality of watershed implementation or at most to rainfall variations (if any). This clearly indicates the need for considering the biophysical aspects while designing and implementing the watersheds. Such integration of designing, implementation, and assessment becomes convenient and comprehensive when watersheds are placed in the context of an HUN (Chapter 7).

Parametric and semiparametric analyses of farmers' perceived drought survival responses were performed to assess the role of the five capitals as well as the households' characteristics in making farmers resilient to repeated droughts. Drought resilience with and without WSD intervention as well as the identified variables that influenced farmers' nonagricultural incomes such as employment programs, dependence on common pool resources, and migration incomes were tested. It was found that households with a significant source of nonagricultural income could either come from vulnerable or resilient categories. Further, it is observed that the role of human capital such as health and education in influencing drought resilience becomes crucial. Healthy individuals are not only found to show higher participation in labor force and employment programs, they also have higher income from common pool resources because they can put in more effort. However, all the healthy households are not necessarily drought resilient. Similarly, a larger number of educated members in the household also made the household more resilient. However, households that spend more on education indicate a marginally lower drought survival. This highlights the trade-offs between accumulating higher human capital (which could provide long-term resilience) at the cost of reducing current or short-term resilience (Chapter 8).

Equity has been the most difficult objective to achieve in any developmental intervention. This is more so with WSD, as the technology is land based; i.e., landless households automatically fall outside the set of beneficiaries. To overcome this bias other interventions targeting the landless, such as supporting nonagricultural activities, self-help groups, etc., have been introduced under the livelihoods support component. Apart from this, the inequity within the landed households is the most controversial as it tends to increase with the interventions due to structural anomalies like access to groundwater. Addressing the structural issues calls for major policy changes, apart from proper planning of WSD interventions. This is seen in both India and Australia; that is, long-term management of WSD requires consideration of how collective decision making and action can be maintained at an appropriate hydrological scale.

While property rights and markets can assist, there is no natural "evolution" to sustainability through these vehicles. Although both can be helpful, there is a need for them to be underpinned by concerted community action based on distributive and procedural justice. It is clear that landholders are motivated by individual profit needs and rely largely on their own judgment or follow lead farmers when choosing the crop type. In the long term, this will lead to the ongoing deterioration of the quantity and quality of the resource.

Communal approaches to groundwater management do exist, but are not a great priority to the community in Andhra Pradesh. However, the move to mesoscale WSD will require careful attention to how justice principles can be used to promote sustained community action and appropriate property rights. In this regard, the eight "rationalities" or criteria for the successful delivery of the WSD identified by Crase et al. (see Chapter 2) will provide a very useful evaluative tool. These rationalities include social, political, organizational, and government, all of which are highly pertinent to the achievement of justice and cooperation at the local level and crucial if meso-institutions are required to be designed and created (Chapter 10).

13.4 THE APPROACH TO INTEGRATION

Integrated modeling methodologies have a greater potential compared with purely disciplinary approaches to support comprehensive assessment of social, economic, and biophysical aspects of a complex natural resource management such as WSD. Climate and recharge estimates drive predictions and assessment of the availability of surface and groundwater resources as impacted by

WSD, climate, and land use (i.e., water extractions). Water availability and land use together influence crop productivity for households that have access to the available water resources, depending on how they use these resources, and, consequently, their decisions and resilience.

The critical aspect of research is to integrate the various technical aspects of WSD interventions and their impact on the socioeconomic fabric and livelihoods of the communities. The BN theory has been used to achieve this. BNs have been applied within the mesoscale project to relate the stocks of the livelihood capitals to the capacity of the households to survive consecutive droughts. The utility of the BN models in analyzing social datasets and how scenario analyses can be implemented using the approach has already been demonstrated. Hence, the BN models form the basis of the integrated model described and are used to link the biophysical and livelihood outcomes to alternative policy scenarios.

The scenarios presented in Chapters 9 and 12 are only a demonstration, although they are seen as feasible "futures" by the study team. They do provide a template, however, against which decision makers and implementers can approach their planning and policy formulation. Since the amount of data available for this study is not always possible, this approach provides a framework for systematic thinking. The BN approach was chosen because it can incorporate expert judgment when necessary and highlight where biophysical or socially based data collection will be of most value for policy formulation. Thus, we have demonstrated the efficacy of a systematic approach to WSD policy and planning, which should benefit the planning and delivery of sustainable catchment management.

13.5 PUTTING SCIENCE TO PRACTICE

Converting good science into practice is critical for achieving the stated objectives for any developmental intervention, especially when technical aspects are involved. Of late, engaging with stakeholders is becoming more important to research programs. Stakeholder engagement needs to be an integral part of a research program from the beginning, but often researchers interact with the stakeholders only when they have something substantial to convey or share. Such an approach, however, does not appeal to the stakeholders, especially policy makers, who often treat the interactions as a formality. This is more evident in integrated research where transdisciplinarity makes technical aspects more difficult to convey. Moreover, achieving integration among the team members often takes a considerable amount of time and effort. This project is no different, although the Department of Rural Development (DRD; Andhra Pradesh), which is the nodal agency for implementing the WSD program, has been a formal partner in the project from the beginning. The real challenge was to differentiate between integrated framework and approach. The project could achieve the shift from the framework to approach quite effectively; it took a focused and concerted effort from the project team to get the right messages to the policy makers at the state (DRD) and national levels (National Rainfed Area Authority).

The stakeholder engagement was targeted at three levels: policy makers (state and national level), implementing agencies at the field level (government and nongovernmental agencies), and the farming communities in the sample villages. The intention was to integrate the priorities of policy makers and the farming communities into the research and enable the middle-level implementing agencies to understand and adopt this integrated approach. Understanding the hydrogeology and managing groundwater accordingly has been the main concern at the policy level, whereas optimizing groundwater use in a sustainable manner has been the priority of the communities. At the community level, the hydrogeology along with soils and land use in their villages and its linkages with the existing groundwater conditions and watershed interventions were presented and discussed. This awareness was expected to change their obsession with on-stream interventions (check dams) as against on-farm interventions. It was also expected to help improve the acceptability of the new design and implementation of watershed interventions. Similarly, at the policy level, although the policy makers are aware of the importance of hydrogeology in watershed design and implementation, they were not clear on how to take it to the implementation level. The simplified and practical approach adopted in the project has attracted the interest of the policy makers at the state as well as national level, which translated into the demand for customized training for their implementing agencies. Most important, the team was requested to help them prepare the Detailed Project Reports (DPRs), which form the basis for watershed implementation. The first pilot training provided by the team to the district-level implementing agencies was well received, and there were requests for more such training with an enhanced focus on hydrogeology (Chapter 11). This is a clear sign of the benefit of stakeholder engagement and putting science into practice.

13.6 THE WAY FORWARD

In a way this research is futuristic, at least when it was initiated in the year 2009 and designed as per the priorities of the DRD and Government of Andhra Pradesh. At that stage, the concept of meso-watershed was just introduced and the implementation was about to start. The priority of the department was to understand the major concerns and the likely livelihood impacts of the WSD program. To achieve this, the research team had to find a prototype meso-watershed in a hydrological context. Thus, the study is not typically an impact assessment of the meso-WSD program, although it provides all the necessary insights into the likely impacts and concerns. On the whole, the assessment of the impacts validates the hypotheses that meso-watersheds could generate differential benefits at scale (upstream/downstream). Given the lag between implementation and assessment, the impacts are most conspicuous in terms of perceived household resilience.

However, it is clear that there is some mismatch of perceptions of the outcome of WSD in terms of the benefits and costs to other parties that relate to hydrogeology and the biophysical characteristics of the location. Therefore, these aspects, which were previously not considered, need to be directly addressed to optimize as well as sustain the watershed impacts. Listed below are some important concerns and challenges that need policy attention while implementing the Integrated Watershed Management Program (IWMP) watersheds:

- A hydrology-based approach of placing IWMP watersheds within the hydrological coordinates would help in understanding the upstream/downstream linkages better.
- Technical inputs need to be used for assessing surface and groundwater hydrology and their linkages in the context of biophysical attributes, while designing the watersheds. Contributions from this project have clearly demonstrated that these models and tools can be simplified with limited data demands. With training, these simple technical tools can be used at the implementation level.
- Based on the hydrogeology and biophysical aspects of the location, WSD interventions can be rationalized with the type and density of interventions— it is not necessary for the entire area to be treated. Some portion of the watershed could be left untreated as a buffer (donor) for the rest of the watershed. Using different scenarios pertaining to biophysical aspects such as variations in rainfall, hydrogeology, soils, and land use, user-friendly decision support tools for watershed design could be developed.
- There is a need for differential allocations within and between watersheds with fixed per hectare allocations. However, the components to which these funds should be allocated need to be location-specific, depending on agroclimatic and hydrogeological factors.
- Household resilience is an important indicator of the socioeconomic impacts of WSD, which needs to be considered while assessing the impacts.
- Achieving equity remains a difficult issue to be resolved and demands judicious planning of interventions with justice principles in mind.
- BNs proved to be useful in the process of integration. Using the simulation exercises, BNs can be developed into a decision-support tool.
- Stakeholder engagement should be an integral part of any research project trying to change policies. Although the research is well received at the policy level, translating it into effective policy is a time-consuming and difficult process. Thus, finding support for stakeholder engagement remains a big challenge in the post-project phase. Funding agencies should identify and support any potential follow-up activities that may add substantial value.

Index

Note: Page numbers followed by "f" indicate figures; "t", tables; "b", boxes.